METHODOLOGICAL APPROACHES TO DERIVING ENVIRONMENTAL AND
OCCUPATIONAL HEALTH STANDARDS
Edward J. Calabrese

NUTRITION AND ENVIRONMENTAL HEALTH—Volume I: The Vitamins
Edward J. Calabrese

NUTRITION AND ENVIRONMENTAL HEALTH—Volume II: Minerals and Macronutrients
Edward J. Calabrese

SULFUR IN THE ENVIRONMENT, Parts I and II
Jerome O. Nriagu, Editor

COPPER IN THE ENVIRONMENT, Parts I and II
Jerome O. Nriagu, Editor

ZINC IN THE ENVIRONMENT, Parts I and II
Jerome O. Nriagu, Editor

CADMIUM IN THE ENVIRONMENT, Parts I and II
Jerome O. Nriagu, Editor

NICKEL IN THE ENVIRONMENT
Jerome O. Nriagu, Editor

ENERGY UTILIZATION AND ENVIRONMENTAL HEALTH
Richard A. Wadden, Editor

FOOD, CLIMATE AND MAN
Margaret R. Biswas and Asit K. Biswas, Editors

CHEMICAL CONCEPTS IN POLLUTANT BEHAVIOR
Ian J. Tinsley

RESOURCE RECOVERY AND RECYCLING
A. F. M. Barton

QUANTITATIVE TOXICOLOGY
V. A. Filov, A. A. Golubev, E. I. Liublina, and N.A. Tolokontsev

ATMOSPHERIC MOTION AND AIR POLLUTION
Richard A. Dobbins

INDUSTRIAL POLLUTION CONTROL—Volume I: Agro-Industries
E. Joe Middlebrooks

BREEDING PLANTS RESISTANT TO INSECTS
Fowden G. Maxwell and Peter Jennings, Editors

GROUND WATER QUALITY

GROUND WATER QUALITY

Edited by

C. H. WARD
Rice University

W. GIGER
Swiss Federal Institute for Water
Resources and Water Pollution Control

P. L. McCARTY
Stanford University

A WILEY-INTERSCIENCE PUBLICATION
JOHN WILEY & SONS
New York · Chichester · Brisbane · Toronto · Singapore

Library of Congress Cataloging in Publication Data:

Main entry under title:

Ground water quality.

   (Environmental science and technology)
   "A Wiley-Interscience publication."
   Papers presented at the First International Conference
on Ground Water Quality Research, Oct. 7–10, 1981 at Rice
University in Houston, Tex., sponsored by the National
Center for Ground Water Research.
   Includes index.
   1. Water, Underground—Pollution—Congresses.
2. Water Quality—Congresses. I. Ward, C. H. (Calvin
Herbert), 1933–   . II. Giger, W. (Walter)
III. McCarty, Perry L.   IV. National Center for Ground
Water Research (U.S.)   V. Title.   VI. Series.

TD426.G72   1985   363.7'394   84-25661
ISBN 0-471-81597-7

CONTRIBUTORS

ANDREA B. ARQUITT   Department of Biochemistry, Oklahoma State University, Stillwater, Oklahoma

D. L. BALKWILL   University of New Hampshire, Durham, New Hampshire

K. M. BAXTER   Water Research Centre. Medmenham Laboratory, United Kingdom

PHILIP B. BEDIENT   National Center for Ground Water Research, Rice University, Houston, Texas

GÖRAN BENGTSSON   Laboratory of Ecological Chemistry, University of Lund, Lund, Sweden

ROBERT C. BORDEN   National Center for Ground Water Research, Rice University, Houston, Texas

L. W. CANTER   National Center for Ground Water Research, University of Oklahoma, Norman, Oklahoma

JENQ C. CHANG   Department of Biochemistry, Oklahoma State University, Stillwater, Oklahoma

CAROL CURRAN   National Center for Ground Water Research, Rice University, Houston, Texas

C. J. DOWNES   Chemistry Division, Department of Scientific and Industrial Research, Petone, New Zealand

ELIZABETH R. DOYEL   Department of Biochemistry, Oklahoma State University, Stillwater, Oklahoma

W. van DUIJVENBOODEN   National Institute for Public Health and Environmental Hygiene, Leidschendam, The Netherlands

WILLIAM J. DUNLAP   Robert S. Kerr Environmental Research Laboratory, Ada, Oklahoma

ROBERT H. FINDLEY   Department of Biological Science, Florida State University, Tallahassee, Florida

MICHAEL J. GEHRON   Department of Biological Science, Florida State University, Tallahassee, Florida

CHARLES P. GERBA   Departments of Microbiology and Immunology and Nutrition and Food Science, University of Arizona, Tucson, Arizona

W. C. GHIORSE   Cornell University, Ithaca, New York

WALTER GIGER   Swiss Federal Institute for Water Resources and Water Pollution Control, Dübendorf, Switzerland

GINGER J. HAMPTON   Department of Biochemistry, Oklahoma State University, Stillwater, Oklahoma

GARY D. HOPKINS   Department of Civil Engineering, Terman Engineering Center, Stanford University, Stanford, California

A. W. HOUNSLOW   Department of Geology, Oklahoma State University, Stillwater, Oklahoma

JEFFREY L. HOWARD   Department of Biochemistry, Oklahoma State University, Stillwater, Oklahoma

S. R. HUTCHINS   National Center for Ground Water Research, Rice University, Houston, Texas

JERALYN Z. JACKSON   Department of Biochemistry, Oklahoma State University, Stillwater, Oklahoma

JACK W. KEELEY   Ground Water Research Branch, Robert S. Kerr Environmental Research Laboratory, Ada, Oklahoma

RUSSELL F. LANG   Drinking Water Research Center, Florida International University, Tamiami Campus, Miami, Florida

FRANKLIN R. LEACH   Department of Biochemistry, Oklahoma State University, Stillwater, Oklahoma

M. D. LEE   National Center for Ground Water Research, Rice University, Houston, Texas

JAY H. LEHR   National Water Well Association, Worthington, Ohio

DAVID I. LEIB   National Center for Ground Water Research, Rice University, Houston, Texas

ROBERT F. MARTZ   Department of Biological Science, Florida State University, Tallahassee, Florida

G. MATTHESS   Geological–Paleontological Institute, Kiel University, Kiel, West Germany

JAMES F. MCNABB   Robert S. Kerr Environmental Research Laboratory, Ada, Oklahoma

ROSALEE MERZ   Department of Biochemistry, Oklahoma State University, Stillwater, Oklahoma

DAVID W. MILLER   Geraghty & Miller, Inc., Syosset, New York

M. M. MORTLAND   Department of Crop and Soil Sciences, Michigan State University, East Lansing, Michigan

JANET S. NICKELS   Department of Biological Science, Florida State University, Tallahassee, Florida

MICHAEL J. NOONAN   Lincoln College, Canterbury, New Zealand

PHYLLIS T. NORTON   Department of Biochemistry, Oklahoma State University, Stillwater, Oklahoma

JEFFREY H. PARKER   Department of Biological Science, Florida State University, Tallahassee, Florida

IRIS L. PAYAN   Drinking Water Research Center, Florida International University, Tamiami Campus, Miami, Florida

A. PEKDEGER   Geological–Paleontological Institute, Kiel University, Kiel, West Germany

WAYNE A. PETTYJOHN   Department of Geology, Oklahoma State University, Stillwater, Oklahoma

G. J. PIET   National Institute for Public Health and Environmental Hygiene, Chemical Biological Division, Leidschendam, The Netherlands

MARTIN REINHARD   Department of Civil Engineering, Terman Engineering Center, Stanford University, Stanford, California

PAUL V. ROBERTS   Department of Civil Engineering, Terman Engineering Center, Stanford University, Stanford, California

RENÉ P. SCHWARZENBACH   Swiss Federal Institute for Water Resources and Water Pollution Control, Dübendorf, Switzerland

J. G. M. M. SMEENK   National Institute for Public Health and Environmental Hygiene, Leidschendam, The Netherlands

GLEN A. SMITH   Department of Biological Science, Florida State University, Tallahassee, Florida

R. SCOTT SUMMERS   Department of Civil Engineering, Terman Engineering Center, Stanford University, Stanford, California

M. B. TOMSON   National Center for Ground Water Research, Rice University, Houston, Texas

GORDON WAGGETT   National Center for Ground Water Research, Rice University, Houston, Texas

C. H. WARD   National Center for Ground Water Research, Rice University, Houston, Texas

JoANN J. WEBSTER   Department of Biochemistry, Oklahoma State University, Stillwater, Oklahoma

C. C. WEST   National Center for Ground Water Research, Rice University, Houston, Texas

DAVID C. WHITE   Department of Biological Science, Florida State University, Tallahassee, Florida

JOHN T. WILSON   Robert S. Kerr Environmental Research Laboratory, Ada, Oklahoma

PAUL R. WOOD   Drinking Water Research Center, Florida International University, Tamiami Campus, Miami, Florida

C. P. YOUNG   Water Research Centre, Medmenham Laboratory, United Kingdom

B. C. J. ZOETEMAN   National Institute for Public Health and Environmental Protection, Leidschendam, The Netherlands

SERIES PREFACE

Environmental Science and Technology

The Environmental Science and Technology Series of Monographs, Textbooks, and Advances is devoted to the study of the quality of the environment and to the technology of its conservation. Environmental science therefore relates to the chemical, physical, and biological changes in the environment through contamination or modification, to the physical nature and biological behavior of air, water, soil, food, and waste as they are affected by man's agricultural, industrial, and social activities, and to the application of science and technology to the control and improvement of environmental quality.

The deterioration of environmental quality, which began when man first collected into villages and utilized fire, has existed as a serious problem under the ever-increasing impacts of exponentially increasing population and of industrializing society. Environmental contamination of air, water, soil, and food has become a threat to the continued existence of many plant and animal communities of the ecosystem and may ultimately threaten the very survival of the human race.

It seems clear that if we are to preserve for future generations some semblance of the biological order of the world of the past and hope to improve on the deteriorating standards of urban public health, environmental science and technology must quickly come to play a dominant role in designing our social and industrial structure for tomorrow. Scientifically rigorous criteria of environmental quality must be developed. Based in part on these criteria, realistic standards must be established and our technological progress must be tailored to meet them. It is obvious that civilization will continue to require increasing amounts of fuel, transportation, industrial chemicals, fertilizers, pesticides, and countless other products, and that it will continue to produce waste products of all descriptions. What is urgently needed is a total systems approach to modern civilization through which the pooled talents of scientists and engineers, in cooperation with social scientists and the medical profession, can be focused on the development of order and equilibrium in the presently disparate segments of the human environment. Most of the skills and tools that are needed are already in existence.

We surely have a right to hope a technology that has created such manifold environmental problems is also capable of solving them. It is our hope that this Series in Environmental Sciences and Technology will not only serve to make this challenge more explicit to the established professionals, but that it also will help to stimulate the student toward the career opportunities in this vital area.

Robert L. Metcalf
Werner Stumm

PREFACE

The First International Conference on Ground Water Quality Research sponsored by the National Center for Ground Water Research stressed the chemical and biological aspects of ground water and was intended to (1) focus attention on the scientific and technological challenges and accomplishments in ground water quality research, (2) provide a forum for assessing the state-of-the-art of knowledge in this area on an international scale, and (3) result in a major technical reference on ground water quality. The Conference was held October 7–10, 1981 on the campus of Rice University in Houston, Texas.

The Directors of the Conference and editors of this volume organized the program into four symposia dealing with (1) sources, types, and quantities of contaminants in ground water, (2) methods for ground water quality research, (3) subsurface characterization in relation to ground water pollution, and (4) transport and fate of subsurface contaminants. Each symposium contained five to seven substantive research or review papers presented, by invitation, by internationally known experts in their respective areas. The Conference was introduced by J. W. Keeley, Chief of the Ground Water Research Branch at the U.S. Environmental Protection Agency's Robert S. Kerr Environmental Research Laboratory at Ada, Oklahoma. Jay H. Lehr, Executive Director of the National Water Well Association, provided a provocative perspective for the Conference.

This volume is not a proceedings of the First International Conference on Ground Water Quality Research. Rather, it brings together chapters in four subject areas which together contribute significantly to our understanding of ground water quality. Collectively, the chapters have considerable breadth in both subject matter and the geographical areas covered. Several papers presented at the Conference have not been included; others have been greatly expanded to ensure depth of coverage or were written specifically for this volume.

C. H. WARD
W. GIGER
P. L. McCARTY

March 1985
Houston, Texas
Zurich, Switzerland
Stanford, California

ACKNOWLEDGMENTS

The multidisciplinary conference which provided the basis for this volume was held with the encouragement and support of Jack W. Keeley, Chief of the Ground Water Research Branch at the U.S. Environmental Protection Agency's Robert S. Kerr Environmental Research Laboratory, Ada, Oklahoma.

The editors and authors gratefully acknowledge the expert editorial assistance of Ms. Maurine Lee of Rice University and the near infinite patience of the staff at John Wiley & Sons. Without both, publication of this book would surely have been abandoned.

CONTENTS

PART TWO   METHODS FOR GROUND WATER QUALITY RESEARCH

GROUND WATER QUALITY

INTRODUCTION

1

NEW DIRECTIONS IN INTERNATIONAL GROUND WATER RESEARCH

Jack W. Keeley

Ground Water Research Branch
Robert S. Kerr Environmental Research Laboratory
Ada, Oklahoma

It was not until the middle of the last decade that water resources experts began to pay proper attention to our underground water resources. This was a consequence of severe droughts in the western states, reports of public water supply contamination with toxic organic substances in the northeastern states, and the determination that past waste-disposal practices throughout the country were threatening ground water quality. Of course, much of our current awareness of these problems is due to the last decade's advances in analytical techniques, which allow us to make observations not possible earlier. This pattern has been much the same for many of the developed countries of the world. There is little doubt that water shortages, ground water contamination, and increased reliance on underground water resources will add substantially to the demands for knowledge from researchers of ground water quality through this decade and beyond.

The science of ground water traditionally has been concerned with pump tests, flow through porous media, partially penetrating wells, leaky aquifers, and the development of mathematical models to describe these and other phenomena. Such efforts are critically important because our considerations of ground water quality must be based on a knowledge of the flow of water in the subsurface environment. Pollutants rely on a moisture flux to transport them through the unsaturated zone to the water table and through ground water.

In order to place our research activities in perspective, it seems advisable to define the goals toward which they should be directed. Areas for attention must include studies of the transport and transformation of contaminants in the subsurface environment as well as the traditional studies of flow through porous media. The problems of ground water contamination are here and have been for many years. Therefore, our efforts must be directed, in the main, toward developing practical, economical solutions.

The first goal is to gather information which allows us to manage waste sources in such a way as to eliminate or at least minimize ground water contamination. Included would be information about waste-disposal practices and about facilities where waste is applied to the land for treatment.

The second goal is to develop means of making accurate damage assessments. We are often charged with assessing damages to ground water from a variety of sources for a variety of reasons. In these assessments, we must determine not only the extent of current contamination but also what it has been in the past and what it is likely to be in the future.

The last goal is to develop technology which will allow us to take the most feasible remedial action necessary to restore an aquifer to some predetermined level of quality, to prevent the contaminants from spreading, or to provide treatment for the contaminated water before use.

It is apparent that developing an understanding of the behavior of pollutants in the subsurface and of the processes that take place in this environment is paramount to the accomplishment of these goals. It is also apparent that we are faced with a crippling need for new methods with which to carry out our research, including the basics of drilling, completing, and sampling observation wells. We must improve ways to obtain undisturbed samples from the subsurface. Adequate techniques in both laboratory and field are needed to ensure that our observations of transport and transformation phenomena accurately duplicate that which occurs in nature.

In recent years, the amount of research directed toward ground water has increased dramatically. Unfortunately, it takes time for much of this work to be completed and more time for it to reach technical journals and become available to other researchers. In addition, there have not been many forums where we can get together and discuss our work and our findings on a collective and individual basis.

A great deal of serious work has been going on in the area of ground water research. For instance, scientists in the United States have been involved with developing new tracer techniques and modifying others for conducting advanced ground water investigations. Other work has been directed toward the development of methods for predicting the effects of mining activities on ground water quality and the technology of sampling the unsaturated zone. On an international basis, we are finding that research is shifting from the traditional posture of dealing with inorganic substances to trace organic compounds that may be of a toxic nature. Researchers in Poland have done work using resins to isolate sources of contamination; some in France are developing seismic techniques applicable to shallow geology, and scientists in the United Kingdom and Israel are working on methods to isolate viruses. Hundreds of predictive models have been developed by many researchers. Most of the models deal with water movement, but a few present heat transfer and mass transfer considerations. Unfortunately, only a small number of these models are generally usable because of the lack of documentation.

Traditional research in pollutant transport and transformation has been directed toward inorganic contaminants, with a significant emphasis on metals. There has also been a great deal accomplished in understanding the movement of nitrogen and phosphorus compounds. Current information permits only generalized conclusions concerning the movement of a few organic compounds in the subsurface environment. Mathematical expressions capable of providing a first-generation predictive ability for these selected groups of organics have been developed. Researchers in Sweden, The Netherlands, Switzerland, the United States, and the United Kingdom are active in organic transport and transformation studies.

The amount of information concerning the specific sources of ground water contamination precludes extensive discussion here, but certain generalities can be made. For example, even though a great deal has been done concerning environmental effects of septic tanks, most efforts have been directed toward such contaminants as coliform bacteria, nitrogen, phosphorus, and total dissolved solids. Little information has been provided on viruses, trace organics, or nitrogen losses to the atmosphere. Research also is needed to define allowable septic tank densities. Most work associated with the development and production of oil and gas deals with the environmental impact of salt-water brines, while little information is available about drilling fluids, chemicals used in treating wells, corrosion inhibitors, or other common compounds associated with drilling practices. Although leachates from municipal landfills and dumps generally have been characterized, the presence of industrial wastes in municipal dumps makes accurate characterizations difficult. There have been many studies of leakage from lagoons, but the study of organics has been limited. Specific sources of contamination are being studied throughout the world; these studies are predominantly associated with waste dumps, agriculture, ground water recharge, land application of wastes, and well construction. Scientists in Belgium have worked with industrial hydrocarbon spills while those in New Zealand are concerned with agriculture, septic tanks, and land application projects. The USSR has done considerable work associated with mining.

Aquifer rehabilitation studies conducted to date have been directed toward prevention, including procedures for control, and methods for monitoring. Case histories of ground water contamination incidents reveal that remedial action is complex, time-consuming, and expensive. About the only large-scale work in this area, outside the United States, is being done in Switzerland and Germany. Swiss researchers have established emergency procedures for pumping and treating contaminated water in spill areas, and scientists at a university in Germany have established a research and demonstration project for pumping, treating, and recharging contaminated ground water.

There is yet another area of ground water research that is just now emerging, that being an attempt to understand the abiotic and biological processes involved with pollutant transport and transformation. These processes appear to be of the utmost importance in characterizing the subsurface environment as a receptor of pollutants. Through the efforts of researchers in such fields as agriculture, petroleum engineering, geology, geochemistry, and hydrology, we know a great deal about various aspects of the composition and structure of the earth's crust and how water moves through it. Unfortunately, available information is inadequate to describe what occurs when pollutants are introduced into the subsurface. Little work of this nature is currently under way in the United States, although some work is being done to define the role of soils in determining the success of land application projects in Australia, Germany, Poland, and The Netherlands. There is some evidence that this type of work is also under way in the USSR, but specifics are difficult to obtain.

It is all too clear that we have just made a beginning in the development of technical information needed to protect one of the world's most valuable natural resources—ground water. We must continue our efforts to provide the tools necessary to protect and restore the integrity of ground water. In the coming decade, for example, we will require the development of modeling capabilities in order to address economic and management problems in addition to those of mass transport, particularly of organic and biological contaminants.

The ultimate destination of contaminants generated by the activities of man is either land or water. It is incumbent upon us to determine how to use the land in such a way as to eliminate or reduce the threat of ground water contamination. A great deal of work is required in the areas of drilling, completing, and sampling of monitoring wells and in the proper use of tracers. New technology is needed in remote sensing. Extensive research will be required to categorize contaminants and select indicators for each of the categories in our pursuit of transport and transformation information, particularly with respect to organics, bacteria, and viruses.

Characterizing the subsurface environment as a receptor of contaminants is an area that is virtually unexplored. This is an extremely important area, which ultimately will lead to developing much-needed technical criteria for waste-disposal site selections. Initial efforts should be directed toward defining the nature of the subsurface environment in terms of its innate physical, chemical, and biological characteristics and toward developing a unified theory correlating these findings with contaminant transport.

We are witnessing a metamorphosis in ground water research, which, in many respects, is late in coming. We are facing ground water pollution problems in all parts of the world, but we would be doing ourselves a great injustice to use scare tactics to heighten public interest in our cause. Though time is growing short, we are far from losing the battle to protect and enhance the vast ground water resources of the world. The late C. L. McGuinness of the U.S. Geological Survey said it best more than a decade ago: "Now we don't know whether to laugh or cry, because the world suddenly wants more from us than we have to give—more knowledge than was ever demanded before and more than we ever dreamed would be needed."

PERSPECTIVE

2

CALMING THE RESTLESS NATIVE: HOW GROUND WATER QUALITY WILL ULTIMATELY ANSWER THE QUESTIONS OF GROUND WATER POLLUTION

Jay H. Lehr

National Water Well Association
Worthington, Ohio

Those of us who count ourselves as ground water scientists have spent the better part of our careers laboring in anonymity. The public did not know the difference between ground water and water lying on the ground until just a few years ago. Most hydrogeologists and hydrologists did little to educate John Q. Public to the wonderful world of the underground and the natural water resources that exist within. I know myself that 10 years ago, when asked what I did for a living, I would often as not make up an exciting war story or answer in Spanish rather than attempt the long explanation that was required to make the listener understand my profession. By default, we have allowed ignorance and misunderstanding to prevail in regard to the public's attitude toward protection of the ground water resource.

Today, recognition of ground water as an important resource has galloped to the forefront. Stories of water shortages and water pollution abound on the covers of our magazines and newspapers and on prime time radio and television news programs. Jobs in hydrogeology are now found at the top of the "help wanted" list. Salaries have risen to a parity with the most respected of scientific professions. Today when a hydrogeologist tells a new acquaintance what he does for a living, he or she asks pertinent questions about ground water pollution, a subject they are reading about so frequently. Thus, while our thoughts and opinions on public issues of water supply once fell on deaf ears, today they have the potential to shape public opinion and eventually bear heavily upon solutions to the water problems we now face.

In the past 8 months, feature articles on ground water problems have been the front page stories in *Newsweek, Time, U.S. News & World Report,* and even the revitalized *Life* magazine. Newspapers across the country have featured week-long series on the subject and all three television networks have focused attention on water in their evening news shows. Without exception, the media have done an excellent job in taking a comprehensive look at the problems of both quantity and quality. Equally without exception, they have overstated these problems, leading the public toward a crisis reaction when a more sober stance might eventually be more productive. The question for us then becomes, "Should we spend any energy calming the restless natives or should we simply cash in on the widespread concern, and get down to work solving our problems?" We want support and we want action. We can no longer afford apathy, but at the same time we do not need knee-jerk responses based on insufficient data and ill-defined questions.

THE QUANTITY MYTH

The overstatement of the water quantity problems can be set aside rather easily. A study of America's water use patterns clearly indicates serious water shortages over 15% of the country. This is the result of the failure to match water use with water availability. Such is the case through the Ogalala areas of Texas, Oklahoma, Colorado, Kansas, and Nebraska, portions of southern California, and limited areas in other parts of the United States. Other water quantity problems can be blamed entirely on mismanagement of clearly anticipated drought and flood cycles throughout our nation, where people continue to be surprised that we do not always receive "normal" rainfall year after year after year. Conjunctive management of ground and surface water resources throughout America on no more than a common sense basis will handle 85% of the nation's periodic water shortages; strict, foresighted management programs will ultimately handle the remaining 15%.

THE QUALITY FACTS

I cannot address ground water pollution problems in nearly so cavalier nor simplistic a fashion. There is little doubt that each of us could discover a polluted well in 75% of the counties in these United States, but there is also little disagreement that the total of our available ground water resources that has been polluted to date is less than 1%. Herein lies the problem. The public tends to believe that ground water pollution spreads within an aquifer in much the same way it would spread in a surface stream; people do not recognize the relatively distinct boundaries of a plume of pollution and its slow movement in predictable directions. The slow, inexorable, but well-defined nature of ground water movement is both a blessing and a curse to us. Once there, pollution cannot readily be removed, but we do have time to plan ahead and avoid compounding the negative outcome of advancing pollution.

There is indeed a great deal of education yet to be completed regarding the public's view of the threat to our ground water resources. The problem is how zealously we wish to approach this task. We could mount a major campaign to calm the restless natives by pointing out that we can restrain ground water pollution in the next decade before it destroys much more of our valuable resource. This course could set the public back on its inactive duff, no longer willing to take the limited action that is required. If, on the other hand, we do nothing to dampen the flames of concern ignited by the media, we will likely encourage some inappropriate action to be taken or turn the nation away from further development of our still underdeveloped ground water resources. As I travel the country talking to hundreds of ground water scientists, I find them amazingly split on this issue. Many fully support the media-generated concern, and a like number feel we must straighten out the fourth estate before it is too late.

Our country can no longer afford fence-straddling silence. But on whichever side of the heated ground water pollution issue you choose to stand, it is most important that you recognize your oral comments and written words are desired by and are important to the public. Your views will be listened to as never before. We therefore have taken on a public responsibility of which we had only dreamed a decade ago. Are we equal to the task that the media have laid at our feet, or would we prefer to return to our former anonymity? Frankly, the latter is a rhetorical question, as the choice is no longer ours. We are now a part of the dialogue, and how well we respond will determine how intelligently this nation will develop solutions to its ground water pollution problems. We are called on daily to influence decisions made in groups ranging from town meetings to the U.S. Environmental Protection Agency and from small, family-owned businesses to the boardrooms of the Fortune 500.

We must reduce existing ground water pollution as best we can. We must protect potable waters from pollution by existing waste sources and disposal activities, we must eliminate the construction of new waste-disposal sites that threaten ground water, and we must alter other activities with potentially dangerous consequences to the underground.

Whether you are optimistic about the future reduction of ground water pollution and thus believe the media have overstated the case, or pessimistic about the future and thus grateful for the attention the problem has received, you should agree on one point: we are still sadly deficient in data to prove either side of the argument. For this reason, we can all be confident regarding the future of ground water monitoring. Whether the new administration takes the high or the low road toward environmental regulation, we can be sure that it will continue the efforts to collect the data necessary to make meaningful decisions in the future.

The data we are beginning to obtain through ground water monitoring wells will decide the future regulation and management of our nation's ground water resources. Thus, those of us involved in this important data collection effort have a secure future. The call for our equipment, services, and expertise will continue to grow for many years to come. No matter where we stand philosophically, we can appreciate the fact that the new recognition of ground water as a resource and of its pollution as a problem can be seen as an indication that our collective professional cups are half full rather than half empty. It is imperative that we learn to cope with our newfound place in the sun. Our words and deeds will likely shape a significant portion of the management of our nation's vast ground water resources.

THE WORLD'S WATER SUPPLY

As we progress toward a specific focus of ground water quality, let us first consider the broad picture of the earth's water resources. Everybody is aware that most of the world's water is in the ocean. The rest is either frozen as ice or in the ground. Half of this ground water is in the upper half mile of the crust and the other half is in the next 2 miles of the crust. We shall deal with what is in the top half mile of crust. Half a million cubic kilometers move through the system each year. Fortunately, we have a net gain on land each year of 40,000 km^3, which is more than 10 times the 3000 km^3 used annually. However, most of it runs off as flood flow, and in spite of every economical dam we have built in the United States, we cannot stop the flow. We will not build another major dam in this country that has a positive cost/benefit ratio. Additionally, the world has controllable water in inhospitable places like the Amazon River basin, where we are not about to go to get it. But we still have 9000 km^3 of controllable water in habitable areas, which is three times what the world uses. What the world uses is 1800 km^3 in reservoirs or dams and another 1200 in well fields; essentially all unused available water in the world is in the ground. Worldwide we do not have a water problem, which of course does not help anybody in Kansas or Nebraska whose water table has dropped 800 ft and who cannot afford to pump it to grow their crops nor does it help those whose water is polluted. But we still have a lot of undeveloped ground water. Ground water has numerous advantages over surface water. Nobody will quarrel with the fact that it is low in cost of operation; you get to use the land for other things, boating excepted, without paying any additional taxes; rising and falling water levels do not create great mud flats; and there is very little loss of storage space. We once thought there was also less chance of pollution, but we found out that the organic chemicals used in industry today persist through our filtering formations.

World prices in 1977 indicated that it cost about five times as much to develop a water supply with dams and reservoirs as with wells. And in this country the average is about 10 to 1. Yet ground water versus surface water use in the United States in 1975 was a 1 to 3 ratio. Texas uses 18 billion gallons of surface water a day and 11 billion gallons of ground water. East of the Mississippi, only Mississippi uses more ground water than surface water. West of the Mississippi, Kansas and Nebraska use much more ground water than surface water because they have inefficient irrigation over the whole countryside. California, Arizona, and New Mexico use a lot of ground water. But in general very few states overuse their ground water. In the east we use it hardly at all. Nationally, we use three times as much surface water as ground water, and there are several reasons why.

SURFACE WATER INCENTIVES

In general the average person on the street is not going to invest in a resource that cannot be seen. Engineers would much rather build dams than drill wells, and if you figure out 7% of a dam versus 7% of a well as your commission, you will understand why. There are also many federal programs that promote dams, but none that promote wells. Bankers like to keep that federal money in their banks. Realtors are all smart enough to know land values increase around reservoirs and hardly ever around wells. Singularly, you would probably rather have your name on a brass plaque on the side of a dam than on the side of a black pipe. Do not take water-witching too lightly, either. As I write, the cover of one of our most popular farm magazines features a fabulous picture of a water witch. The magazine is *Country Living,* one of the most popular rural magazines in America. As long as people give any credence to water-witching, ground water development will always have at least one strike against it.

Five percent of our available water is in the rivers, but the freshest water is the 95% in the ground. The large amount of ground water is the premise on which I am going to state the case that we do not have that much ground water pollution, and that we have a chance to heal our wounds and get on top of the problem. I want to teach you a little bit of optimism.

Before we take a look at some grocery store arithmetic showing how many gallons of ground water have been polluted, let us take a brief visual sojourn through the world of a ground water flow.

THE PHYSICS OF GROUND WATER FLOW

It has taken hydrogeologists many years to turn the tables on the surface water hydrologists, who used mathematics in an attempt to overwhelm us. In reality their mathematics were crude by any means of comparison to what we are using today in ground water. In fact, I do not believe you can really solve the Chezy–Manning Equation for the velocity of a surface stream. A primary variable is the roughness of the stream, which you must locate in a picture book. You look for pictures that look like your stream and plug in the number assigned to it; it is really quite absurd. Ground water is a much neater physical system than surface water flow. It is all moving out of the ground, even though getting out might take a few years, a few decades, or many centuries. When I was at the University of Arizona, we measured the age of the water in the wells on campus. We had a very good geo chem lab and used carbon-14 dating. We found that the water we were drinking from our wells was 25,000 years old. It had traveled 25 miles from the hills in which the rain had fallen and then moved underground, so it was traveling 1 mile every 1000 years, or about 5 ft/yr.

Figure 2.1 is a very common ground water flow net showing a cross section through a minor valley and a major valley. Recharge goes vertically down, reaches the water table, and then moves toward a surface water body. All ground water ends up going into a lake, stream, or ocean. We who study ground water, in fact, consider surface water to be rejected ground water.

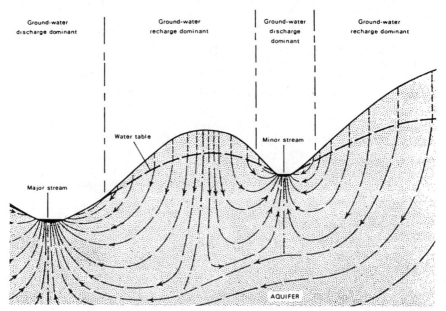

Figure 2.1. Typical cross-sectional view of ground water flow net in vicinity of major and minor stream valleys.

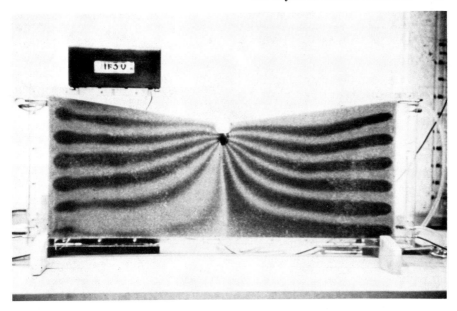

Figure 2.2. Hydraulic sand tank model illustrating flow to a perennial gaining system.

Figure 2.2 shows a sand model with a stream in the middle and the ground water moving toward the stream. The stream flow is perpendicular to the viewer. Flow lines are produced in the system by burying a perforated tube in the rock into which dye is injected. It shows laminar flow moving toward a stream, a very orderly process. It is a saturated system flowing everywhere, but you only see it where dye appears. You may imagine the flow lines to be plumes of pollution, working their way into a surface stream.

Variations in geology will alter the flow directions. Figure 2.3 shows the variation from moderately permeable sandstone to highly permeable. The permeability ratio between the two is about 8 to 1. The picture resembles the illustration in your high school physics book of Snell's Law, with light refracting as it passes from water to air. A similar kind of refraction characterized by an equation developed by a scientist at the U.S. Geological Survey, M. King Hubbert, takes place in this sandstone. Hubbert determined that ground water refracts as it crosses boundaries at other than a right angle and that it refracts according to a tangent law.

Figure 2.4 shows a clay layer on the right and a gravel lens on the left. The flow lines are affected by the impermeable clay layer on the right and the highly permeable zone in the middle. Lines 1 and 5 are drawn toward the high permeability. There is a refraction pattern going in and out of the gravel ellipse. The flow is slowed by the clay layer but not stopped. If we imagine these to be plumes of pollution, we can see why we must understand the hydrogeology.

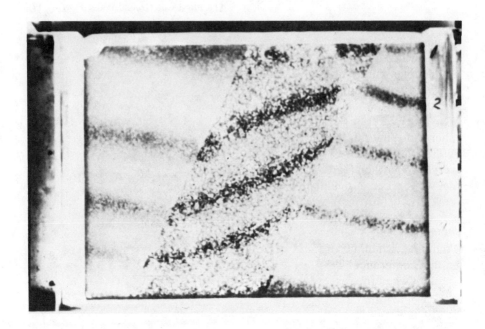

Figure 2.3. Hydraulic sand tank flow model illustrating the refraction of laminar ground water flow across boundaries between rocks of differing permeability.

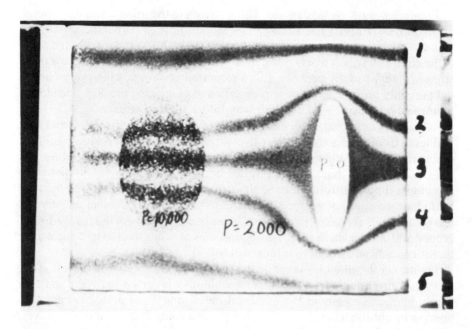

Figure 2.4. Hydraulic sand tank flow model illustrating flow aroiund and through lenses of varying permeability.

One of the things in our favor is the fact that there is little dispersion in ground water so that ground water mixing is limited. We can see in Figure 2.4 that lines 4 and 5 are very close to each other but do not converge. This is particularly important when we try to educate people about ground water pollution. If your well in the middle of the community is polluted, it does not mean that everybody else's well will become polluted but the public does not understand that at all. In Gray, Maine, a few years ago they had a ground water pollution scare with a few wells polluted, and the newspapers told everybody in the area to expect their wells to become polluted. How could one well become polluted without everybody's becoming polluted? We know the absurdity of that question, but we are not doing a very good job of educating the public in that area.

On that point, I really must pay an accolade to the U.S. EPA Lab at Ada, Oklahoma, which is responsible for the ground water research center holding this conference. Ten years ago the Ada lab was studying ground water pollution when the rest of the hydrogeology profession was still dealing with pump tests and step-drawdown tests, how to calculate transmissibility, and how to figure out for the umpteenth time an equation for a leaky aquifer. As Editor of *Ground Water* I still get an article every month from somebody in India who has figured out a new equation for calculating the transmissibility of a leaky aquifer. I do not want to read another one of those papers. Frankly, the day of research in quantitative hydrogeology is waning; we have enough tools to do that job. Tranpsort and fate of pollutants are our current areas of concentration, but the Ada laboratory was studying them 10 years ago. At that time I wanted to tell them they did not know what they were doing, but they were paying the bills. Now I can pay them a compliment for having had a lot more foresight than I and most other hydrologists. On the other hand, it is possible that the attention ground water pollution is now receiving needs modifying by means of some realistic estimate of the total impact of the problem.

Figure 2.5 shows a model of a hazardous-waste site, a pit we are dumping our garbage into. Whether it is tin cans, leaking or liquid waste put into a pond with the idea that it will just sit there, the pollution will move out of the ground as shown. It is not uncommon to have the water table rise above the bottom of a pit, pond, or lagoon, in which case we get saturated flow out of the pit and not vertical flow downward. The flow follows standard flow net theory and moves at right angles to the equal/potential lines. Moving in all directions as well as down, eventually it will change direction and move with the hydraulic gradient.

Figure 2.6 shows a hazardous-waste site, represented by the little circle in the middle of this plan view. The pollutant is moving into the ground and is flowing in all directions backward, right, left, and forward until it reaches some kind of equilibrium in the direction of the flow path. The dimensions of that plume of pollution are normally measured in hundreds of feet in width

Figure 2.5. Hydraulic sand tank flow model illustrating flow from a waste-disposal pit in which the water table has risen to the base of the pit.

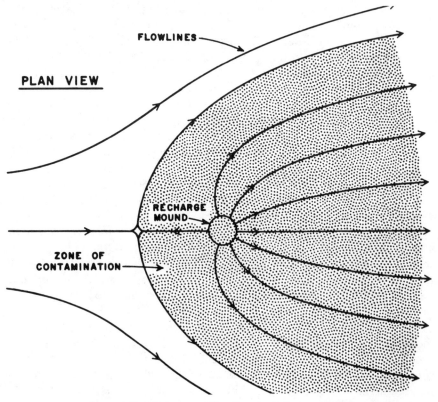

Figure 2.6. Plan view of flow from hazardous-waste site.

CROSS SECTION

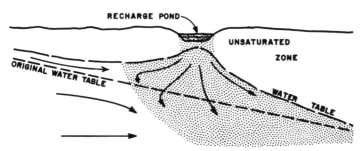

Figure 2.7. Cross-sectional view of flow from hazardous-waste site.

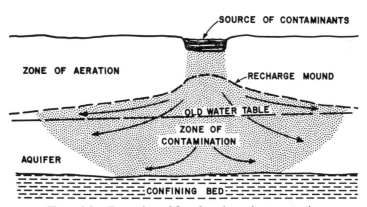

Figure 2.8. Front view of flow from hazardous-waste site.

and in tens of feet in thickness. Figure 2.7 shows what a cross section in the direction of flow might look like and Figure 2.8 shows what it might look like coming at you. Assuming a fairly thin permeable aquifer sitting on a less permeable zone, your plume is going to do one of three things. If it is lighter than the ground water that it is going into, it will either float or merge with the upper zone of the ground water system. If it is more dense, it will move downward until it stops at a less permeable zone; then it will start moving with the direction of flow. It may have some intermediate destination and find neutral buoyancy somewhere between the bottom and top of the aquifer.

THE BOTTOM LINE

You have heard and read from a variety of sources that less than 1% of the nation's ground water supply is polluted. This figure has been used by numerous independent hydrogeologists, and even officials of the U.S. Geological Survey. I will now prove that they are absolutely right.

POLLUTION ASSUMPTIONS

First I will make some assumptions. Let us assume that if we take all point sources of pollution—lumping together the septic tanks and the landfills, the pits, ponds, and lagoons, and other such nefarious improperly designed waste disposal techniques—there are 200,000 sources of pollution oozing pollutants into the ground. The EPA and the news media talk only in terms of 75,000 to 100,000 industrial sites, but I am including all other more innocuous point sources of pollution and calling the total 200,000. Let us further assume that each is creating a plume of pollution 1000 ft wide and 100 ft thick. Let us assume that each plume moves about 4 in./day or roughly 125 ft/yr (faster than most hydrogeologists would believe), and that each has been traveling for 40 years since its initial emplacement (a longer history than most have). Forty years at 125 ft/day is 5000 ft. Thus, each of these 200,000 plumes has moved about a mile from its source. Let us assume that they are polluting aquifers that are 25% porous.

ADDING IT UP

How much water have we polluted? 200,000 times width (= 1000 ft) times thickness (= 100 ft) times length (= 5000 ft) times porosity (= 0.25 equals 25,000,000,000,000 ft^3). Converting cubic feet to gallons (7.5 gal/ft^3, but rounding to 8 for grocery store simplicity) we arrive at 200,000,000,000,000 gallons—200 trillion gallons of polluted water.

AVAILABLE GROUND WATER

How much water do we have? Any grocery store clerk can figure out how much ground water we have in the United States, but few have done it. The area of the conterminous United States which is 3000 miles wide and about 1000 miles long is 3,000,000 mi^2. How much ground water is contained therein depends on how deep we are willing to drill. We will make it simple and say one-third of a mile (1760 ft) is as deep as we will drill in the future. You may not want to drill a well one-third of a mile deep, but what choice are you going to have 100 years from now? We did not want to pay a buck and a half for a gallon of gas either. On the other hand, we will ignore all water below one-third of a mile, even though we recognize it represents 60% of the earth's ground water reserves. Now, how much ground water can we get out of the upper third of a mile on the earth's crust? That depends on its specific yield, or quantity of water expressed as a percentage of the total rock volume that will drain out under the influence of gravity. While it varies widely between the sedimentary, igneous, and metamorphic rocks that make up the earth's crust, I will assume a national average of 10%, which most hydrogeologists would agree is a conservative figure. Thus 3,000,000 mi^2 in area times one-third of a mile in depth is 1,000,000 mi^3 of rock, times 10% specific yield, gives us 100,000 mi^3 of water. Using a rounded mile of 5000 ft, we get 125,000,000,000 ft^3/mi^3. Thus 100,000 mi^3 contain 12,500,000,000,000,000 ft^3 of ground water. Then converting to gallons at 8 gal/ft^3, we arrive at 100,000,000,000,000,000 or 100 quadrillion gallons of water. Thus, since we have 2×10^{14} (200 trillion) gallons of ground water polluted, the percentage of ground water already polluted is $2 \times 10^{14} \div 10^{17}$, which equals 0.002, or two-tenths (0.2) of 1%.

ASSUMPTIONS CHALLENGED

If you insist that the average plume is 1500 ft wide, then we have polluted three-tenths (0.3) percent. If you also want to say that it is 150 ft thick, then we have polluted 0.45%. If you believe that average plume has gone 1.5 mi instead of 1 mi, then we have polluted 0.675%. If you insist the specific yield is closer to 5% than 10, then we have polluted more than 1%, and if you further want to decide that there are 300,000 of these plumes, then the pollution does reach 2%. But no matter how liberally or conservatively we consider the problem we are dealing with a very small fraction of our ground water.

ALTERNATE CALCULATIONS

Attempting the calculation from another direction, EPA estimates of known pollution sources leaking into the ground average about 1.5 trillion gallons per year. If we double that to account for unknown sources, we get 3×10^{12} (three trillion) gallons leaking in every year. Let us assume it has been leaking in for $33\frac{1}{3}$ years (a nice round number, clearly in the ballpark). If 3 trillion gallons have been leaking into the ground for $33\frac{1}{3}$ years, then we have 1×10^{14} (100 trillion) gallons of polluted ground water, which is only half as much as we calculated using our 200,000 sites. Similarly, if we use USGS estimates of ground water availability (in the neighborhood of 200 quadrillion gallons), the percentage of polluted ground water drops even further.

AVERAGES DO NOT ALWAYS COUNT

I believe I have been conservative on all my numbers. It is therefore reasonable to say we have not polluted even 1% of our ground water system. The part you can argue with is whether we have polluted the ground water unevenly and concentrated our misdeeds where our dependence on it is greatest. In the industrial sections of New York, New Jersey, and New England, we have indeed polluted far more than 1%. They have very serious problems, but there is still a considerable amount of unpolluted ground water available in these areas as well.

Additionally, we will likely uncover many more yet unknown locations of severe ground water pollution. However, one should not be led to a conclusion that they are ubiquitous, for they are not.

WHAT DOES IT ALL MEAN?

My primary point is that we have yet to pollute 1% of our total water supply. We can eliminate the initiation of new sources within 10 years and create a public ethic and morality which will put an end to all such ill-conceived practices. The existing plumes will not have moved much further or polluted much more. The odds are in our favor to keep 98% of our ground water still unpolluted at the end of this century.

It will frequently cost us more to get because we may have to drill at greater distances from the point of use and to greater depths, but that is still a small price to pay for such a precious resource.

If we start work now, we can stop polluting ground water; we can keep the polluted percentage low. Additionally, we can restore some aquifers, treat some of the water, and reduce the movement of some hazardous plumes, even though we may have to abandon other areas.

THE BOTTOM LINE

Finally, we have to tell this story to the public before it says "What a shame we didn't discover ground water before it was polluted." Ten years ago the public did not know there was such a thing as ground water. They drilled a well only on the judgment of the court of last resort.

It is in the best interest of the public to understand the situation better, specifically that ground water pollution, while serious, is not quite the crisis some lead us to believe. We have a problem and we want action; we want the press and the public to keep arguing for ground water cleanup. But we do not need any Chicken Littles telling us that the sky is falling, because it most certainly is not.

PART ONE

SOURCES, TYPES, AND QUANTITIES OF CONTAMINANTS IN GROUND WATER

3

OVERVIEW OF CONTAMINANTS IN GROUND WATER

B. C. J. Zoeteman

National Institute for Public Health and Environmental Protection, Leidschendam, The Netherlands

Increased research activities in the field of ground water pollution have resulted in a recent growing awareness of the large number of cases of soil contamination. It seems that only very recently the total magnitude and impact of the worldwide soil pollution problem is becoming visible.

To deal with these soil problems, research programs and legislative instruments similar to those developed during the last decade for water contamination problems are needed. After the present phase of collecting data on the occurrence of chemicals in soil and ground water and on the sources of contamination, the next step inevitably is to answer the question of which cases need to be handled with the highest priority and for what reasons. For this latter purpose a better insight into the possible effects of soil contamination on the soil ecosystem and on public health has to be developed in conjunction with methods to predict the transport and behavior of ground water contaminants.

Many symposium papers discuss in detail the analytical aspects of detecting contaminants in ground water and the behavior in the soil. This paper will cover these items briefly and will also focus on effects on soil ecology and health. Finally, indications for future research needs are delineated.

TYPES AND QUANTITIES OF CONTAMINANTS

Pathogenic Organisms

Contamination of ground water by pathogenic organisms most frequently causes problems in situations where private wells and poorly constructed septic tanks are in proximity (Craun et al., 1976). Because of the filtering capability of soil, most of the pathogenic organisms present in sewage, including bacteria, viruses, protozoa, and parasitic worms, are effectively removed. Due to their small size, viruses can penetrate into the ground water most extensively, and their removal from the water phase is dependent on adsorption (Gerba and Keswick, 1981). Adsorbed viruses, however, are not devitalized and may create health risks later on. Data on levels of enteric bacterial pathogens and viruses in contaminated ground waters are very scarce.

Inorganic Contaminants

Contaminated ground water generally shows increased levels of chloride, sulfate and sodium ions. Depending on the redox potential, high levels of nitrate or ammonia are also reported. Elevated nitrate levels up to 50–100 mg NO_3/L are not exceptional in contaminated ground waters (Csáki and Endrédi, 1981; Zoeteman et al., 1981). Leachate of domestic waste tips may contain concentrations in the order of g/L of sodium, potassium, ammonia, chloride, and carbon dioxide/bicarbonate (Kooper et al., 1981).

Levels of metals in contaminated soil and ground water can also reach very high values near dumps of specific waste materials and under conditions of low pH and low redox potential. Matthess (1981) reported high arsenic levels up to 56 mg/L in contaminated ground water with a low redox potential. Many waste tips contain cyanides in combination with heavy metals; an example has been reported by Kakar and Bhatnagar (1981) in India, where levels up to 12 mg/L chromium (VI) and up to 2 mg/L of cyanide were found. Disposal of pyritic coal wastes in Illinois resulted in acidic leachates containing up to 30 g/L of sulfate, 1400 mg/L of iron, 1500 mg/L of zinc, and 20 mg/L of cadmium (Schubert and Prodan, 1981). Typical levels of metals in domestic waste dump leachate are as illustrated in Table 3.1. Unlike most of the organic and the microbiological contaminants, many metals tend to accumulate in the top layer of the soil, which aggravates their effect on the soil ecosystem, as will be pointed out later. Complexation of metals, by fatty acids for example, generally mobilizes them only for a short period, after which the complexing agent is degraded and the metals precipitate as carbonates or sulfides, or are strongly adsorbed (Loch et al., 1981).

Organic Contaminants

The detection of organic contaminants in ground water has been mainly responsible for the growing awareness of soil contamination. Like chlorides and related inorganic ions, organic contaminants are often easily transported through the soil. Another factor of interest is that organic compounds that travel most rapidly through the unsaturated or the saturated zone are often the compounds suitable for detection by routine gas chromatography-mass

Table 3.1   Heavy Metals Detected in Leachate of a Domestic Waste Dump at Noordwijk, The Netherlands[a]

Element	Concentration Range (μg/L)
Arsenic	10–60
Cadmium	2–8
Cobalt	10–80
Copper	20–90
Iron	50,000–60,000
Lead	50–100
Manganese	1,000–2,000
Nickel	40–80
Zinc	100–600

[a] Bom, 1981.

spectrometry techniques. Finally, the detectability by the sense of smell and the toxic properties of a number of organic chemicals contributed to the widespread interest in these chemicals, since they were found to be ubiquitously present.

Compounds like chlorinated hydrocarbons (chloroform, carbon tetrachloride, trichloroethylene, dichlorobenzenes), and aromatic hydrocarbons (benzene, alkylbenzenes, naphthalene), have been reported most frequently near contaminated sites (Giger and Schaffner, 1981; Zoeteman et al., 1981). Maximum concentrations detected in ground water of The Netherlands range from 100–600 μg/L for the alkylbenzenes and from 10–3000 μg/L for halogenated hydrocarbons. Due to their high water solubility and resistance to biological or chemical degradation, these volatile chemicals remain in ground water for many years, whereas they evaporate rapidly from surface water. The behavior of these chemicals in the soil has been studied near sites for ground water recharge with reclaimed water (Roberts and Valocchi, 1981), and during bank filtration (Piet and Zoeteman, 1980; Schwarzenbach and Westall, 1981). It is generally found that sorption is the main process controlling the behavior of these volatile non-polar compounds in the soil. Retardation of these chemicals during ground water movement is related to the carbon content of the solid phase and the octanol-water partition coefficient of the chemical.

Besides these compounds a large group of more polar, halogenated substances can be present in ground water, as detected by the AOCl and EOCl. The identity of these substances is not completely known. Known substances of this category, which are easily transported in ground water, are bis(-2-chloroisopropyl)ether and trifluoromethylaniline (Zoeteman et al.,

1980). Also, ground water contamination by relatively non-polar pesticides has been reported, such as a case of atrazine contamination (up to 2 μg/L) in Platte Valley, Nebraska, USA (Wehtie et al., 1981), and of 1,2-dibromo-3-chloropropane (up to 60 μg/L) in California (Nelson et al., 1981). In contaminated anaerobic ground waters organic sulfides have been found, such as dipropyldisulfide at 1 μg/L in leachate of a waste tip (Zoeteman et al., 1981), and chlorinated dialkylsulfides, which are supposed to be reaction products of alkylbromides and hydrogen sulfide (Giger and Schaffner, 1981).

SOURCES OF CONTAMINANTS

Although common sources of contamination, like uncontrolled leakage from septic tanks, agricultural land use, surface water infiltration, and contaminants present in rainwater, need our full attention, the major concern of soil protection authorities goes at present to the overwhelming number of sites where chemical wastes of an affluent society have been carelessly dumped.

In a small country like The Netherlands an inventory of existing waste sites has revealed the astonishing number of 3857 registered locations of chemical waste dumps. About 1% of these sites need immediate clean-up

Table 3.2   Classification of Types and Sources of Soil Contamination in The Netherlands Based on a Sample of 100 Cases[a]

Source of Contamination	Type of Contamination	Frequency (%)
Gasworks	Aromatic hydrocarbons, phenols, CN^-	45
Waste dumps and land fills	Halogenated hydrocarbons, alkyl-benzenes; metals like As, Pb, Cd, Ni; CN^-; pesticides	26
Chemicals production and handling sites (including painting industries and tanneries)	Halogenated hydrocarbons, alkyl-benzenes; metals like Pb, Cr, Zn, As	13
Metal plating and cleaning industries	Tri- and tetrachloroethylene, benzene, toluene, Cr, Cd, Zn, CN^-	9
Pesticide manufacturing sites	Pesticides, Hg, As, Cu	4
Automobile service facilities (including gasoline storage tanks)	Hydrocarbons, Pb	3

[a] Luijten, 1981.

and soil protection measures to reduce the risks of further damage to the ecosystem and to human health. About 30% of the sites need to be investigated in more detail to identify the types of chemicals present and to establish the priority for future clean-up activities. A sample of 100 well-defined cases showed the important role of old gas works, as indicated in Table 3.2.

Serious cases of soil contamination are found near pesticide manufacturing sites and certain chemical waste dumps where elevated levels of polychlorinated benzenes, chlorophenols, and TCDD have been detected (Wegman et al., 1981). Among the diffuse sources of ground water pollution, the use of fertilizers in agriculture deserves special attention, as their use is of increasing importance to soil pollution on one hand while a better optimization of the dose applied is often achievable. Not only are salt and nitrate levels increasing in agricultural areas, but pesticide and pesticide metabolites can also be found in ground water. According to a study of Bignoli and Sabbioni (1981), metals like Cd, Se, Mo, and U (present as contaminants in fertilizers), may ultimately also create ground water quality problems.

PRIORITY ASSESSMENT OF SOIL POLLUTION CASES

Natural Levels of Soil Constituents

To assess the importance of soil pollution cases, one first has to establish whether the type and concentration of water constituents is indeed different from natural levels, which vary depending on the nature of the soil. Ranges of the content of various elements in uncontaminated soil have been compiled by Lindsay (1979). Values of interest in relation to soil pollution are shown in Table 3.3. Natural levels of these elements in ground water are, roughly speaking, lower by a factor of 1000. Levels of organic substances in uncontaminated ground waters generally are below 0.1 μg/L, as illustrated in Table 3.4. If higher levels than indicated in Tables 3.3 and 3.4 are present, the seriousness of the contamination must be evaluated on the basis of the following aspects:

1. Effects on the soil ecosystem.
2. Direct or indirect effects on human health.
3. Effects on agriculture and cattle breeding.
4. Effects on the aquatic environment and related usages.

Based on the earlier mentioned survey in The Netherlands, different types of usage affected by soil contamination could be identified as important aspects in the priority assessment (see Table 3.5).

Table 3.3   Content of Some Elements in Uncontaminated Soils[a]

Element	Common Range in Soils (mg/kg)
Ag	0.01–5
As	1–50
B	2–100
Ba	100–3,000
Be	0.1–40
Br	1–10
Cd	0.01–0.7
Co	1–40
Cr	1–1,000
Cu	2–100
Hg	0.01–0.3
Mo	0.2–5
Ni	5–500
Pb	2–200
Se	0.1–2
Sn	2–200
Ti	1,000–10,000
V	20–500
Zn	10–300

[a] After Lindsay, 1979

Effects on the Soil Ecosystem

The area of soil toxicology is at present practically unexplored. It is not unlikely that the application of standards to protect soil organisms will be sufficiently strict to result also in an effective protection of man against unacceptable exposure to soil contaminants.

People have long been aware of effects of fertilizers and irrigation water on crops. Apart from the direct effects on agricultural production, very little is known of the effects of chemicals on soil organisms such as bacteria, fungi, protozoa, worms, and higher organisms in the soil food chain. As stated before, unlike many organic compounds, metals are strongly bound in the top layer of the soil, and once introduced, may stay there forever. Their presence, in combination with other stress factors such as organic chemicals, extreme temperatures, and extreme drought, will affect soil ecology in a number of ways. At increased levels they reduce the activity of enzymes involved in soil respiration, N-mineralization, and C-mineralization apart from the other toxic effects on individual organisms. Doelman (1981), using

Table 3.4   Content of Some Organic Compounds in Eight Uncontaminated Ground Waters in The Netherlands[a]

Compound	Odor Threshold Concentration (μg/L)	Range in Ground Water (μg/L)
Cyclohexane	200,000	<0.005–0.1
Benzene	10,000	<0.01–0.03
Toluene	1,000	0.01–0.1
Xylenes	1,000	0.01–0.1
Ethylbenzene	100	0.005–0.03
Styrene	50	<0.005
Naphthalene	5	<0.005–0.03
Fluoranthene	>100	<0.005–0.03
Pyrene	>1,000	<0.005–0.1
3,4-Benzopyrene	>1,000	<0.005
Carbon tetrachloride	300	<0.01–0.1
Tetrachloroethylene	300	<0.01–0.1
Dichlorobenzenes	1	<0.005–0.01
Hexachlorocyclohexanes	1	<0.01–0.01
p,p'DDE	300	<0.01–0.01

[a] After Zoeteman, 1980.

five soil types and an exposure period of 1 year, determined the metal concentration that reduced the activity of five soil enzymes by 10%. Using the most sensitive criteria and applying a safety factor of 10 to account for additional stress factors, he arrived at the following limits for the six metals studied (Table 3.6). Comparison of these data with those of Table 3.3 shows that these levels are similar to those occurring in most uncontaminated soils.

Table 3.5   Soil Functions Affected by Contamination in The Netherlands Based on a Sample of 100 Cases

Soil Function	Percentage of Cases
Use for housing	45
Planned use for housing	21
Ground water catchment area	20
Camping, recreation	4
Gardening	1
Grassland used for dairy cattle	3
Drainage to surface water	6

Table 3.6 Acceptable Concentrations of Six Metals in Soil, Based on a Reduction of Soil Enzyme Activity of 10% or Less and a Safety Factor of 10[a]

Metal	Concentration (mg/kg soil)
Cd	1
Cr	10
Cu	50
Ni	25
Pb	100
Zn	50

[a] Doelman, 1981

This means that for many soils, any increase of the metal contamination is likely to reduce the microbiological activity.

Similar data for organic chemicals are needed, in combination with a more complete battery of soil toxicity tests. Furthermore, the impact of chemical speciation of metals on their behavior and biological effect is yet unclear.

Effects on Human Health

Ultimately, soil and water ecology requirements should be essential criteria in soil protection measures. However, it is the risk of exposure of the population to dangerous chemicals that presently plays the primary role in setting priorities for the further handling of soil pollution sites. From this point of view, mobile toxic chemicals are of the most concern. The possible routes of exposure include skin contact, inhalation, and ingestion via drinking water contaminated by substances such as agricultural products (Figure 3.1). The risk of skin contact, which may occur with children when waste materials have been dumped near the surface in residential areas, of course necessitates immediate remedial action.

Inhalation may occur when waste chemicals evaporate. Generally this is first noticed by an offensive odor in the area. Sometimes such odors have been noticed in houses that may have been built on former illegal waste-disposal sites. Generally chemicals can be smelt at much lower concentrations than concentrations which are acutely toxic (Zoeteman, 1980). Therefore, a general rule should be that in those areas where people recreate or live, the odor of volatile chemicals should not be noticeable. Concentrations of chemicals in ground water should be at least below their odor threshold concentration in such areas. For some organic contaminants the available odor threshold concentrations (Zoeteman, 1980) have been additionally

EFFECT ON HUMAN HEALTH

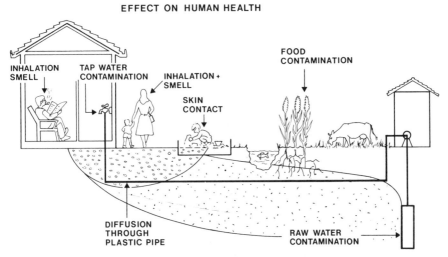

Figure 3.1.   Effect on human health.

listed in Table 3.4. From these data it is evident that odor problems can be related to compounds like styrene, naphthalene, and chlorinated benzenes.

Existing criteria-documents for acceptable daily intakes can be used to assess the acceptability of the presence of a number of chemicals in contaminated drinking water or agricultural food products destined for human consumption. Obviously, the presence of a water catchment area for a public water supply or the use of a contaminated site for food crops or cattle breeding requires special precautions and monitoring procedures, including application of an integrated assessment of the various routes of exposure to the chemical.

Generally, highest priority will be given to the clean-up of those contaminated sites where chemicals are relatively toxic (or odorous) and chemicals are easily transported (e.g., due to complexing agents, high permeability of soil, etc.). Local factors that might raise the priority are (1) presence of drinking water distribution pipes (e.g., PVC) that are permeable for certain waste chemicals in the contaminated soil, (2) proximity of a ground water catchment area that is or may be affected, and (3) (future) presence of houses on the dump site or landfill.

RECOMMENDATIONS FOR FUTURE RESEARCH

Based on the foregoing considerations the following recommendations for future research can be derived.

Research in the area of the *occurrence* of soil and ground water contamination should focus on:

1. Metal speciation.
2. Identification and quantification of more polar, halogenated contaminants which escape present analytical methods.
3. Application of overall-effect-detecting techniques such as odor assessment and mutagenicity screening tests, on the basis of which advanced analytical techniques can be used to identify the compounds of greatest concern.

Research in the area of *transport and transformation* of soil contaminants seems to have advanced well in relation to adsorption processes of inorganic and relatively non-polar organic compounds. Future emphasis is needed on areas such as:

1. Microbial degradation of organic chemicals under aerobic and anaerobic conditions.
2. Production of metabolites (e.g., from certain pesticides).
3. Prediction techniques, based on mathematical models for the half-life and/or retardation factor of chemicals introduced with the infiltrating rain water or surface water into the soil.

Both soil research areas are at present undergoing rapid development and should in the near future be expanded up to the point where exposure data can be linked to data on human and ecological *effects*. In my opinion, the study of effects on the soil ecosystem is the most important research area for the coming years, as I briefly indicated in the paragraph on effects on the soil ecosystem.

After sufficient ecotoxicity data have been obtained for the soil ecosystem, research can explore emission of contaminants to the soil surface in relation to exposure levels, and biological effects on the most critical soil organisms. It is evident from this brief outlook into the future that we stand at present only at the beginning of a complex and fascinating field of research.

ACKNOWLEDGMENT

The author wishes to thank Dr. Ir. J. A. Luyten and Dr. Ir. J. P. G. Loch for critically reviewing the text.

REFERENCES

Bignoli, G. and Sabbioni, E. (1981). Long term prediction of the potential impact of heavy metals on groundwater quality as a result of fertilizer use. In: W. van Duijvenbooden, P. Glasbergen, and H. van Lelyveld (Eds.), *Quality of Groundwater*. Elsevier, Amsterdam, pp. 857–862.

Bom, C. M. (1981). Evaluation of different analytical procedures in the trace element analysis of a landfill leachate. In: W. van Duijvenbooden, P. Glasbergen, and H. van Lelyveld (Eds.), *Quality of Groundwater*. Elsevier, Amsterdam, pp. 495–499.

Craun, G. F., McCabe, L. J., and Hughes, J. M. (1976). Waterborne disease outbreaks in the U.S. 1971–1974. *J. Am. Water Works Assoc.* **68**:420–424.

Csáki, F. and Endrédi, I. (1981). Nitrate contamination of underground waters in Hungary. In: W. van Duijvenbooden, P. Glasbergen, and H. van Lelyveld (Eds.), *Quality of Groundwater*. Elsevier, Amsterdam, pp. 89–94.

Doelman, P. (1981). The effect of heavy metals on soil microflora. Internal report of the National Institute for Nature Conservation, Leersum, The Netherlands.

Gerba, C. P. and Keswick, B. H. (1981). Survival and transport of enteric viruses and bacteria in groundwater. In: W. van Duijvenbooden, P. Glasbergen, and H. van Lelyveld (Eds.), *Quality of Groundwater*. Elsevier, Amsterdam, pp. 511–515.

Giger, W. and Schaffner C. (1981). Groundwater pollution by volatile organic chemicals. In: W. van Duijvenbooden, P. Glasbergen, and H. van Lelyveld (Eds.), *Quality of Groundwater*. Elsevier, Amsterdam, pp. 517–522.

Kakar, Y. P. and Bhatnagar, N. C. (1981). Groundwater pollution due to industrial effluents in Ludhiana, India. In: W. van Duijvenbooden, P. Glasbergen, and H. van Lelyveld (Eds.), *Quality of Groundwater*. Elsevier, Amsterdam, pp. 265–272.

Kooper, W. F., van Duijvenbooden, W., and Peeters, A. A. (1981). Invloed van vuilstortpercolaat te Noordwijk op de bodem en het oppervlaktewater. Internal Report of the National Institute for Water Supply, The Netherlands, No. HyH 81-13:11.

Lindsay, W. L. (1979). *Chemical Equilibria in Soil*. Wiley, New York, p. 7.

Loch, J. P. G., Lagas, P., and Haring, B. J. A. M. (1981). Behaviour of heavy metals in soil beneath a landfill: Results of model experiments. In: W. van Duijvenbooden, P. Glasbergen, and H. van Lelyveld (Eds.), *Quality of Groundwater*. Elsevier, Amsterdam, pp. 545–555.

Luijten, J. A. (1981). Unpublished data.

Matthess, G. (1981). In site treatment of arsenic contaminated groundwater. In: W. van Duijvenbooden, P. Glasbergen, and H. van Lelyveld (Eds.), *Quality of Groundwater*. Elsevier, Amsterdam, pp. 291–296.

Nelson, S. J., Iskander, M., Volz, M., Khalifa, S., and Haberman, R. (1981). DBCP contamination of groundwater in California. In: W. van Duijvenbooden, P. Glasbergen, and H. van Lelyveld (Eds.), *Quality of Groundwater*. Elsevier, Amsterdam, pp. 169–174.

Piet, G. J. and Zoeteman, B. C. J. (1980). Organic water quality changes during sand bank and dune filtration of surface waters in The Netherlands. *J. Am. Water Works Assoc.* **72**:400–404.

Roberts, P. V. and Valocchi, A. J. (1981). Principles of organic contaminant behavior during artificial recharge. In: W. van Duijvenbooden, P. Glasbergen, and H. van Lelyveld (Eds.), *Quality of Groundwater*. Elsevier, Amsterdam, pp. 439–450.

Schubert, J. P. and Prodan P. F. (1981). Groundwater pollution resulting from disposal of pyritic coal wastes. In: W. van Duijvenbooden, P. Glasbergen, and

H. van Lelyveld (Eds.), *Quality of Groundwater*. Elsevier, Amsterdam, pp. 319–327.

Schwarzenbach, R. P. and Westall, J. (1981). Transport of non-polar organic compounds from surface water to groundwater: Laboratory sorption studies. In: W. van Duijvenbooden, P. Glasbergen, and H. van Lelyveld (Eds.), *Quality of Groundwater*. Elsevier, Amsterdam, pp. 569–574.

Wegman, R. C. C., Bank, C. A., and Greve, P. A. (1981). Environmental pollution by a chemical waste dump. In: W. van Duijvenbooden, P. Glasbergen, and H. van Lelyveld (Eds.), *Quality of Groundwater*. Elsevier, Amsterdam, pp. 349–357.

Wehtje, G., Leavitt, J. R. C., Spalding, R., Mielke, L., and Schepers, J. (1981). Atrazine contamination of groundwater in the Platte Valley of Nebraska: Non-Point sources. In: W. van Duijvenbooden, P. Glasbergen, and H. van Lelyveld (Eds.), *Quality of Groundwater*. Elsevier, Amsterdam, pp. 141–145.

Zoeteman, B. C. J. (1980). *Sensory Assessment of Water Quality*. Pergamon Press, Oxford.

Zoeteman, B. C. J., Harmsen, K. H., Linders, J. B. H. J., Morra, C. F. H., and Slooff, W. (1980). Persistent organic pollutants in river water and ground water of The Netherlands. *Chemosphere* **9**:231–249.

Zoeteman, B. C. J., de Greef, E., and Brinkman, F. J. J. (1981). Persistency of organic contaminants in groundwater, lessons from soil pollution incidents in The Netherlands. In: W. van Duijvenbooden, P. Glasbergen, and H. van Lelyveld (Eds.), *Quality of Groundwater*. Elsevier, Amsterdam, pp. 465–476.

4

CHEMICAL CONTAMINATION OF GROUND WATER

David W. Miller

Geraghty & Miller, Inc.
Syosset, New York

Federal and state laws designed to protect ground water have focused on landfills and surface impoundments at industrial facilities where chemicals have been stored or disposed of on the land. However, landfills and waste water impoundments may be only the most obvious landmarks, not the only sources of ground water contamination. Other industry-related sources include chemical leaks from storage areas, accidental spills, and vapor condensate from solvent-recovery systems. Increasing regulation by local, state, and federal agencies has effectively barred the use of many traditional disposal facilities that were available to industry, and has led in turn to poor housekeeping and dependence on unsuitable sites within plant boundaries.

Nonindustrial sources of ground water pollutants include road runoff, municipal landfills, junkyards, and domestic waste water. Household products contain many soluble organic chemicals that find their way into septic tanks, cesspools, and leaky sewer lines and eventually migrate to the water table. In fact, some products used for unclogging septic tank drain-fields contain industrial solvents suspected of being human carcinogens. Common commercial operations, such as automotive service, auto body repair, dry-cleaning, and printing, are often unsuspected contributors to ground water contamination. These less obvious sources can contribute substantially to local and regional ground water contamination.

Ground water is the source of drinking water for approximately half of the population of the United States. Except for chlorination, it is rarely treated, presumed to be naturally protected, and considered free of the impurities associated with surface waters because it comes from deep within the earth.

There are tens of thousands of community water-supply wells and millions of domestic wells in the nation. Until the Federal Safe Drinking Water Act (SDWA) was passed in 1972, there was no formalized national program for analyzing potentially toxic elements in public water supplies, including ground water. The Act requires a program of regular testing of public water supplies and wells, and establishes drinking-water standards to protect public health. Regulations under the Act still do not require analysis for most of the synthetic organic compounds associated with hazardous-waste sites, and there is no testing program for the domestic wells that serve some 40 million people in the United States.

The lack of monitoring at tens of thousands of sites where there is a potential for contamination, along with the lack of comprehensive analysis of water quality at hundreds of thousands of wells, rules out the possibility for a reliable determination of the extent and severity of ground water degradation and associated health risks in the United States. Segments of important aquifers are known to be degraded and may be essentially lost forever as sources of drinking water. Most important, some portion of our population may have been exposed to chemical contamination for an unknown period of time.

Most ground water contamination incidents are local, and affect only the uppermost aquifers. The area over which ground water quality is significantly degraded is typically less than a mile long and one-half mild wide, with pollutants moving at an average rate of less than 1 ft/day. On the other hand, the resource itself is huge. At almost any location, ground water can provide a supply sufficient for single-family domestic use, and more than one-third of the nation is underlain by aquifers capable of yielding at least 100,000 gal/day to an individual well. Although the specific volume of contaminated ground water may be only a very small percentage of the nation's ground water resource, the impact of this contamination can be very large. There is a critical need to identify and manage contamination sites both from the standpoint of protecting public health and preserving the resource.

MOVEMENT OF CONTAMINANTS IN THE GROUND WATER ENVIRONMENT

The wide range of contamination sources is one of many factors contributing to the complexity of a ground water quality assessment. It is important to know the geochemistry of the chemical–soil–ground water interactions in order to assess the fate and impact of chemicals discharged onto the ground. Chemicals will pass through several different hydrologic zones (Figure 4.1) as they migrate through the soil to the water table.

The water table (the level at which water stands in a shallow well) is the upper surface of the ground water system. The pore spaces between soil particles above the water table are occupied by both air and water (the

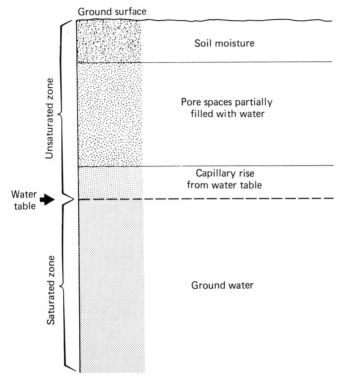

Figure 4.1. Relationship between unsaturated and saturated zones.

unsaturated zone). Flow in this zone is vertically downward, as liquid contaminants or solutions of contaminants and precipitation move under the force of gravity.

The uppermost region of the unsaturated zone (the soil zone) is the site of important processes leading to pollutant attenuation. Some chemicals are trapped in this zone by adsorption onto organic material and chemically active silt and soil particles; there they are decomposed through oxidation and microbial activity. Many end-products are taken up by plants or released into the atmosphere.

Below the soil zone, the pore spaces are also unsaturated, and as chemical-bearing precipitation percolates through this zone, oxidation and aerobic biological degradation continue to take place. Some chemicals are also adsorbed in this zone, and precipitates may be filtered out.

In the capillary zone, spaces between soil particles may be saturated by water rising from the water table under capillary forces. Certain chemicals that are lighter than water will "float" on top of the water table in this zone. These floating chemicals may move in different directions and at different rates than contaminants dissolved in the percolating recharge.

Once dissolved contaminants reach the water table, they enter the ground water flow system—the direction of which depends upon the hydraulic gradients. All pore spaces between soil particles below the water table are saturated. The relative unavailability of dissolved oxygen in the saturated zone limits the potential for oxidation of chemicals. Varying levels of attenuation may take place, depending on the geologic conditions.

Unlike the turbulent flow of surface water systems, ground water flow is laminar; particles of fluid move along distinct and separate paths, with little mixing occurring as the ground water moves (Figure 4.2). Dissolved chemicals in the saturated zone will flow with the ground water. The direction of flow is governed by hydraulic gradients, and ground water will move in response to differences in hydrostatic head.

The major components of the flow system are the recharge area (where flow is generally downward) and the discharge area (where flow may be generally upward) (Figure 4.3). The direction of flow in a shallow, local flow system could in some cases be opposite to flow in a deeper flow system. The ability of a monitoring well to detect the presence of a plume is therefore based on the location and depth at which the well is set (Figure 4.4). Knowledge of the flow system is an essential precondition in assessing chemical contamination problems, and a monitoring program implemented without adequate hydrogeological information can be very misleading.

Ground water flow rates in aquifers generally range from a few inches to a few feet per day. A body of contaminated ground water may contain the accumulation of decades of leachate discharge; and it may take many years for contaminants to be detected in a nearby water supply.

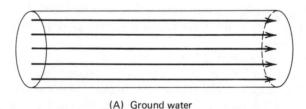

(A) Ground water

(B) Surface water

Figure 4.2.   (A) Flow paths of molecules of water in laminar flow. (B) Flow paths of molecules of water in turbulent flow.

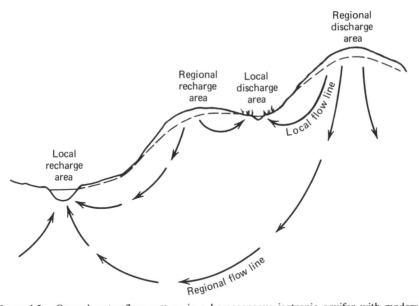

Figure 4.3. Ground water flow pattern in a homogeneous isotropic aquifer with moderate relief.

Because ground water flows in a laminar fashion, dissolved chemicals will follow ground water flow lines and form distinct plumes. Plumes of contaminated ground water have been traced from a few feet to several miles downstream of the pollution source.

The shape and size of a plume depends on a number of factors, including the local geologic framework, local and regional ground water flow, the type

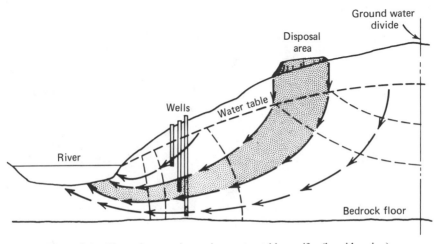

Figure 4.4. Flow of contaminants in a water table aquifer (humid region).

and concentration of contaminants, and variations in the rates of leaching. Figure 4.5 illustrates the shapes of two plumes of contamination in different geologic settings and the lengths of time it took for them to develop.

The fact that chemicals are attenuated in the soil through adsorption and chemical interaction with other organic and inorganic constituents of the aquifer (Figure 4.6) makes it difficult to predict the movement and fate of chemicals in the ground water. Volatile organic chemicals in ground water are extremely mobile while other chemicals are not so mobile. There are differences in attenuation through sorption, and some chemicals are less changed in the ground water environment than are others.

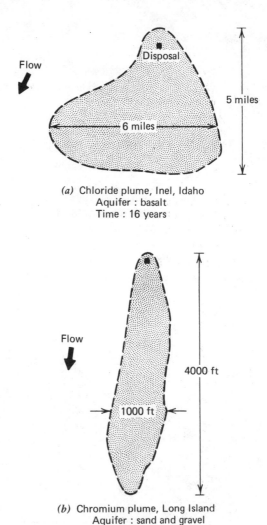

(a) Chloride plume, Inel, Idaho
Aquifer : basalt
Time : 16 years

(b) Chromium plume, Long Island
Aquifer : sand and gravel
Time : 13 years

Figure 4.5. Effect of differences in geology on shapes of contamination plumes.

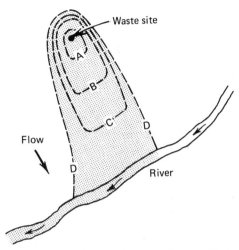

Figure 4.6. Schematic diagram showing areal extent of contamination by different specific contaminants A, B, C, and D in a mixed-waste plume in a water table aquifer.

The density of contaminated fluids is another important factor in the formation and movement of a plume. Some chemicals will tend to flow on top of the water table, while others will tend to sink to the bottom. Multiple discharges of different kinds of chemicals have led to a complex pattern of plumes at the site illustrated in Figure 4.7. Note that the figure shows a

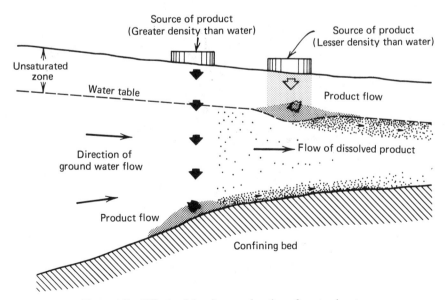

Figure 4.7. Effects of density on migration of contaminants.

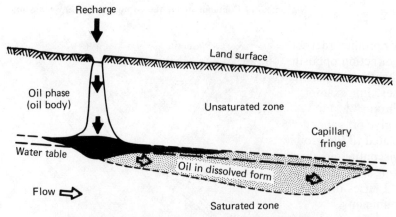

Figure 4.8. Petroleum product reaching ground water.

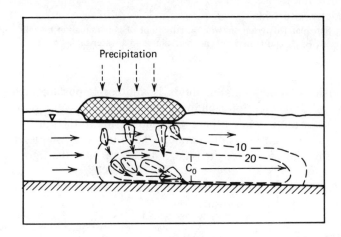

Explanation

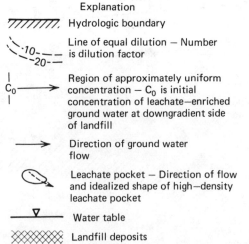

Hydrologic boundary

Line of equal dilution — Number is dilution factor

Region of approximately uniform concentration — C_0 is initial concentration of leachate—enriched ground water at downgradient side of landfill

Direction of ground water flow

Leachate pocket — Direction of flow and idealized shape of high—density leachate pocket

Water table

Landfill deposits

Figure 4.9. Leachate movement in ground water beneath a landfill.

heavy product (denser than water) flowing down the slope of a confining bed in a direction opposite to the flow of a dissolved and floating product.

Slightly soluble materials that are lighter than water flow in multiple phases. For example, oil will move as a body, its components flowing with the ground water system (Figure 4.8). In addition, the undissolved phase may give off vapors which migrate through the unsaturated zone in patterns unrelated to the ground water flow system.

Discontinuous discharges may result in "slugs" of contaminated water, causing wide spatial and temporal fluctuations in well-water quality (Figure 4.9). Lenses of sand and clay can cause other variations by stratifying the contaminants.

Surveys of ground water contamination are complicated by operating practices at waste disposal facilities as well as by natural phenomena. There can be numerous distinct plumes of contamination moving independently under a site where a variety of materials has been discharged from multiple sources. Pumping from wells can modify ground water flow patterns and consequently alter the movement of a contaminant plume (Figure 4.10). Furthermore, detailed monitoring of sites more than 5 years old has revealed fluctuations in the concentration of some constituents while other constituents have remained relatively constant. Thus the factors influencing movement of ground water and contaminants within aquifers are complex, and the investigation of ground water contamination can require extensive and costly work over a considerable time.

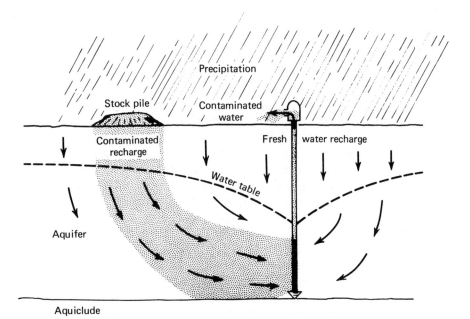

Figure 4.10.   Influence of pumping on plume migration.

INFORMATION GENERATED BY GEOHYDROLOGIC INVESTIGATIONS

The first step in a typical assessment of ground water quality at a waste facility suspected of causing contamination is a thorough survey of potential discharges into ground water in the vicinity of the site. This involves compiling a detailed history of the facility, including records of spills, leaks, and abandoned operations. Background information is extremely important because contaminants that may have entered the ground water system many years before can still be in the study area and be intercepted by monitoring wells installed as part of the investigation.

This preliminary work is generally followed by a hydrogeologic investigation, which would include soil borings and the installation of monitoring wells to yield information about water-table elevations, horizontal and vertical flow, and the extent and degree of ground water contamination. The final phase of a typical investigation will evaluate the project data to identify pollution sources, evaluate any threat to nearby water supplies, and recommend control or abatement alternatives.

Typical costs for determining ground water quality at an industrial disposal site where contamination has occurred can range from $25,000 to several hundred thousand dollars, depending on the nature and extent of the work. This effort may determine only the extent of contamination and its rate of movement. The investigation may not provide data sufficient to allow a confident prediction of the future movement of the plume or concentrations of particular chemicals. This analysis may require thousands of dollars more in field work, computer time, and specialized labor.

PROBLEMS WITH GEOHYDROLOGIC INVESTIGATIONS

The predominant reason that geohydrologic investigations fail is the lack of a clearly defined objective. The fault is shared by the regulators and the regulated, both of whom tend to embark on directionless studies amassing inconclusive data at tremendous cost. This problem is compounded by the fact that the state of the geohydrologic art is such that some crucial questions regarding the risk to public health cannot be answered with certainty. Plume shapes, lengths, rates of movement, and concentrations over time cannot always be evaluated accurately, even after extensive study. Lack of adequate information on the history of disposal and discharge concentration makes it impossible to determine how long the public has been exposed to a contaminant at a particular site.

Where multiple sources contribute to ground water degradation, the development of cause–effect relationships is extremely difficult. Finally, water-supply wells and monitoring wells tend to dilute contaminated water with large volumes of clean water from unaffected portions of the aquifer (Figure

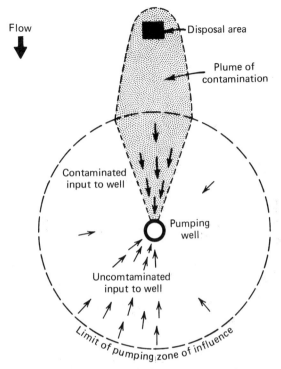

Figure 4.11.   Dilution effects of natural ground water on contaminants at a pumping well.

4.11). The resulting monitoring data may lead the investigator to underestimate public health threats.

Because of the extremely slow rate of ground water flow, contaminants may infiltrate a water supply aquifer for years before the pollution is revealed by monitoring wells. By the time contamination is detected in a well, the source may be modified or abandoned, and it will be difficult to establish a useful and cost-effective analytical protocol. Different analytic approaches may be required to identify the products of breakdown of an organic chemical in the ground water rather than in the material initially buried. Selection of drilling methods, well-screen and casing materials, and sampling techniques are also governed by the nature of the constituents of concern. For example, use of glues and plastic casings may interfere with analyses of volatile organics, and improper well construction may cause cross-contamination of aquifers.

The combined influences of geology and pumping can result in erratic variations in pollutant concentrations. In one case, nitrate concentrations have been found to fluctuate and decline with longer pumping periods (Figure 4.12). Initial concentrations were higher as pollutants were drawn from more permeable and more contaminated layers. Over longer periods of

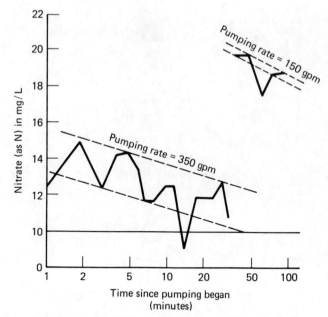

Figure 4.12.   Change in nitrate concentration at different pumping rates.

pumpage, uncontaminated water was drawn from zones of lower permeability. This explanation was confirmed when the well exhibited higher levels of nitrate under reduced pumping rates.

In another situation, organic chemical concentrations in a supply well were reduced substantially when the well was pumped for more than 10 days (Figure 4.13). In this case the well was receiving recharge from another (uncontaminated) source—a nearby river—after pumping for this period of time.

Sampling procedures and data analysis must be based on knowledge of pumping effects. The correct sampling procedure will depend upon the objective of the sampling program. Different approaches are required to determine what is in the aquifer at a specific site as opposed to what people are drinking out of water-supply wells.

Thus, the significance of a water sample from any particular well must be carefully evaluated because it may represent a segment of a very limited plume, or a slug of pollutant, or more widespread contamination. Fluctuations due to sampling methods (such as length of time the well is pumped prior to sampling), geohydrologic conditions, or intermittent discharge of contaminants into ground water, must be subjected to appropriate statistical analysis. Critical levels of many toxic pollutants are measured in parts per billion; therefore quality controls and carefully thought-out protocols for both field and laboratory procedures must be stringently enforced.

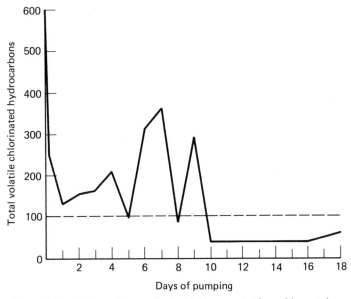

Figure 4.13.   Change in organic chemical concentration with pumping.

To improve the success of geohydrologic investigations, additional research and information exchange are required. Research should be diverted toward development of better surface investigative techniques, such as improved geophysical surveying, to avoid the high cost of drilling to obtain subsurface information. More information is needed on the fate and transport of toxic materials in the unsaturated zone and in ground water. Statisticians must provide better guidance on the nature and applicability of statistical analysis in ground water investigations.

REFERENCES

Braids, O. C., Wilson, G. R., and Miller, D. W. (1977). Effects of industrial hazardous waste disposal on the ground-water resource. In: *Drinking Water Quality Enhancement Through Source Protection,* Ann Arbor Science, Ann Arbor, MI.

Deutsch, M. (1963). Groundwater contamination and legal controls in Michigan. U.S. Geological Survey Water-Supply Paper 1691.

Fetter, C. W. Jr. (1980). *Applied Hydrogeology.* Charles E. Merrill, Columbus, OH.

Edward E. Johnson, Inc. (1966). *Ground Water and Wells.* Edward E. Johnson, Inc., St. Paul, MN.

Kimmel, G. E. and Braids, O. C. (1980). Leachate plumes in ground water from Babylon and Islip landfills, Long Island, New York. U.S. Geological Survey Professional Paper 1085.

LeGrand, H.E. (1965). Patterns of contaminated zones of water in the ground. *Water Resour. Res.* **1**:83–95.

Miller, D. W. (Ed.) (Jan. 1977). The Report to Congress on Waste Disposal Practices and Their Management Program. Office of Water Supply and Office of Solid Waste Management Program, U.S. Environmental Protection Agency, Washington, D.C. EPA-570/9-77-001.

Miller, D. W., DeLuca, F. A. and Tessier, T. L. (June 1974). Ground-water contamination in the northeast states. U.S. Environmental Protection Agency, Washington, D.C. EPA-660/2-74-056.

Schmidt, K. D. (1977). Water quality variations for pumping wells. *Ground Water* **5**(2):130–137.

5

MICROBIAL CONTAMINATION OF THE SUBSURFACE

Charles P. Gerba

Departments of Microbiology and Immunology and
Nutrition and Food Science
University of Arizona
Tucson, Arizona

It is estimated that septic tanks and cesspools contribute over 800 billion gallons of domestic wastes to ground water each year (Keeley, 1977). Leakage from municipal sewer systems and treatment lagoons contributes another 268 billion gallons (Keeley, 1977). In addition to these sources are direct injection and percolation of domestic wastewater from purposeful land application for ground water recharge and crop irrigation.

Domestic sewage almost always contains bacteria, viruses, protozoa, and helminths pathogenic to man. The most common bacterial species associated with sewage are *Salmonella* sp. (the cause of typhoid), *Shigella* sp., and *Vibrio* sp. (both associated with severe gastroenteritis). Over 100 different types of pathogenic viruses have been isolated from fecal material, including poliovirus, coxsackievirus, Norwalk agent(s), adenovirus, rotavirus, and hepatitis A virus. These viruses have been associated with a wide variety of diseases in man, including paralysis, gastroenteritis, hepatitis, meningitis, and eye infections. Parasitic protozoa and helminths are not a problem associated with ground water contamination because their large size results in highly efficient removal by filtration.

Bacterial and viral contamination of ground water is a serious problem that can result in large outbreaks of waterborne disease. From 1971 to 1978, there were 224 reported outbreaks of waterborne disease affecting 48,193 individuals (Craun, 1981). A pronounced increase in outbreaks of water-borne disease has occurred since 1970, and the United States now has an average of 34 reported outbreaks per year compared with 15 per year reported during the period from 1966 to 1970 (Craun, 1981). Almost half of these outbreaks occur because of contaminated ground water. Overflow or seepage of sewage from septic tanks and cesspools was believed responsible for over 40 percent of the outbreaks. However, the actual source of ground water contamination was proved in only 17 of the outbreaks (Wilson et al., 1983). Of the ground water-related disease outbreaks, almost 70% could be attributed to illness of probable viral etiology (hepatitis A, Norwalk agent, etc.) according to Wilson et al. (1983). Because of limitations in surveillance, the number of reported waterborne disease outbreaks is believed to be only a fraction of the total number that occur each year.

SURVIVAL AND TRANSPORT OF MICROORGANISMS IN THE SUBSURFACE

The fate of pathogenic bacteria and viruses in the subsurface will be determined by their survival and their retention by soil particles. Both survival and retention are largely determined by the three factors listed in Table 5.1. Climate will control two important factors in determining viral and bacterial survival: temperature and rainfall. The survival of microorganisms is greatly prolonged at low temperatures; below 4°C they can survive for months or even years (Gerba et al., 1975). At higher temperatures, inactivation or die-off is fairly rapid. In the case of bacteria, and probably viruses, the die-off rate is approximately doubled with each 10°C rise in temperature between 5 and 30°C (Reddy et al., 1981). Above 30°C, temperature is probably the dominant factor determining virus survival time. Adsorbed viruses are protected to some degree against thermal effects, so their survival is prolonged at higher temperatures (Liew and Gerba, 1980). Another principal factor determining survival of bacteria and viruses is soil moisture and drying.

Table 5.1   Factors Affecting Microbial Fate in the Subsurface

Effectiveness of pathogen removal is determined by:

1.   Survival
2.   Retention

which are largely determined by:

1.   Climate
2.   Nature of the soil
3.   Nature of the microorganism

Table 5.2   Factors Affecting Survival of Enteric Bacteria in Soil

Factor	Remarks
Moisture content	Greater survival time in moist soils and during times of high rainfall
Moisture-holding capacity	Survival time is less in sandy soils with greater water-holding capacity
Temperature	Longer survival at low temperatures; longer survival in winter than in summer
pH	Shorter survival time in acid soils (pH 3–5) than in alkaline soils
Sunlight	Shorter survival time at soil surface
Organic matter	Increased survival and possible regrowth when sufficient amounts of organic matter are present
Antagonism from soil microflora	Increased survival time in sterile soil

Survival is greater under saturated conditions; marked inactivation occurs during drying near the soil surface. Results of studies by Yaeger and O'Brien (1979) suggest that poliovirus and perhaps other viruses are inactivated by different mechanisms in moist soils versus drying soils.

The nature of the soil will also play a major role in determining survival. Soil properties influence moisture-holding capacity, pH, and organic matter—all of which will control the survival of bacteria in the soil (Table 5.2). Virus survival will also be influenced by these factors, especially pH, which will affect virus adsorption to soil. Results of recent studies indicate that adsorption of viruses to the soil surface plays a major role in determining viral survival time. Apparently, adsorption of virus to soil particles offers protection against factors responsible for viral inactivation in the soil. Soil pH, exchangeable aluminum, and resin-extractable phosphorus can be correlated with virus survival (Table 5.3), but all of these factors appear to be related to the degree of virus adsorption. Using these factors, a linear regression equation has been developed for comparing virus survival in different soil types (Hurst et al., 1980).

The nature of the microorganism will also play a role in survival. Resistance of microorganisms to environmental factors is dependent upon the species as well as the particular strain. Most enteric bacterial pathogens die off very rapidly outside the human gut, whereas indicator bacteria such as *Escherichia coli* will persist for longer periods of time. Survival times among different types of viruses vary greatly and are difficult to assess without studying each type individually.

Table 5.3   Factors That Affect Virus Survival

As virus adsorption to soil increases, virus survival is prolonged

Virus survival increases with increasing levels of exchangeable aluminum

Virus survival decreases with increasing pH and resin-extractable phosphorus

As temperature increases, survival decreases

Aerobic soil microorganisms adversely affect virus survival while anaerobic microorganisms have no effect

In general, virus survival is less at lower moisture levels

Fulvic and humic acids may mask virus infectivity (Sobsey, 1981)

Soil moisture, temperature, pH, and availability of organic matter can also indirectly influence the survival of enteric bacteria by regulating the growth of antagonistic organisms. Numerous workers (Rudolfs et al., 1950) have found a longer survival time of enteric bacteria after inoculation into sterilized soil as compared to non-sterilized soil, which indicates that antagonism is an important factor.

Soil microorganisms may also play a role in virus survival (Hurst et al., 1980). In a comparative study of the survival of enteric viruses in soil under sterile and non-sterile conditions and aerobic and anaerobic environments, it was found that virus survival was much longer under sterile conditions in aerobic environments, but not under anaerobic conditions. This is indicative that aerobic, but not anaerobic, soil microorganisms are antagonistic to virus survival. Thus, viruses retained near the soil surface would be expected to be more rapidly inactivated than those that penetrate the soil surface.

In summary, survival of those enteric microorganisms retained near the soil surface would be expected to die off at a fairly rapid rate due to the combined effects of sunlight, antagonism, and drying. On the other hand, those organisms that penetrate the aerobic zone could be expected to survive for a more prolonged period of time.

A simple conceptual model based on the current state of knowledge on indicator and pathogen die-off has been described by Reddy et al. (1981). Microbial die-off was described by assuming first-order kinetics. First-order die-off rate constants (k) were calculated from the literature for enteric microbial die-off in soil-water systems. Correction factors were presented to adjust constants for changes in temperature, moisture, and pH of the soil. Average die-off rate constants for selected microorganisms are shown in Table 5.4. In the article by Reddy et al. (1981), data on die-off of viruses during anaerobic digestion were used; only data on virus die-off in soil systems are shown in Table 5.4. These values were obtained from various experiments and represent an average value of several soil and environmental variables. Such an approach could prove useful for estimating microbial survival in soil-water systems, but a greater data base is needed, especially

Table 5.4   Die-Off Rate Constants (day^{-1}) for Enteric Microorganisms in Soil[a]

Microorganism	Average	Maximum	Minimum	Number of Observations
Escherichia coli	0.92	6.39	0.15	26
Fecal coliforms	1.53	9.10	0.07	46
Fecal streptococci	0.37	3.87	0.05	34
Salmonella sp.	1.33	6.93	0.21	16
Shigella sp.	0.68	0.62	0.74	3
Enteroviruses	0.10	0.16	0.04	4

[a] Based on table by Reddy et al. (1981).

for viruses and pathogenic bacteria. Also, most of our data base on microbial survival is in soil-water systems and not in ground water. For example, information on virus survival in ground water is almost nonexistent.

OCCURRENCE OF INDICATOR BACTERIA IN GROUND WATER

A few extensive bacteriological surveys on ground water quality in the United States have been conducted and are summarized in Table 5.5. In

Table 5.5   Microbial Surveys of Drinking-Water Wells

Survey	Number of Samples	Percent Positive for Coliforms[a]	Percent Positive for Fecal Coliforms[a]	Reference
South Carolina rural supplies	460	84.8	75	Sandhu et al. (1979)
Colorado rural supplies	164	41.3	—	Ford et al. (1980)
Community water supply study	621	9.0	2.0	Allen and Geldreich (1975)
Tennessee-Georgia rural water supplies	1257	51.4	27.0	Allen and Geldreich (1975)
Interstate highway drinking-water systems	241	15.4	2.9	Allen and Geldreich (1975)
Umatilla Indian Reservation	498	35.9	9.0	Allen and Geldreich (1975)

[a] One or more organisms per 100 mL.

these studies, 9 to 85% of the samples examined contained coliforms, and 2 to 75% of these same waters were positive for fecal coliforms. These studies indicate that the microbial quality of ground water cannot be taken for granted, especially in rural areas.

Several surveys have indicated that rainfall and well depth are related to the microbial quality of ground water. Studies in Washington indicated that shallow drinking-water wells average median coliform values of 8 MPN/100 mL with an average depth of 9.4 m (31 ft), while deep wells with an average depth of 153.3 m (503 ft) average 4 MPN/100 mL (DeWalle et al., 1980). They also noted that virtually all bacterial contamination coincided with the periods of heaviest rainfall. Brooks and Cech (1979) observed in rural eastern Texas that practically all wells with depths of 50 ft (15 m) or less were positive for either fecal coliforms or fecal streptococci. While presence of fecal bacteria was much less common in deeper wells, some wells as deep as 250 ft (80 m) were positive. Increased levels of bacterial contamination of drinking-well water after periods of rain have been noted in several studies (Lewis et al., 1980; Barrell and Rowland, 1979; Lamka et al., 1980; DeWalle et al., 1980). In one study it was noted that while an increase in coliform bacteria appeared almost immediately after periods of heavy rainfall in shallow wells, the increase did not occur in deeper wells until two weeks later (Loehnert, 1981). Thus, any satisfactory study of well water quality should include sampling during periods of highest rainfall.

Sandhu et al. (1979) found that basic well design had little effect on the extent of microbial pollution in their study area; drilled, covered, artesian, and open wells were compared. It is possible, however, that the degree of care during installation and operation within each design might have a significant impact on the water quality.

OCCURRENCE OF VIRUSES IN GROUND WATER

Only in the last decade have adequate methods become available for the concentration and detection of human enteric viruses in large volumes of water. Such methods are believed necessary for adequately assessing the occurrence of enteric viruses in water. One of the first applications of this technology was the sampling of ground waters beneath sites of land application of wastewater (Wellings et al., 1974). The prime concern with the presence of viruses in ground water is the transmission of infectious disease. Because of previous limitations in concentration methodology and the sudden occurrence of outbreaks, few reports exist on the isolation of viruses from waters associated with waterborne disease outbreaks. In addition, no methodology currently exists for the detection of the Norwalk agent in water that may be responsible for many gastroenteritis outbreaks. Although the Norwalk agent cannot be isolated from ground water, it certainly occurs there, as evidenced by the growing list of ground water disease outbreaks associated with this virus (Table 5.6).

**Table 5.6   Ground Water-Borne Outbreaks of
Gastroenteritis Associated with Norwalk-Like Agents**[a]

Location	Year	Water Source	Number of Cases
Colorado	1976	Well	418
Pennsylvania	1978	Well	350
Pennsylvania	1978	Well	120
Washington	1978	Well	467
California	1979	Ground water	30
North Carolina	1979	Well	146
Pennsylvania	1979	Well	151
Maryland	1980	Well	139

[a] Modified from Melnick and Gerba (1982).

Table 5.7 lists the reported isolation of enteric viruses from drinking-water wells during outbreaks of waterborne disease.

Mack et al. (1972) were the first to isolate a virus from ground water associated with waterborne disease. Coliform levels in the well ranged from 0 to 16/100 mL, but no *Salmonella* or *Shigella* were found. The source was a waste drain field that allowed sewage to enter a 30.5 m-deep well by passing through 5.5 m of clay, 2.5 m of shale, and 22.5 m of limestone. Although poliovirus was isolated from the well, it was not suggested as the causative agent of the gastroenteritis outbreak. That outbreak was probably attributable to some other virus (such as rotavirus or the Norwalk agent) that could not be detected by the methods available at that time. Also, it is evident that soil type, structure, and proximity of a well to a source of sewage played an

**Table 5.7   Isolation of Viruses from Drinking-Water Wells
Associated with Waterborne Disease Outbreaks**

Location	Virus Type	Reference
Florida	Echo 22/23	Wellings et al. (1977)
Michigan	Polio 2	Mack et al. (1972)
Texas	Coxsackie B3, B2 Hepatitis A	Hejkal et al. (1982)
Maryland	Polio 1 Echo 27, 29 U[a]	Woodward and Sobsey (1981)
Israel	Polio 1 U[a]	Shuval (1969)

[a] U = unidentified.

important role in the contamination of ground water sources, as the virus in this case penetrated 30.5 m through several soil profiles and traveled 91.5 m laterally.

Wellings et al. (1977) reported the isolation of an echovirus 22/23 complex in 100-gal samples from a 12.2 m-deep well during an outbreak of gastrointestinal illness at a migrant labor camp in Florida. The well was located 30.5 m from a solid-waste field and was in the middle of an area bordered by septic tanks. The echovirus was isolated from sewage, from potable well water containing 0.4–0.6 mg/L residual chlorine, and from stools collected from individuals living in the camp. It is also interesting to note the occurrence of 15 cases of hepatitis A in the camp some six weeks later.

Thus, the virus was probably present because of well water contamination from the septic tanks and existed in chlorinated water in the absence of evidence of bacterial contamination. It was surmised that the chlorine was ineffective against the virus, which was probably associated with solids introduced with the sewage. Even though this level of chlorine treatment reduced bacterial counts to undetectable levels, the virus survived. This raises concern about the efficacy of chlorine disinfection of contaminated ground water, based on currently accepted bacterial standards.

In June 1980, an outbreak of gastrointestinal illness and infectious hepatitis was found to be associated with drinking-water wells in the city of Georgetown, Texas. The attack rate of gastroenteritis among the 10,000 individuals living in the areas supplied by the wells was almost 80%. An increased number of cases of hepatitis A began to appear early in July 1980. Georgetown received its drinking water from wells with no water treatment other than chlorination. At the time of the peak of the gastroenteritis outbreak, 400- to 1100-L samples of both well water and chlorinated drinking water were taken for virological analysis. Some well samples were found to be heavily contaminated with coliform bacteria before receiving chlorination. Potable water samples taken for virological analysis contained 0.8 mg/L residual chlorine. Coxsackieviruses B2 and B3 were isolated from the well water. Coxsackievirus B3 was also isolated from a chlorinated tap water sample. Hepatitis A antigen was identified by radioimmunoassay in one of the well water samples. This was the first demonstration of the occurrence of hepatitis A in water used as a source for potable water prior to a suspected waterborne disease outbreak.

Also in 1980, enteric viruses were isolated from iodinated ground water at a summer camp in western Maryland during an outbreak of gastroenteritis (Woodward and Sobsey, 1981). The water supply for the camp was a 95-ft-deep well which, during the outbreak, contained 0.7–1.0 ppm iodine (recommended 0.3–0.5 ppm). The coliform count in the raw well water was 5000/100 mL, but all drinking water was within allowable standards, during and after the epidemic. Preliminary evidence has indicated that the Norwalk agent may have been responsible for the outbreak. To date, poliovirus 1 and echoviruses 27 and 29 have been isolated from the well water.

Unfortunately, few systematic studies on the occurrence of viruses in ground water over a region have been conducted (Table 5.8). The largest to date was done by Marzouk et al. (1979), who collected 20- to 440-L samples of ground water from 3-m-deep wells in Israel; 20 of the 99 were found to be positive for viruses. The isolated viruses included coxsackievirus B6, echoviruses 6 and 7, poliovirus 1, and several unidentified types. Total bacteria, fecal coliform, and fecal streptococcal concentrations ranged from 0–10^4, 0–200 and 0–100, respectively, per 100 mL. In at least one case, viruses were detected in samples that contained no standard plate count bacteria, while 12 of 17 samples that were negative for fecal coliforms and fecal streptococci were positive for viruses. No statistical correlation could be demonstrated between indicator bacteria and the presence of viruses, which suggests the failure of bacterial indicators to represent viral contamination adequately.

Even though there have been no reports of disease outbreaks associated with land treatment of sewage wastes, there is a growing number of studies concerning the detection of viruses in ground water after waste water application to land or direct ground water recharge. These studies have previously been reviewed in detail (Keswick and Gerba, 1980). In summary, these studies indicate that depending on the nature of the soil and other site-specific factors, viruses can travel as far as 67 m vertically and 408 m horizontally from land application sites. These field studies have shown that viruses can travel long distances in sand and gravel soils and that rainfall can play an important role in their penetration into the subsurface (Wellings et al., 1974).

Overflow from septic tanks and cesspools has been estimated to be responsible for 42% of the outbreaks and 71% of the illness caused by using untreated ground water in non-municipal systems (Craun, 1981). Yet, almost no field work has been done on the occurrence of enteric viruses in ground water near septic tank systems. Recent studies using "seeded" marker viruses added to septic tanks indicate that viruses can travel significant distances in the subsurface from these sources (Gerba, 1981). Two recent stud-

Table 5.8 Factors That Influence the Movement of Viruses and Bacteria in Soil

1. Rainfall
2. pH
3. Soil composition
4. Flow rate
5. Soluble organics
6. Cations
7. Adsorption characteristics of the virus and bacteria
8. Degree of saturation

ies in progress have reported the isolation of enteric viruses from wells near septic tank systems (Wang et al., 1981; Vaughn and Landry, 1981).

In the study in Texas, an enteric virus was isolated from an abandoned well 25 m distant from a septic tank system serving a mobile home park. No viruses were isolated from another well 300 m distant from the same drain field (Wang et al., 1981). In a survey of septic tank systems on Long Island, New York, enteric viruses have been detected in sampling wells as far as 65 m from a known septic tank source and at depths as great as 45.7 m (Vaughn and Landry, 1981).

MOVEMENT OF BACTERIA AND VIRUSES THROUGH THE SUBSURFACE

Retention of microorganisms by the soil is, of course, a paramount consideration in protecting ground water from contamination during the land application of sewage. Filtration plays a major role in bacterial removal during this process, because of their large size, although adsorption is also involved. Virus removal is believed to be almost totally dependent on adsorption (Lance, 1978).

Factors currently believed to play a role in viral and bacterial removal are listed in Table 5.8. Viruses have been shown to travel as far as 408 m horizontally in the ground water from sewage infiltration basins (Keswick and Gerba, 1980). Coliform bacteria have been observed to be transported in loamy sand aquifers for more than 1 km, and in fissured karstic aquifers for several kilometers (Matthess and Pekdeger, 1981). Factors involved in bacterial and viral removal have recently been reviewed by Hagedorn (1983) and Sobsey (1983). In his review on bacterial movement through ground water, Hagedorn (1983) concluded that (1) coliforms and other microorganisms move only a few dozen centimeters with the percolating waters in unsaturated soil layers although much greater distances are possible under saturated flow conditions, (2) with all soil conditions the degree of bacterial retention by the soil is inversely proportional to the size of the component particles in the unstructured matrix, (3) the physical straining or filtration of bacteria by soil particles is the main limitation to travel through soils, and sedimentation of bacterial clusters occurs through the zone of saturated flow (Krone, 1968), and (4) adsorption is a factor in the retention of bacteria by soil and becomes more effective in soils with higher clay content.

Virus adsorption to soil surfaces is believed to be largely governed by both electrostatic double-layer interactions and van der Waals forces (Gerba et al., 1975). Thus, the surface charge on both the virus and the soil, and factors controlling the net charge, are important in determining the efficiency of virus adsorption.

A number of studies using batch reactors to evaluate virus adsorption to soils has been conducted (Gerba et al., 1975; Sobsey, 1983). In these studies a given amount of soil was mixed with virus suspended in a solution, and adsorption was determined after a given period of time (Goyal and Gerba, 1979). The results of such studies indicate that virus adsorption increases with increasing cation exchange capacity, clay content, exchangeable aluminum, and low flow rate; and that it decreases with decreasing pH (Burge and Enkiri, 1978; Gerba et al., 1975; Sobsey, 1983). Unfortunately, firm predictive correlations between virus adsorption and these factors have not been established. Establishment of predictive relationships between soil factors and virus adsorption is further complicated by genetic variability among the different types and strains of viruses. In recent studies we found that efficiency of virus adsorption was dependent on not only the particular type of virus but also the particular strain. For example, in batch studies the adsorption of different echovirus 1 strains to a given weight of soil was found to vary from 0 to higher than 99%. Differences in adsorption among different strains of the same virus probably result from variability in the configuration of proteins on the outer capsid of the virus, since this configuration will influence the net charge of the virus, which would in turn affect the electrostatic potential between virus and soil and thereby influence the degree of interaction between the two particles.

Batch studies have also shown that the pH of the suspending media, soluble organics, and the presence of cations will influence virus adsorption (Table 5.8). All of these factors act to influence the electrostatic potential between the virus and the soil.

Unfortunately, viruses cannot be considered permanently immobilized after adsorption onto a soil particle. This became clear in the study of a land application site in Florida by Wellings et al. (1974). Viruses remained undetected in wells approximately 3 m (10 ft) and 6 m (20 ft) below the soil surface until after periods of high rainfall. Subsequent laboratory studies confirmed that viruses previously adsorbed near the soil surface desorb and migrate further through the soil column (Lance et al., 1976). The degree of elution which occurs during a rainfall event also appears to be dependent on both the type and specific strain of virus (Landry et al., 1980). Viruses eluted near the soil surface will eventually readsorb further down a soil column (Lance et al., 1976), but it has been speculated that viruses could continue to travel vertically through the soil by a chromatographic effect controlled by periodic rainfall events.

SUMMARY AND RECOMMENDATIONS

Current information on bacterial indicators and potentially pathogenic microorganisms in ground water is limited. More is currently known about the bacteriological quality of ground water than of its virological quality, but there are still areas in need of further research. For example, studies are needed on the adequacy of bacteriological indicators for judging ground water quality and on the occurrence of opportunistic pathogens and bacterial pathogens of emerging significance such as *Campylobacter* and *Yersinia*. Rainfall events appear to play a key role in the transport of both bacteria and viruses in the subsurface, and detailed studies are needed to define the significance of these events on long term ground water quality.

A significant amount of information is available on the survival of indicator bacteria and specific bacterial pathogens in soil and ground water, but information on virus survival is almost nonexistent. Such information would be useful in the development of predictive models on microbial survival in ground water.

Results of previous surveys on the bacteriological quality of ground water indicate that viral pollution is more widespread than might be supposed. Furthermore, many of the previous systems used for virus concentration had efficiencies of 2–3% and were limited to the detection of enteroviruses (Sobsey et al., 1980). Now, however, systems with efficiencies of 30% and greater are available (Sobsey and Glass, 1980).

Viruses are known to enter ground water under certain conditions, but these conditions have not been adequately defined. It is not yet possible to determine exactly what factors affect the retention and mobility of viruses in soils. Such an understanding is essential so that predictions may be made and methods developed for estimating virus migration.

The current body of information on the occurrence of viruses in ground water is too limited to develop guidelines on safe distances between waste sources and drinking-water wells. Field studies are sorely needed to define such guidelines. The last decade has seen major developments in technology for virus detection in water, and it would appear time to apply this technology to answer some of the major questions concerning viral contamination of ground water.

REFERENCES

Allen, M. J. and Geldreich, E. E. (1975). Bacteriological criteria for ground-water quality. *Ground Water* **13**:45–51.

Barrell, R. A. E. and Rowland, M. G. M. (1979). The relationship between rainfall and well water pollution in a West African (Gambian) village. *J. Hyg. Camb.* **83**:143–150.

Brooks, D. and Cech, I. (1979). Nitrates and bacterial distribution in rural domestic water supplies. *Water Res.* **13**:33–41.

Burge, W. D. and Enkiri, N. K. (1978). Virus adsorption by five soils. *J. Environ. Qual.* **7**:73–76.

Craun, G. F. (1981). Outbreaks of waterborne disease in the United States: 1971–1978. *J. Am. Water Works Assoc.* **73**:360–369.

DeWalle, F. B., Schaff, R. M., and Halten, J. B. (1980). Well water quality deterioration in central Pierce County, Washington. *J. Am. Water Works Assoc.* **72:**533–536.

Ford, K. L., Schoff, J. H. S., and Keefe, T. J. (1980). Mountain residential development minimum well protective distances—well water quality. *J. Environ. Health* **43:**130–133.

Gerba, C. P. (1983). Virus occurrence in groundwater. In: Microbial Health Considerations of Soil Disposal of Domestic Wastewaters. U.S. Environmental Protection Agency publication EPA-600/9-83-017, pp. 240–263.

Gerba, C. P., Wallis, C., and Melnick, J. L. (1975). Wastewater bacteria and viruses in soil. *J. Irrig. Drain. Div. ASCE* **101:**157–174.

Goyal, S. M. and Gerba, C. P. (1979). Comparative adsorption of human enteroviruses, Simian rotavirus and selected bacteriophages to soils. *Appl. Environ. Microbiol.* **38:**241–247.

Hagedorn, C. (1981). Transport and fate: Bacterial pathogens in ground water. In: Microbial Health Considerations of Soil Disposal of Domestic Wastewaters. U.S. Environmental Protection Agency publication, no. EPA-600/9-83-017, pp. 153–172.

Hejkal, T. W., Keswick, B. H., LaBelle, R. L., Gerba, C. P., Sanchez, Y., Dressman, G., and Hafkin, R. (1982). Viruses in a community water supply associated with an outbreak for gastroenteritis and infectious hepatitis. *J. Am. Water Works Assoc.* **74:**318–321.

Hurst, C. J., Gerba, C. P., and Cech, I. (1980). Effects of environmental variables and soil characteristics on virus survival in soil. *Appl. Environ. Microbiol.* **40:**1067–1079.

Keeley, J. W. (1977). Magnitude of the groundwater contamination problem. In: W. R. Kerns (Ed.), Public Policy on Ground Water Quality Protection. Virginia Water Resources Center, Blacksburg, VA, pp. 2–9.

Keswick, B. H. and Gerba, C. P. (1980). Viruses in groundwater. *Environ. Sci. Technol.* **14:**1290–1297.

Krone, R. B. (1968). The movement of disease producing organisms through soils. In: C. W. Wilson and F. F. Beckett (Eds.), *Municipal Sewage Effluent from Irrigation.* The Louisiana Technical Alumni Foundation, Ruston.

Lamka, K. G., LeChevallier, M. W., and Seidler, R. J. (1980). Bacterial contamination of drinking water supplies in a modern rural neighborhood. *Appl. Environ. Microbiol.* **39:**734–738.

Lance, J. C. (1978). Fate of bacteria and viruses in sewage applied to soil. *Trans. Am. Soc. Agric. Eng.* **21:**1114–1119.

Lance, J. C., Gerba, C. P., and Melnick, J. L. (1976). Virus movement in soil columns flooded with secondary sewage effluent. *Appl. Environ. Microbiol.* **32:**520–526.

Landry, E. F., Vaughn, J. M., and Penello, W. F. (1980). Poliovirus retention in 75-cm soil cores after sewage and rainwater application. *Appl. Environ. Microbiol.* **40:**1032–1038.

Lewis, W. J., Farr, J. L., and Foster, S. S. D. (1980). The pollution hazard to village water supplies in eastern Botswana. *Proc. Instn. Div. Engrs.* **69:**281–293.

Liew, P. and Gerba, C. P. (1980). Thermostabilization of enteroviruses by estuarine sediment. *Appl. Environ. Microbiol.* **40:**205–308.

Loehnert, E. P. (1981). Hohe Nitratgehalte in einem ländlichen Gebiet in Nigeria verursacht durch ungeordnete Ablagerung hävslicher Abtälle und Exkremente. International Symposium on the Quality of Groundwater, Amsterdam, The Netherlands.

Mack, W. N., Lu, Y., and Coohon, D. B. (1972). Isolation of poliomyelitis virus from a contaminated well. *Health Services Rept.* **87:**271–274.

Marzouk, Y., Goyal, S. M., and Gerba, C. P. (1979). Prevalence of enteroviruses in groundwater in Israel. *Ground Water* **17:**487–491.

Matthess, G. and Pekdeger, A (1981). Concepts of a survival and transport model of pathogenic bacteria and viruses in groundwater. International Symposium on the Quality of Groundwater, Amsterdam, The Netherlands.

Melnick, J. L. and Gerba, C. P. (1982). Viruses in water and soil. *Public Health Rev,* **9:**185–213.

Reddy, K. R., Khaleel, R., and Overcash, M. R. (1981). Behavior and transport of microbial pathogens and indicator organisms in soils treated with organic wastes. *J. Environ. Qual.* **10:**255–266.

Rudolfs, W., Frank, L. L., and Ragotzkie, R. A. (1950). Literature review on the occurrence and survival of enteric, pathogenic and relative organisms in soil, water, sewage and sludges, and on vegetation. *Sew. Indust. Wastes* **22:**1261–1281.

Sandhu, S. S., Warren, W. J., and Nelson, P. (1979). Magnitude of pollution indicator organisms in rural potable water. *Appl. Environ. Microbiol.* **37:**744–749.

Shuval, H. I. (1969). Detection and control of enteroviruses in the water environment. In: H. I. Shuval (Ed.), *Developments in Water Quality Research.* Ann Arbor Science, Ann Arbor, MI, pp. 47–71.

Sobsey, M. (1983). Transport and fate of viruses in soils. In: *Microbial Health Considerations of Soils Disposal of Domestic Wastewaters.* U.S. Environmental Protection Agency publication no. EPA-600/9-83-017, pp. 174–213.

Sobsey, M. D. (1981). Personal communication.

Sobsey, M. D. and Glass, J. S. (1980). Poliovirus concentration from tapwater with electropositive adsorbent filters. *Appl. Environ. Microbiol.* **40:**201–210.

Sobsey, J. D., Glass, J. S., Carrick, R. J., Jacobs, R. R., and Rutala, W. A. (1980). Evaluation of the tentative standard method for enteric virus concentration from large volumes of tapwater. *J. Am. Water Works Assoc.* **72:**292–299.

Vaughn, J. and Landry, E. F. (1981). Personal communication.

Walter, R. and Rudiger, S. (1977). Significance of virus isolation in terms of municipal hygiene. *Zeit. Ges. Hyg.* **23:**461–469.

Wang, D. S., Keswick, B. H., and Gerba, C. P. (1981). Personal communication.

Wellings, F. M., Lewis, A. L., and Mountain, C. W. (1974). Virus survival following wastewater spray irrigation of sandy soils. In: J. F. Malina and B. P. Sagik (Eds.), *Virus Survival in Water and Wastewater Systems.* Center for Research in Water Resources Systems, Austin, TX, pp. 253–260.

Wellings, F. M., Mountain, C. W., and Lewis, A. L. (1977). Virus in ground water.

In: *Proceedings of the 2nd National Conference on Individual On-site Wastewater Systems*. Ann Arbor Science, Ann Arbor, MI, pp. 61–66.

Wilson, R., Haley, C. E., Relmar, D., Lippy, E., Craun, G. F., Morris, J. G., and Hughes, J. M. (1983). Waterborne outbreak, related to contaminated ground water reported to the Centers for Disease Control, 1971–1979. In: *Microbial Health Considerations of Soil Disposal of Domestic Wastewaters*. U.S. Environmental Protection Agency, publication no. EPA-600/9-83-017, pp. 264–288.

Woodward, W. E. and Sobsey, M.D. (1981). Personal communication.

Yaeger, J. E. and O'Brien, R. T. (1979). Enterovirus inactivation in soil. *Appl. Environ. Microbiol.* **38**:694–701.

6

EFFECTS OF LOCAL SOURCES OF POLLUTION ON GROUND WATER QUALITY IN THE NETHERLANDS

W. van Duijvenbooden

National Institute for Public Health and Environmental Hygiene
Leidschendam, The Netherlands

This chapter deals with the effects of local sources of pollutants on ground water quality in The Netherlands. As background it should be mentioned that The Netherlands is a small, densely populated and heavily industrialized country in western Europe. The population is about 14 million people. The total area is about 40,000 km²; the United States is about 280 times the area of The Netherlands, whose mean population density is 350 persons/km².

Only 12% of the area of The Netherlands can be characterized as "natural." But even in relatively undisturbed areas, ground water is at times heavily polluted by camping, parking places, recreational facilities, and so on. The other 88% of the country is cultivated, with a high degree of urbanization. Nearly 8% of the whole land area is urbanized. As a result, nearly all shallow ground waters in The Netherlands are qualitatively influenced by human activities. Organohalogen compounds are widespread in the ground waters, and the registered areas with waste disposal sites total about 25 km² for more than 4000 sites. In addition, there are large industrial areas with significant pollution potential, many illegal disposal sites, and areas polluted by accidents and spills. In rural areas dumping of manure can be an important source of ground water pollution (Table 6.1). The responsibility for the quality of soil and ground water in The Netherlands is mainly delegated to

Table 6.1 Nitrate Concentrations in Rural Water Wells[a]

Concentration Range (mg/L)	Percent of Wells
0–45	32.4
45–90	15.4
90–225	28.9
225–1350	23.3

[a] Based on 1000 rural wells (after Trines, 1952, in *Verslagen en Mededelingen voor de Volksgezondheid*); more recent research indicates similar values.

the regional authorities. On the national level the Ministry of Housing, Physical Planning and Environment is responsible for the protection of the environment, including soil and ground water.

GROUND WATER QUALITY RESEARCH NEAR LANDFILL SITES

Until about 1970 there was no structured research in The Netherlands on the effects of local sources of pollution on ground water quality. Nevertheless, incidental research was and in many cases is still being done. Recently attention has been given to the effects on ground water quality of fly ash, low level radioactive waste, harbor sludge, cyanide waste, organic micropollutants, and so on. Leaching of several trace elements from fly ash has been detected (Table 6.2). Several coal-fired power plants that will soon generate enormous quantities of fly ash are currently under construction. Emissions of sulfur and nitrogen into the air can cause ground water quality problems. In fact, in several industrial areas significant acidification of rain water and elevated concentrations of sulfate in ground water have been detected (Figure 6.1).

Table 6.2 Trace Element Contamination from a Fly Ash Disposal Site (μg/L)

Element	Control	Leachate Below Site	Ground Water
Ba	100	2850	300–600
Mo	2	1750	200–800
V	<5	230	<5

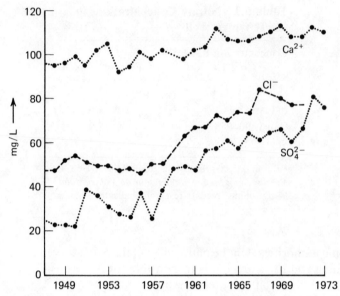

Figure 6.1.   Effects of air pollution on ground water quality (Duijvenbooden, 1979, 1981).

In harbor sludge very high concentrations of nitrogen, trace elements, polyaromatic hydrocarbons, and even pesticides have been detected. In one case, sulfate concentrations up to 1680 mg/L coming from a normal rubbish dump could be detected in ground water. Insufficiently screened chemical waste dumps can cause serious ground water pollution. Table 6.3 shows the results of a leaky screen on ground water quality. The effects of leaky pipes and tanks on ground water quality are generally known. The quantities of wastes generated in The Netherlands are given in Table 6.4 (Duijvenbooden, 1979).

Table 6.3   Quality of Ground Water Near an Industrial Dump

Contaminant	Concentration (mg/L)	
	Control	Downstream
CN	0.004	1.6
CNS^-	<0.5	4.4
S^{2-}	<0.1	11.8
NH_4^+	0.19	775
SO_4^{2-}	190	3100

Table 6.4   Waste Production in The Netherlands[a,b]

Type	Quantity (10^6 tons)
Chemical waste	1
Household waste	4
Industrial waste	3.7
Construction waste	6.7
Sewerage sludge	5
Coal ash	0.7
Harbor sludge	20
Manure	40

[a] Institute for Waste Research (SVA, 1977; personal communication).

[b] Duijvenbooden (1979).

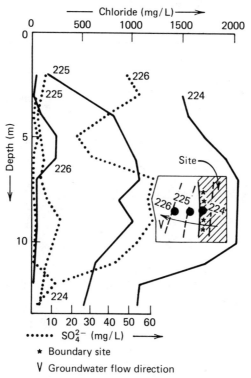

Figure 6.2. Cl^- and SO_4^- profiles in the ground water near the Delden landfill.

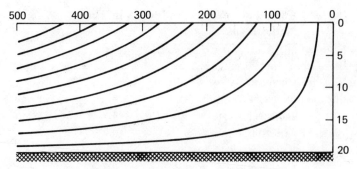

Figure 6.3. Flow tubes, vertical profile.

In 1970 more structured research on the effects of local sources of pollution, especially on normal waste disposal sites, started. Initially, little attention was given to the presence of organic micropollutants and trace elements
or to the effects of local hydrogeological parameters. Table 6.5 and Figure
6.2 present data on chemical contamination resulting from a waste disposal
site in Delden (Kooper, 1983).

In 1976 the National Institute for Water Supply started intensive research

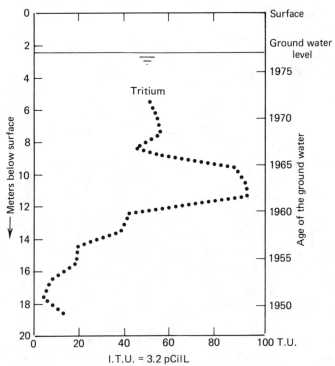

Figure 6.4. Effects of atmospheric nuclear tests on ground water quality (Duijvenbooden,
1980, 1981).

on the effects of local sources of pollution on ground water quality. The research was concentrated on two waste disposal sites. Limited attention also has been given to several other sites. Results obtained to date clearly indicate the important effects of the local hydrogeology, including such things as density currents caused by the landfill itself, on transport of polluted ground waters (Kooper, 1983). Often the effects are restricted to narrow flow tubes surrounded by nonpolluted ground water, which hampers detection of the pollutants. Figure 6.3 shows in a vertical profile a schematic view of such a flow tube. Field data indicate that there is a real risk of sampling outside the pollution plume (Figure 6.4). Figure 6.4 also shows the effect of atmospheric nuclear tests in the early 1960s on the tritium content of ground water (Duijvenbooden, 1980, 1981). The vertical transport of ground water in the cental part of the dune region is illustrated in Figure 6.5 (Duijvenbooden, 1981).

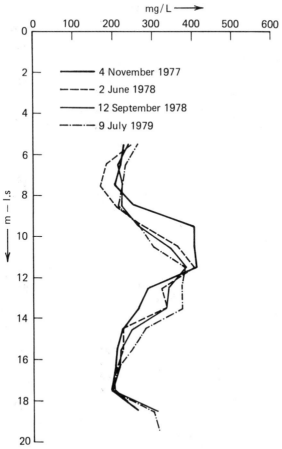

Figure 6.5. Vertical transport of ground water as observed by measurements of HCO_3^- (Duijvenbooden, 1981).

Table 6.5 Leachate Characteristics and Ground Water Contamination at the Delden Waste Disposal Site[a]

	Concentration (mg/L)	
Contaminant	Leachate[a]	Ground Water Downstream
NH_4^+ (N)	190–2300	212
Cl^-	1100–3700	2,000
SO_4^{2-}	80–1470	18
COD	19000–74000	—

[a] Institute for Waste Research (personal communication).

THE NOORDWIJK LANDFILL

One of the landfill sites investigated intensively is a waste disposal site at Noordwijk (Kooper et al., 1981; Kooper, 1983). The site is situated in a former sandpit near the dunes (Figure 6.6). The average depth of the sandpit is about 13 m. In the central part of the sandpit, the maximum depth is 16–17 m. The level of the ground water is about 1 m below the surface. From 1960 to 1973, the sandpit was used as a waste disposal site. The wastes deposited reached a height of about 3 m above the former level of the surface. After some preliminary investigations carried out in 1975 and 1976, more intensive

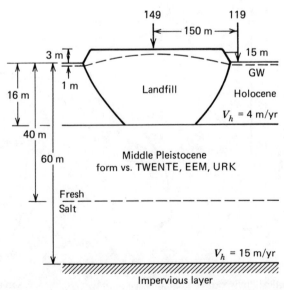

Figure 6.6. Cross section of the Noordwijk landfill (Kooper et al., 1981).

research was started in 1977 on the influence of the waste disposal site on ground water flow and ground water quality. For this research 23 monitoring wells were drilled, six of which had a depth of about 60 m. Two boreholes were drilled through the central part of the waste disposal site. Normal and mini-screens were placed in the wells. At least once a year the ground water around, below, and inside the landfill was monitored. Ground water samples and leachate were analyzed for macro- and trace elements, organic micro-pollutants, and tritium. In addition, electrical resistivity and electromagnetic measurements around the landfill were made along with temperature measurements.

Based on information gathered during the research, the geohydrological situation around the landfill can be described as follows:

0–18 m below land surface (m − l.s.)	A medium permeable phreatic aquifer with clayey fine to medium coarse sand, and thick clay layers (Holocene)
18–57 m − l.s.	Permeable aquifer with coarse sand (Pleistocene)
57–105 m − l.s.	Impervious clay layers of the Formation of Kedichem (base of the aquifer)

In the immediate vicinity of the landfill a clayey, loamy layer is present between 1.5 and 4–5 m below the land surface.

At about 40 m below land surface a fresh–salt-water transition zone is found. Generally the ground water flow is toward a lower situated polder area southeast of the landfill, as can be seen from the equipotential lines of the ground water in the aquifers (Figure 6.7).

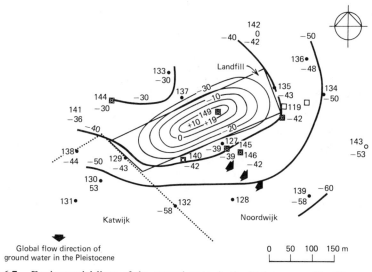

Figure 6.7. Equipotential lines of the ground water in the Holocene aquifer (Kooper et al., 1981).

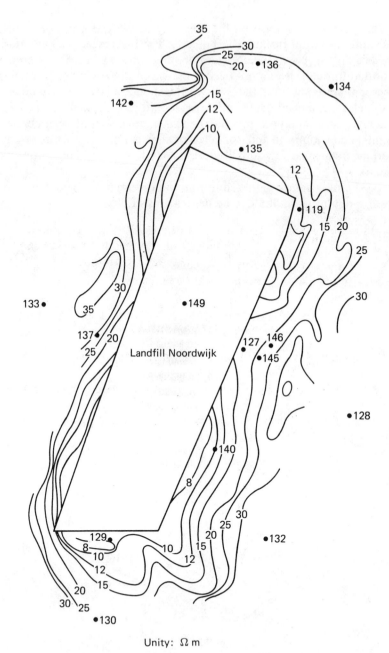

Unity: Ω m

Figure 6.8. Lines of equal resistance (in Ω m) around the Noordwijk landfill (Kooper et al., 1981).

On the average, the velocity of the ground water in the Holocene aquifer is about 4 m/yr and in the Pleistocene aquifer, which has higher permeability, about 15 m/yr. The area upstream and directly downstream of the landfill is an infiltration area. Further downstream, in the polder area, there is upward seepage. Locally, the ground water in the Holocene aquifer is strongly influenced by the landfill while, as will be seen later, the ground water in the Pleistocene aquifer is influenced by effects of density flow, due to infiltration of contaminated water from the landfill. During the excavation of the former sandpit, a silty layer was deposited around the landfill site. The resistance of this layer against waterflow, together with the absence of drainage systems inside or below the landfill, probably caused the elevated ground water level inside the landfill (Figure 6.7). This elevated ground water level results in an all-sided migration of leachate into the Holocene and Pleistocene aquifers. For this reason no ground water coming from the aquifer flows through the landfill. The outflowing leachate is replenished by precipitation infiltrating into the landfill. The dispersion of the pollutants in the ground water in the Holocene aquifer can be illustrated by lines of equal resistance based on electrical resistivity measurements (Figure 6.8). Pollution was found as far as 30 m upstream and about 80 m downstream of the landfill. The results of the electromagnetic measurements and the results of analyses of water samples are very similar. Lines of equal conductivity based on the analytical results are given in Figure 6.9. Pollution near the central part of the landfill is comparatively less than on the periphery, probably due to a lower flow velocity of the ground water.

From electrical resistivity measurements (equal resistance lines in the vertical profile), contamination of ground water could be detected in a zone

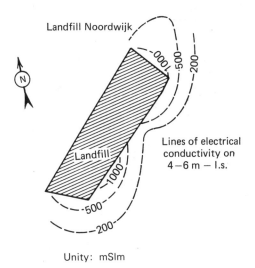

Unity: mSlm

Figure 6.9.   Lines of equal electrical conductivity 4 to 6 m − l.s. (in mSlm).

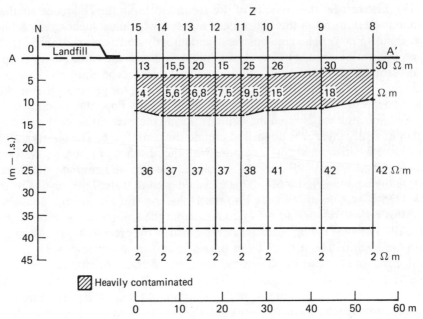

Figure 6.10. Results of measurements of electrical resistance (in Ω m) in a vertical profile (Kooper et al., 1981).

between 5 and 13 m below land surface (Figure 6.10). Hydrogeological conditions and the water analyses indicate that no pollution was present in the upper 5 m, and only a slight influence on ground water quality could be found in the deeper part of the aquifer. Due to the disturbing effects of the brackish water pollution, the fresh–salt-water boundary could not be detected by the electrical resistivity measurements. (The same holds true for the contaminated ground waters on a clayey or loamy layer.)

To obtain more insight into leachate migration in the Pleistocene aquifer just beneath the landfill, boreholes were sunk through the waste disposal site after numerous pressure soundings were made through the landfill in order to find suitable drilling sites. A vertical profile based on these soundings is given in Figure 6.11. In one of the boreholes, a mini-screen was placed every 1.5 m (28 m total). Thirteen electrodes and nine normal well-screens (length 2 m) were placed in the landfill, divided by artificially made clay layers (Figure 6.12). Leachate ran down the boreholes, resulting in temporary contamination of the ground water. This occurred over the whole depth of the Pleistocene aquifer and was probably caused by a short-circuit flow due to a temporary decrease in packing of the grains in the aquifer in the immediate vicinity of the borehole. It took nearly one year before the situation stabilized, as judged by measurements with the electrodes. The upper part of the aquifer beneath the landfill recovered first.

Dispersion of the leachate, based on the HCO_3^- concentration in the ground water, is illustrated in Figure 6.13 and indicates that:

1. The total amount of dissolved solids in the ground water in the landfill increased with depth, which supports the conclusion that replenishment of outflowing leachate takes place only by precipitation.
2. From the bottom of the landfill down to about 27 m below land surface (m − l.s.) the ground water is heavily polluted with almost pure leachate. At the same depth the ground water downstream of the landfill was only slightly polluted.
3. From 27 m − l.s. to about 35 m − l.s., the ground water beneath the landfill was less contaminated. From 35 m − l.s. to about 41 m − l.s. (just above the fresh−salt-water transition zone), a pollution plume appears at about half the concentration of the leachate. This is probably due to density flow combined with dispersion.

Values similar to those observed for HCO_3^- are found for several other parameters such as tritium and K^+ (Figures 6.14 and 6.15). Except for some heavy metals, adsorption does not seem to be important in this coarse sandy

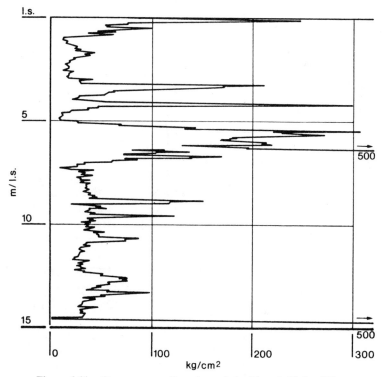

Figure 6.11. Pressure soundings through the Noordwijk landfill.

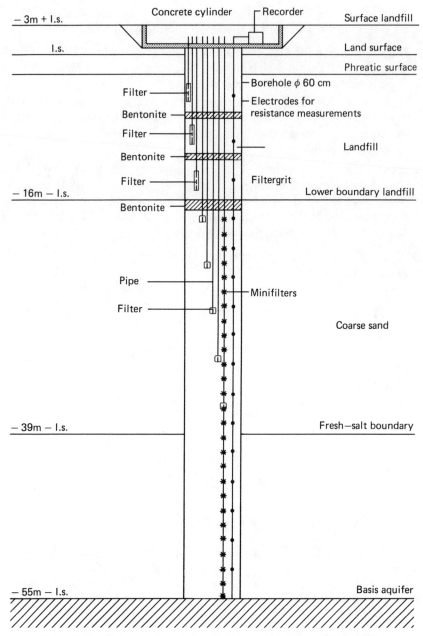

Figure 6.12. Cross section of a borehole through the Noordwijk landfill.

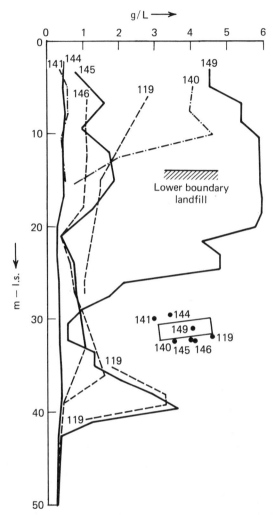

Figure 6.13. HCO$_3^-$ concentration in the ground water near the Noordwijk landfill.

aquifer. It is interesting to note that the measured temperature profiles of the ground water downstream of the landfill do not coincide with the chemical profiles, possibly due to density effects. The highest temperatures were found several meters above the most polluted zone. The contamination at 35 to 41 m − l.s. is remarkable, but can be explained by density flow. The density of the leachate was 1.0094 kg/L while the densities of the polluted ground water just above the salty ground water and the nonpolluted ground water were, respectively, 1.0051 and 1.00027 kg/L. Based on equations developed by Obdam (1979) and Kruijtzer (1980) for the Noordwijk landfill, vertical flow caused by density effects can be calculated to be from 50 to 90

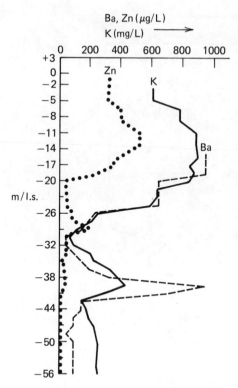

Figure 6.14. Ba^-, Zn^-, and K profiles in the ground water near the Noordwijk landfill.

m/year. This means that pollution of the ground water beneath the landfill can be expected in the fresh–salt-water transition zone.

Until now, attention has been given to the spread of pollution based on the conservative behaviors of the pollutants. But the behavior of many macroparameters, even the positively charged ones, also seem to be almost conservative due to the low adsorption capacity of the soil (<1 meq/100 g) in the middle-deep aquifer. Analytical results on macroelements are given in Table 6.6. The high bicarbonate and ammonium concentrations in the leachate indicate the presence of anaerobic, methanogenic conditions in the landfill, with formation of sulfides. Based on the chemical composition of the leachate and the cation exchange capacity of the soil (about 1 meq/100 g dry wt), transport velocities for Na (0.8) and K and NH_4 (0.65) can be calculated relative to the transport velocity of conservative substances. Significant ion exchange could be detected between K and NH_4 and between Ca and Mg.

If sulfides are formed in the landfill, low concentrations of heavy metals would be expected (Table 6.7). The increase in concentration compared with nonpolluted ground water may possibly be attributed to the presence of complexing organic agents. It is interesting to note the strong retardation of Ni and Zn. On the other hand, Ba and Li seem to behave similarly to the so-called macroelements (Figure 6.13). However, wide ranges in concentra-

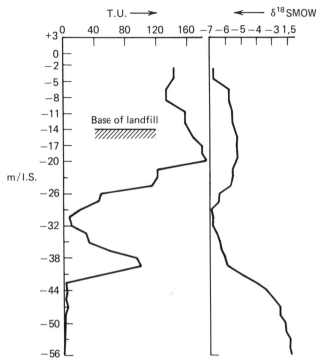

Figure 6.15. ^3H and ^{18}O profiles beneath the Noordwijk landfill.

tions of heavy metals can be found depending on the methods used for sample-handling, sample-destruction, and analysis; use of uniform analytical procedures is therefore essential for quantitative results (Bom, 1981). The mobility of the heavy metals can be influenced by the complexing properties of the organic compounds present. Although about 140 organic compounds (12 carbon atoms or fewer) were detected in the leachate and polluted ground water, fewer than 10 percent of all organic compounds present in the leachate were identified (Morra et al., 1979). Numerous heavier organic compounds were present, the most important of which are presented in Table 6.8. Little is known about the adsorption, degradation, or mobility of these compounds (Zoeteman, 1981). These factors should vary widely, depending on local circumstances and the nature and concentration of the organic compounds. The organic compounds decreased more rapidly in concentration with distance from the landfill than did most of the so-called macroelements.

The tritium concentration beneath and around the landfill is interesting. Besides the effects of the nuclear tests in the 1960s, there are effects of the landfill. The tritium concentrations in the leachate contaminated ground water are clearly higher than in surrounding ground waters of the same age (Table 6.9 and Figure 6.15), indicating that tritium is an acceptable tracer to detect polluted ground water descended from this waste disposal site.

Table 6.6 Macroelement Contamination at the Noordwijk Disposal Site (mg/L)[a]

Contaminant	Control	Leachate	Ground Water Below Site	Ground Water Downstream (38 m − l.s.)
Cl^-	44	3310	3090	1670
HCO_3^-	312	5920	5510	3230
NH_4^+	1.4	700	640	236
Mg^{2+}	38	232	232	153
Na^+	42	2000	1850	1100
K^+	22	880	820	475

[a] Kooper (1983).

Table 6.7 Trace Elements at the Noordwijk Waste Disposal Site (μg/L)[a]

Contaminant	Control	Leachate	Ground Water Below Site	Ground Water Downstream (38 m − l.s.)
Ba	50	900	900	700
Li	5	500	485	270
Ni	10	100	80	20
Zn	10	310	270	20

[a] Kooper (1983).

Table 6.8 Organic Micropollutants at the Noordwijk Waste Disposal Site (μg/L)[a]

Contaminant	Control	Leachate	Ground Water Below Site	Ground Water Downstream (38 m − l.s.)
Benzene	—	100	30	10
Toluene	1	300	100	3
Xylene	1	600	400	300
Ethylbenzene	0.3	300	100	100
C_9H_{12}	3	300	100	30
Phenols	—	200	90	90
Camphor	—	1000	100	100
Org. Cl	—	26	14	16

[a] Morra et al. (1979).

Table 6.9   Tritium Levels at the Noordwijk Waste Disposal Site (T.U.)[a]

Depth (m − l.s.)	Control		Leachate	Polluted Ground Water	
7	109	88	143	112	93
10	42	7	133	116	103
15	8	12	158	63	88

[a] Tritium unit = 3.2 pCi/L.

Based upon the results obtained at Noordwijk, the following conclusions can be made:

1. Density flow can be an important phenomenon near waste disposal sites. Because of the high vertical flow component, it may be necessary to sink observation wells to the base of the aquifer for effective monitoring.

2. When a borehole is drilled through a polluted area, contamination of the aquifer can take place by short-circuit flow. The time for recovery to the original condition may significantly delay the taking of representative samples.

3. In a landfill situated partly in the ground water, a water-divide can be formed in the landfill resulting in an all-sided migration of landfill leachate. No ground water will flow through the landfill, and the ground water upstream may be contaminated. Control wells must be carefully selected.

4. Electrical resistivity and electromagnetic measurements can be used to trace movement of pollutants in both the horizontal and vertical directions. This reduces the number of observation wells required. However, because of interferences, electrical and electromagnetic measurements cannot be used to monitor pollutants above a clay base or in a fresh–salt-water transition zone. Electromagnetic measurements are faster and much cheaper than electrical resistivity, but both are equally effective.

5. Temperature profiles in general do not coincide with profiles of chemical parameters and cannot be used for tracing chemical pollution.

6. Uniform methods for sample-handling, sample-destruction, and analysis are required to obtain reliable data on trace elements in leachate and in heavily polluted ground water.

THE LEKKERKERK AFFAIR

The discovery of several severe cases of ground water pollution at ground water pumping stations in The Netherlands resulted in increased interest in possible risks of ground water contamination by local sources. At some pumping stations very high concentrations (mg/L) of organohalogen compounds such as trichloroethene and tetrachloroethene have been observed. In one case the pumping station was temporarily closed. In other cases the contaminated pumping wells were disconnected and used as screening wells until a purification plant was built.

The so-called Lekkerkerk affair (Brinkmann, 1981) focused attention in The Netherlands on ground water pollution from local sources. In the 1970s in the village of Lekkerkerk—situated in a very wet, peaty, clayey area with many ditches—rubbish, household wastes, and chemical wastes were dumped in ditches to prepare the land as a building site. Within a few years the waste penetrated to the ground water and contaminated tap water. Foul odors were present in the houses. The wastes and polluted soil were excavated and transported to an incineration plant. A total of 87,000 m^3 of contaminated soil and 1650 drums, some still filled with chemical waste, had to be destroyed. A special purification plant was built to reclaim the polluted ground water.

In 1978 the first signs of severe pollution appeared. The chemicals present caused fractures in drinking-water pipes. A foul-smelling floating layer was found on the surface of the ground water; the odor was also detected in several houses. Preliminary research by the National Institute for Water Supply indicated the presence of several volatile aromatic compounds commonly used in the paint industry. Organoleptic and visual mapping indicated that the pollution was mainly restricted to the ditches and adjacent areas (Figure 6.16). For these studies, soil samples were taken by simple handborings to a depth of about 2 m − l.s. in order to determine the most polluted localities. These studies coincided well with the results of a mine-detection survey indicating the possible presence of metal drums containing chemical wastes. Additional sampling of soil and ground water confirmed the results of the preliminary research. In a test excavation of an area 50 m^2 carried out at one of the most polluted locations, 42 rusted drums were found between 1.5 and 3 m − l.s. After removal of the pavement, concentrations of volatile aromatic compounds in the air required the use of gas masks. The extent of ground water contamination discovered during the test excavation is given in Table 6.10. The results of the test excavation, together with the discovery that pollutants had diffused through the water pipes into the tap water, led to a decision by authorities to evacuate the population and to remove the waste and polluted soil. The presence of a ground water pumping station a short distance from Lekkerkerk led to research on the quality of deeper ground water in the area. Contamination of the aquifer was possible via sand-filled creeks and river dunes in the upper Holocene clayey and peaty developed layers. However, because of upward seepage and the absence of heavy fluids that can cause density flow, no pollution in the deeper ground waters was detected. The soils at Lekkerkerk were heavily contaminated with metals and organics (Tables 6.11 and 6.12).

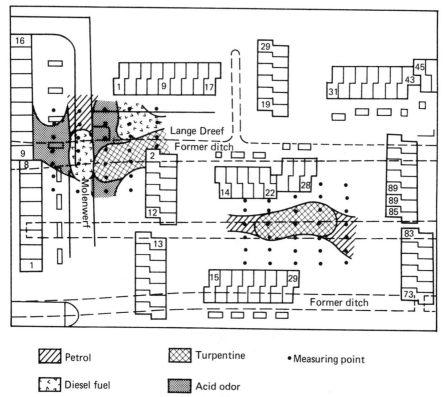

Petrol   Turpentine   • Measuring point

Diesel fuel   Acid odor

Figure 6.16. Results of organoleptic mapping in Lekkerkerk.

**Table 6.10 Ground Water
Quality at the Lekkerkerk
Disposal Site**[a]

Compound	Concentration (mg/L)
Toluene	100
Xylene	200
Ethylbenzene	30
C_3-Benzene	100

[a] Internal report Nat. Inst. for Water Supply.

Table 6.11 Metal Contamination of Soil at the Lekkerkerk Disposal Site (mg/kg dry soil)[a]

Metal	Concentration
Sb	230
As	9
Cd	97
Cr	140
Cu	490
Hg	8
Pb	740
Zn	1670

[a] Brinkmann, 1981.

Table 6.12 Organic Contamination of Soil at the Lekkerkerk Disposal Site[a]

Compound	Concentration (mg/kg)
Benzene	0.3
Toluene	1000
Xylene	400
Ethylbenzene	30

[a] Brinkmann (1981).

Table 6.13 Chemical Waste Dumps in The Netherlands

Description	Number
Dumps (total)	3860
Potential risks	1190
Protected by built-up areas	750
Require short-term clean-up	350

NATIONAL INVENTORY OF WASTE DISPOSAL SITES

According to a recent inventory, there are about 4000 registered waste disposal sites in The Netherlands (Table 6.13). An estimated 1200 of them are a possible risk to human health or the environment. Roughly 750 of these dumps are near ground water pumping stations or in urban areas like Lekkerkerk. Beside waste dumps, polluted industrial areas, especially former gasworks in urban areas and near ground water pumping stations, are serious threats to associated ground waters. Soil and ground water contamination by a gaswork in the city of Schiedam is presented in Table 6.14. The gaswork was closed in 1967. High concentrations of cyanide were widely distributed in the area. Polyaromatic and volatile hydrocarbons were limited to specific areas associated with their use. Heavy metals were detected in relatively low concentrations.

A classification based on 500 cases has been made of types and sources of soil contamination in The Netherlands (Table 6.15).

Short-term remedial action will be necessary at about 350 of the sites; over the long term this concerns 1600 cases. Regulations governing dump-

Table 6.14 Soil and Ground Water Pollution from a Gaswork in Schiedam (measured concentration ranges)

Compound	Dry Soil (mg/kg)	Shallow Ground Water (μg/L)
Benzene	<0.5–3.5	<2.0–440.0
Toluene	<0.5–52.0	<2.0–3300.0
Xylene	7.0–86.0	2.0–110.0
Naphthalene	<1.0–18.0	<200
Phenols	<0.2–5.7	< 20
Cresols	<0.3–12.0	<50.0–280.0
Pyridine	<1.0–33.0	<20.0–1300.0
Fenantrene+	2.8–196.7	100.0–160.0
Fluorantene	2.0–28.7	16.0–310.0
1,2-Benzantracene	0.1–31.2	0.1–88.4
Perylene	0.1–9.9	0.1–30.2
Benz(k)fluorantene	0.03–4.5	0.1–11.9
3,4-Benzpyrene	0.02–13.0	0.1–38.4
1,2-Benzperylene	0.1–9.7	0.7–42.7
CN	<1.0–3185.0	500.0–15100.0
CN (soluble)	<0.1–6.2	100.0–9300.0

Table 6.15 Classification of Types and Sources of Soil Contamination in The Netherlands Based on 500 Cases

Source of Contamination	Type of Contamination	Number	Frequency (%)
Gasworks	Aromatic hydrocarbons, benzene, phenols, CN$^-$	138	28
Waste dumps	Halogenated hydrocarbons, benzene, alkylbenzenes, metals like Hg, As, Pb, Cd, Ni, pesticides, CN$^-$	106	21
Oil pollution	Hydrocarbons, Pb	37	8
Chemical production and handling sites	Halogenated and aliphatic hydrocarbons, alkylbenzenes, phenols, metals like Hg, Pb, As, Cu, Zn, Ni, Cr	33	7
Metal plating industries	Trichloroethylene, CN$^-$, metals like Cr, Ni, Cu, Zn, Cd, Sn	31	6
Metal industries	Tri- and tetrachloroethylene, (aliphatic) hydrocarbons, phenols, metals like Cr, Ni, Cu, Zn, Cd, Sn, Pb	31	6
Painting industries	Alkylbenzenes, (halogenated)hydrocarbons, metals like Ni, Cu, Zn, Cd, Pb, Ti	27	5
Garages	Hydrocarbons	16	3
Vessel cleaning	Benzene, aromatic, aliphatic, and halogenated hydrocarbons, metals like As, Hg, Pb	16	3
Timber industry	Pentachlorophenol, aromatic and aliphatic hydrocarbons, metals like Pb, Sn, Zn	10	2
Dry cleaning	Tri- and tetrachloroethylene	6	1
Textile works	Hydrocarbons, Pb, Cr	6	1
Pesticide manufacturing works	Halogenated hydrocarbons, phenols, As	5	1
Sludge disposal	(Aliphatic) hydrocarbons, Pb, Zn	4	1
Enamel works	Aromatic and aliphatic hydrocarbons, tetrachloroethylene	3	0.5
Tanneries	Hydrocarbons	2	0.5
Various other types	Benzene, alkylbenzenes, phenols, trichloroethylene, (aliphatic) hydrocarbons, metals like Zn, Cd, Sn, Hg, Pb	26	5

site clean-up and risks associated with sites contaminated by hazardous wastes are now under consideration.

EFFECTS OF POLLUTANTS OF DRINKING WATER SUPPLY

The risk of ground water pollution is strongly influenced by hydrogeological conditions. About 35% of all ground water pumping stations in The Netherlands withdraw water from phreatic aquifers. Another 35% withdraw water from shallow aquifers covered by layers of low permeability. Human activities and sandy spots frequently increase the permeability of the layers covering shallow aquifers. Hence, layers of low permeability do not guarantee the protection of underlying aquifers against pollution (Figure 6.17). It appears that only about 30% of the 250 ground water pumping stations in The Netherlands can be characterized as relatively safe against pollution in the short term; most of the pumping stations may be influenced over the long term (Duijvenbooden, 1980). Taking into account that about 80% of the nearly

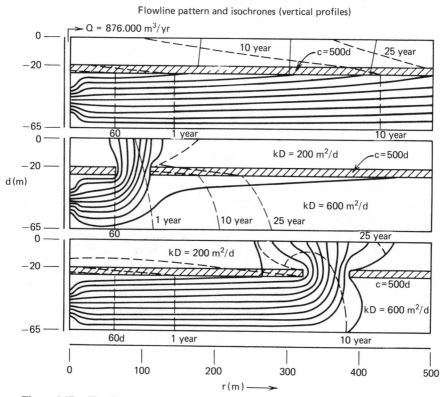

Figure 6.17.   Flowline patterns and isochrones (vertical profiles) (den Blanken, 1979).

1000 million m³ of tap water supplied each year by public water supplies is derived from ground water, it is clear that the presence of hundreds of local sources of pollution in the catchment and even protected areas of ground water pumping stations is of utmost concern. Recent severe cases of ground water pollution reinforce this statement. Without protective measures it should be expected that more and more pumping stations will be influenced by local sources of pollution. Remedial action is needed to prevent unacceptable ground water pollution. In addition, monitoring networks around ground water pumping stations will be required. Detailed knowledge of local flow patterns of ground water and a more or less complete inventory of all local sources of contamination in the catchment areas are necessary. Nevertheless, it can be expected that in the near future it will be necessary to build more and more purification plants to purify ground waters from polluted aquifers (Duijvenbooden, 1980).

REFERENCES

Blanken, M. G. M. den (1979). De beschermende werking van afdekkende lagen voor de grondwaterkwaliteit. H_2O 12(23):514–517.

Bom, C. M. (1981). Evaluation of different analytical procedures in the trace element analysis of a landfill leachate. In: W. van Duijvenbooden, P. Glasbergen, and H. van Lelyveld (Eds.), Quality of Groundwater. Elsevier, Amsterdam, pp. 495–500.

Brinkmann, F. J. J. (1981). Lekkerkerk. In: W. van Duijvenbooden, P. Glasbergen, and H. van Lelyveld (Eds.), Quality of Groundwater. Elsevier, Amsterdam, pp. 1049–1052.

Duijvenbooden, W. van (1979). Diffuse en lokale verontreinigingsbronnen en hun effecten op het grondwater. H_2O 12(23):525–529.

Duijvenbooden, W. van (1980). Verontreiniging grondwater vormt bedreiging drinkwater. De Ingenieur 92(49):10–13.

Duijvenbooden, W. van (1981). Groundwater quality in The Netherlands: Collection and interpretation of data. Sci. Total Envir. 21:221–232.

Kooper, W. F. (1983). Bodem- en grondwaterverontreiniging door vuilstortplaatsen. Internal report of the National Institute for Water Supply, The Netherlands, No. HyH 83-18.

Kooper, W. F., van Duijvenbooden, W., and Peeters, A. A. (1981). Invloed van vuilstortpercolaat te Noordwijk op de bodem en het oppervlaktewater. Internal report of the National Institute for Water Supply, The Netherlands, No. HyH 81-12.

Kruijtzer, G. F. J. (1980). The downward penetration of a spherical foreign fluid substance in an aquifer. L.G.M.-Mededelingen, part XXI, no. 2, pp. 153–160.

Morra, C. F. H., van Duijvenbooden, W., Slingerland, P., and Piet, G. J. (1979). Orienterend onderzoek naar organische microverontreinigingen in het grondwater bij de vuilstortplaats te Noordwijk. RID-mededeling, No. 79-1.

Obdam, A. N. M. (1979). Calculation of the salinity process of partially penetrating wells in a semi-confined anisotropic aquifer. *Proceedings SWIM*, Hannover, pp. 159–175.

Trines, H. (1952). Cyanose bij zuigelingen, als gevolg van het gebruik van nitraathoudend putwater bij de voeding. *Verslagen en Mededelingen Volksgezondheid* **1952**(May, June):481–503.

Zoeteman, B. C. J., de Greef, E., and Brinkmann, F. J. J. (1981). Persistency of organic contaminants in groundwater: Lessons from soil pollution incidents in The Netherlands. *Sci. Total Envir.* **21**:187–203.

7

REDOX REACTIONS, MINERAL EQUILIBRIA, AND GROUND WATER QUALITY IN NEW ZEALAND AQUIFERS

C. J. Downes

Chemistry Division
Department of Scientific and Industrial Research
Petone, New Zealand

Most of the papers presented at this conference documenting changes in ground water quality are concerned with aquifers that have been contaminated with industrial or agricultural effluents. In contrast, the variations in the water quality characteristics for the two aquifers considered in this paper are the result of natural processes.

The first study, which comprises the bulk of the paper, is an investigation of changes in quality of the Hutt Valley ground waters arising from rock–water interactions and oxidation–reduction reactions as the water passes from the recharge to the discharge zone of this greywacke gravel aquifer. A wider range of oxidation–reduction conditions was encountered in this system than in the few reported studies where a similar approach has been adopted. In the past, undesirable features such as high levels of dissolved iron, or the corrosivity of water from a particular bore, have usually been considered as isolated chemical problems. It will be demonstrated that an appreciation of the chemistry of a water in terms of the oxidation–reduction sequence, in combination with the hydrology, is important in deciding on the strategy required for the exploitation and management of a ground water resource. For instance, such an approach allows predictions to be made of possible changes in water quality that would result from increased abstraction rates or relocation of major abstraction points. Knowledge of the oxidation–reduction conditions is likely to be required to interpret chemical transformations in polluted aquifers.

The second study, of the Takaka Valley ground water, is an investigation of the sea water intrusion into a coastal karst aquifer and the unexpected variation of the extent of this intrusion with changes in the flow of the Waikoropupu Springs, the only on-shore discharge point for this ground water system.

CHEMISTRY OF THE HUTT VALLEY GROUND WATERS

Artesian wells in the lower part of the Hutt Valley supply all the potable water requirements for the municipalities of Lower Hutt, Petone, and Eastbourne and are a supplementary source for Wellington City. An average of 66×10^3 m^3/day and a maximum of about 91×10^3 m^3/day is withdrawn from the main aquifer, largely via the Lower Hutt and Petone supply wells, which are situated in the lower part of the Valley (Figure 7.1). In general the quality of the ground water is good, the only treatment required before distribution of public supplies being the addition of lime to suppress corrosion of the reticulation by high levels of dissolved carbon dioxide and the addition of fluoride to the level of 1 g/m^3 as a public health measure.

Although the Hutt River is the ultimate source of most of the water abstracted from the Hutt Valley aquifer, there are important advantages in using the ground water for potable purposes rather than drawing on the river directly. During periods of high precipitation, the Hutt River is turbid, and settling or flocculation would be required to obtain a water of acceptable clarity at these times. Further, chlorination would be required for the destruction of pathogenic bacteria, a process which is accomplished naturally during passage of the water through the aquifer. Another advantage is that the aquifer provides storage, which allows withdrawal rates to exceed average recharge rates, at least for short periods.

Nitrate Levels

The original impetus for this water quality investigation came from the observation (Stevenson, 1975) that nitrate levels in the Lower Hutt supply wells had increased from generally below the detection limit of 0.01 g/m^3 NO$_3$–N in the early 1960's to moderate levels of 0.4–1.5 g/m^3 NO$_3$–N. This concentration is far below the World Health Organization maximum recommended value of 10 g/m^3 NO$_3$–N, but it was important to establish the reason for the increase in order to guard against the possibility of a continuing trend to higher levels. Two possible reasons were advanced to explain

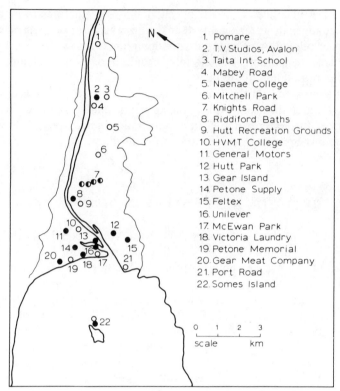

Figure 7.1. Location of bores sampled; filled, half-filled, and open circles are, respectively, production, new production not yet in regular use, and observation bores.

this rise: first, that there had been an increase in the input of nitrogen species to the aquifer in the unconfined region in the upper part of the Valley; second, that there had been a change in the dissolved oxygen levels in the vicinity of the supply wells that was influencing the extent to which denitrification was proceeding. Because the average nitrate level of the Hutt River is only about 0.3 g/m^3 NO$_3$–N, and because urban development had spread further during the period over fertile agricultural land in the unconfined zone, the first alternative appeared unlikely.

Present Investigation

Initially it was planned to determine variations in the concentration of dissolved nitrogen species, dissolved oxygen, iron, and manganese between the recharge zone and the main abstraction wells to test the applicability of the second hypothesis. However, because the preliminary results showed such significant variations in water quality parameters within this relatively small aquifer, the study was extended to characterize fully the chemical changes occurring over the entire length of the aquifer. Additional features, including the extensive array of bores for sampling the ground water and the well-documented geology and hydrology, indicated the Hutt Valley aquifer to be an ideal system for a field study of plausible, but incompletely verified, changes in water quality to be expected from rock–water interactions and oxidation–reduction reactions.

In the following sections a brief outline is given of the stratigraphy and hydrology in order to place the water quality results in context, together with a section dealing with the thermodynamic framework used in the interpretation of these results.

STRATIGRAPHY AND HYDROLOGY

Near the southern end of the North Island of New Zealand, a series of basins has been formed by longitudinal warping about an axis normal to the Wellington Fault. The Lower Hutt-Port Nicholson basin, the most southerly in the series, has been infilled by the alluvial Hutt Formation, which includes a prominent aquifer called the Waiwhetu Artesian Gravels (Stevens, 1956). The ground water system occupies a triangular area (Figure 7.1). Two confining boundaries are the Western Hills, forming the scarp of the Wellington Fault, and the Eastern Hills, converging to an apex at the south end of Taita Gorge, and the third side is the natural discharge zone south of Somes Island in Port Nicholson. Well-log data—most of which are for wells not deeper than the Waiwhetu Artesian Gravels, the main aquifer—supplemented by geophysical investigations have been used to determine the stratigraphy (Figure 7.2). In the northern section the ground water is unconfined, but south of about Mitchell Park (6),* the ground water is confined by a capping structure that includes the Melling Peats and the Petone Silts and Clays.

Below the Wilford Shell Bed, which is in the upper part of the Moera Basal Gravels, there is a fresh-water aquifer with a head 1.5–2 m above that of the main aquifer. It extends across the valley and at least 1.5 km inland from the harbor, but only a limited amount of water is withdrawn from it at Gear Meat Company (20) on the foreshore. At deeper levels, under the same hydraulic head, up to six saline horizons have been identified with reported maximum salinities of 1760 g/m^3 at a depth of 140–161 m (Hutton, 1965). All the chemical results reported in the present paper are for bores tapping the main aquifer.

Of the two possible sources of recharge water—the Hutt River and rainfall infiltrating the unconfined zone—Donaldson and Campbell (1977) concluded, as did earlier workers, that the Hutt River is by far the more important. Under low flow conditions during droughts, gauging of the river has shown that net infiltration, mostly in the vicinity adjacent to bores 2 and 4

---

* Numbers in parentheses correspond to those in Figure 7.1 and are given to aid the reader in locating bores discussed in the text.

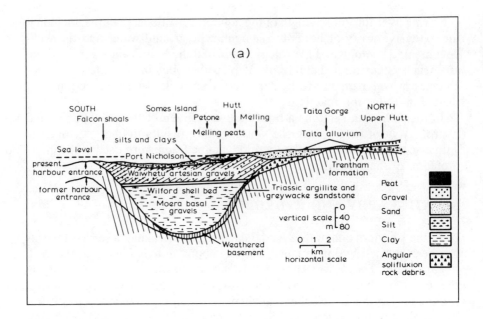

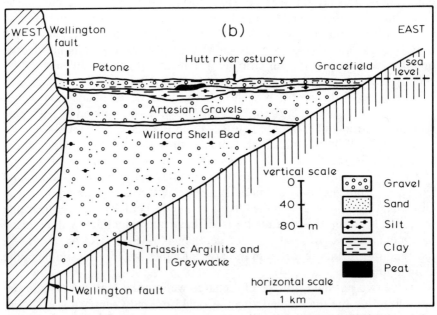

Figure 7.2. Stratigraphy of the Lower Hutt Valley. (*a*) Longitudinal section. (*b*) Cross section near the Petone foreshore. After Stevens (1956) and Donaldson and Campbell (1977).

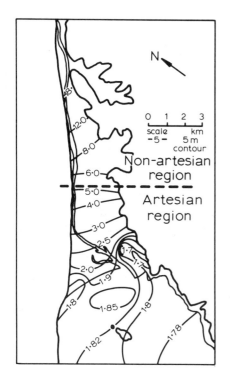

Figure 7.3. Pressure distribution under normal operation, municipal withdrawal at Hutt Park (12), heads in meters above the Lower Hutt City Corporation datum. After Donaldson and Campbell (1977).

(Figure 7.1), amounts to 50×10^3 m³/day while average usage at these times was about 77×10^3 m³/day. These figures indicate that storage is being called upon. They assessed the rainfall contribution as about 14×10^3 m³/day, their estimate being based on rainfall records for the 9 km² catchment of the unconfined section and the relevant part of the Eastern Hills.

Hydrologic models developed by Donaldson and Campbell (1977), who used data provided by an array of observation bores fitted with automatic recorders, are used in the day-to-day management of the aquifer during critical periods to avoid the possibility of sea water intrusion. At present (1981), the main abstraction point is being shifted from Hutt Park (12) near the foreshore to a transect along Knights Road (7) further up the Valley because these models indicate this shift will improve the average yield of ground water by up to 60%. The main abstraction point at Hutt Park (12) strongly influences the flow of water through the aquifer (Figure 7.3).

THERMODYNAMICS

Solution-Mineral Equilibria

To test whether ground waters are undersaturated, at saturation, or supersaturated with respect to possible reactant and product minerals, the observed reaction quotient Q can be compared with the equilibrium constant K. It is convenient to express the reaction state as ΔG_R, the difference in Gibbs free energy between the actual state and the equilibrium state,

$$\Delta G_R = RT \ln (Q/K)$$

where R is the gas constant and T is the absolute temperature. The three possible states, $\Delta G_R < 0$, $\Delta G_R = 0$, $\Delta G_R > 0$, correspond to the solution being unsaturated, saturated, or supersaturated, respectively, with the mineral being considered. The ΔG_R values do not indicate which reactions actually do occur in a finite time. They do, however, allow some reactions to be excluded as impossible; solids cannot dissolve into saturated solutions, nor can solids precipitate from unsaturated solutions. Values of ΔG_R were obtained using the SOLMNEQ program of Kharaka and Barnes (1973), which takes account of complexing and acid–base reactions affecting the activities of the ions considered in the solubility product. A restriction of this approach is that the reaction states pertain only to the water sample as collected and analyzed and do not necessarily define the pathway by which the solutes were acquired.

Oxidation–Reduction Reactions

Although this familiar concept has been applied to ground water chemistry from a theoretical point of view (Stumm and Morgan, 1970), there have been few comprehensive accounts published which use it for the interpretation of field observations (Edmunds, 1973; Champ et al., 1979). This is probably due to the existence of relatively few aquifers where bores are available to permit sampling of ground waters covering other than a restricted portion of the range of possible oxidation–reduction reactions.

The same principles of oxidation–reduction apply to reactions occurring progressively with depth below the sediment–water interface in unconsolidated sediments, but there the relations are obscured by movement of water and solutes due to compaction, diffusion of solutes from the overlying water, inhomogeneity of the sediment, and the activities of burrowing organisms.

The oxidation of organic carbon by increasingly weaker oxidizing agents, as each is depleted in turn, is the basis of the model. In natural aquatic systems these oxidizing agents are, in order of decreasing strength, molecular oxygen, nitrate ions, manganese (IV) and iron (III), oxides, sulfate ions, carbon dioxide, and molecular nitrogen. An equivalent statement is that organic matter reduces each of the species in turn, starting with molecular oxygen.

Qualitatively, it is easy to see that each of the oxidants must be consumed, or nearly consumed, before the reduction of the next strongest oxidant commences. For instance, if reduction of ferric compounds were to proceed before the molecular oxygen had been largely removed, then the ferrous ions produced would be susceptible to being oxidized back to ferric oxides/hydroxides by molecular oxygen.

Following Stumm and Morgan (1970), CH_2O is taken as a model organic compound corresponding to the active reducing agent in natural waters, and its oxidation, either fully to CO_2 or partially to another organic compound, here represented by $HCOO^-$, is given by equations (7.1a) and (7.1b), respectively.

$$\tfrac{1}{4}CH_2O + \tfrac{1}{4}H_2O = \tfrac{1}{4}CO_2(g) + H^+(W) + e \qquad (7.1a)$$

$$\tfrac{1}{2}CH_2O + \tfrac{1}{2}H_2O = \tfrac{1}{2}HCOO^- + \tfrac{3}{2}H^+(W) + e \qquad (7.1b)$$

Apart from the organic matter, most species entering the aquifer will be in an oxidized form, namely O_2, NO_3^-, and SO_4^{2-}; iron and manganese oxides already exist in the aquifer–many of the well logs note brown-staining of the gravels.

The reduction equations to be considered are:

$$\tfrac{1}{4}O_2(g) + H^+(W) + e = \tfrac{1}{2}H_2O \qquad (7.2)$$

$$\tfrac{1}{5}NO_3^- + \tfrac{6}{5}H^+(W) + e = \tfrac{1}{10}N_2(g) + \tfrac{3}{5}H_2O \qquad (7.3)$$

$$\tfrac{1}{8}NO_3^- + \tfrac{5}{4}H^+(W) + e = \tfrac{1}{8}NH_4^+ + \tfrac{3}{8}H_2O \qquad (7.4)$$

$$\tfrac{1}{2}MnO_2(s) + 2H^+(W) + e = \tfrac{1}{2}Mn^{2+} + H_2O \qquad (7.5)$$

$$Fe(OH)_3(s) + 3H^+(W) + e = Fe^{2+} + 3H_2O \qquad (7.6)$$

$$\tfrac{1}{2}CH_2O + H^+(W) + e = \tfrac{1}{2}CH_3OH \qquad (7.7)$$

$$\tfrac{1}{8}SO_4^{2-} + \tfrac{9}{8}H^+(W) + e = \tfrac{1}{8}HS^- + \tfrac{1}{2}H_2O \qquad (7.8)$$

$$\tfrac{1}{8}CO_2(g) + H^+(W) + e = \tfrac{1}{8}CH_4(g) + \tfrac{1}{4}H_2O \qquad (7.9)$$

$$\tfrac{1}{6}N_2(g) + \tfrac{4}{3}H^+(W) + e = \tfrac{1}{3}NH_4^+ \qquad (7.10)$$

These equations are written in a form with unit stoichiometry for the electron, which facilitates the computations of the free energies for the combinations given below. $H^+(W)$ is the hydrogen ion concentration of neutral aqueous solutions ($10^{-7}M$). The equations for the oxidation of organic matter, (7.1a) and (7.1b), can be combined with the reduction equations, (7.2) through (7.10), to describe the processes given in the following list (Stumm and Morgan, 1970).

Process	Combination	$-\Delta G°(W)$ kJ
Aerobic respiration	(7.1a) + (7.2)	125
Denitrification	(7.1a) + (7.3)	119
Nitrate reduction	(7.1a) + (7.4)	82
Manganese (IV) reduction	(7.1a) + (7.5)	81
Iron (III) reduction	(7.1a) + (7.6)	28
Fermentation	(7.1b) + (7.7)	27
Sulfate reduction	(7.1a) + (7.8)	25
Methane fermentation	(7.1a) + (7.9)	23
Nitrogen fixation	(7.1a) + (7.10)	20

$\Delta G°(W)$ is the free energy change for the process considered with all reactants and products in their standard states, except for H^+.

Assuming the presence in the aquifer of organic matter able to act as the reducing agent, there will be a sequence of changes in the concentrations of species participating in oxidation–reduction reactions (Figure 7.4). Proceeding from the recharge area there will be a decline in dissolved oxygen values with a concurrent increase in the level of dissolved carbon dioxide. Some increase in the nitrate concentration is possible due to nitrification of organic nitrogen under aerobic conditions. When dissolved oxygen has fallen to a low level, nitrate values will decrease because of the onset of denitrification, and manganous oxides, followed by iron oxides, will be reduced to yield soluble manganous and ferrous ions in solution. At still lower levels of dissolved oxygen, sulfate reduction will commence and lead to decreasing iron (and manganese, not shown in Figure 7.4) values due to the precipitation of insoluble sulfides. After all the sulfate has been reduced, dissolved iron (and manganese) will again be able to increase. Finally, under extremely

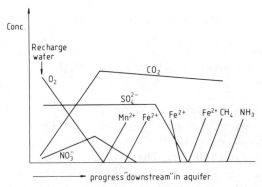

Figure 7.4. Schema of predicted changes in concentrations of species susceptible to oxidation–reduction reactions with progress of water through an aquifer.

reducing conditions, methane and ammonia are to be expected. The decrease depicted for carbon dioxide is expected to arise mainly not from oxidation-reduction reactions, but from the reaction of carbonic acid with the aquifer minerals.

Rates and Mechanism

This thermodynamic reaction sequence assumes there is an approach to an equilibrium state at each stage, but it is possible for some species to exist metastably for very long periods of time because the rates of the reactions which would tend to restore thermodynamic equilibrium are infinitely slow; for example, under conditions prevailing in the Earth's atmosphere, N_2 should be largely oxidized to nitrate. Also, this sequence does not imply that processes such as the reduction of ferric compounds, before molecular oxygen has been largely depleted, are mechanistically unimportant. Indeed, such a pathway, and there is some evidence to support it, has been proposed for the oxidation of organic matter by molecular oxygen (Stumm and Morgan, 1970). Thus

$$Fe(III)-organic\ complex \rightarrow Fe(II) + oxidized\ organic$$

$$Fe(II) + \tfrac{1}{4}O_2 + organic \rightarrow Fe(III)-organic\ complex$$

the *net* reaction being the oxidation of organic carbon by molecular oxygen with the ferrous–ferric system acting as a catalyst. Nevertheless, thermodynamics provides a useful framework for systematizing observations of changes in natural systems, and it identifies reactions which are kinetically controlled.

Biological Mediation

It is generally assumed that there will be microbiological species present in aquifers (McNabb and Dunlap, 1975) to catalyze all or most of the oxidation–reduction reactions. Recent work, including that discussed above for the ferrous–ferric catalysis of the oxidation of organic matter by molecular oxygen, has shown that nonbiological processes operating under ambient conditions may also be important in transformations which were previously thought to be the exclusive domain of microbes. Also, as nitrate is reduced by ferrous ions in acid solution (Chao and Kroontje, 1966), then, at least in principle, the coupling of denitrification with oxidation of organic matter, catalyzed by the ferrous–ferric system, would be possible without the need for the participation of microbes. The investigation of the relative importance of biological as opposed to nonbiological transformations, however, is outside the scope of the present study.

METHODS

Sampling

Because short term variations in the chemical composition of the river water in response to flow and seasonal changes were expected to be greater than that of the ground water, in which such effects are attenuated by mixing of waters from multiple flow paths, the Hutt River was sampled weekly in the infiltration area over a period of 1 year to characterize chemically the main source of recharge for the aquifer. Most of the bores shown in Figure 7.1, on the other hand, were included in two major sampling rounds, separated by 6 months, and, in addition, selected bores in both the confined and the unconfined zones were sampled repeatedly at intervals of 1 to 2 months for up to 4 years to detect long-term changes. There is good coverage of much of the confined zone by production bores; in the unconfined zone, however, more reliance had to be placed on observation bores. A submersible electric pump was used to flush the observation bores and to obtain the samples. Samples for chemical analysis were filtered on-site, through 0.45 μm filters in Swinex holders, into acid-cleaned polyethylene bottles.

Analysis

The pH was usually measured in the field using an Orion 401 portable pH meter, calibrated with appropriate buffer solutions; under adverse weather conditions the pH was measured in the laboratory within an hour or two of collection of the sample. Dissolved oxygen levels were measured in the field using a Yellow Springs instrument, model YSI57, and supplemented with Winkler determinations; particular care was taken to exclude air from the samples.

Filtered samples for the determination of reactive aluminum by the method of Barnes (1975) were extracted in the field. Other species were determined by the standard methods of the Water Laboratory, Chemistry Division, DSIR, which are similar to those of APHA (1976) and Brown et al. (1970).

Parameters to be determined for each sample were selected from the following list: pH, bicarbonate, free carbon dioxide, lithium, sodium, potassium, magnesium, calcium, strontium, iron, manganese, zinc, reactive aluminum, fluoride, chloride, sulfate, dissolved silica, dissolved oxygen, nitrate nitrogen, ammonia nitrogen, soluble organic nitrogen, reactive dissolved phosphorus. (Some of these parameters of lesser intrinsic interest are required in determining the reaction state of minerals. For example, fluoride is required because aluminum complexes contribute to the level of "reactive" aluminum, from which is evaluated the concentration of "free," that is uncomplexed, Al^{3+}.) All these parameters were determined for samples collected in the two major sampling rounds of the ground water. For supplementary samples collected later the analysis was generally restricted to pH, dissolved oxygen, nitrogen species, iron, and manganese. Included in the analyses of the river water were nitrogen species and, for most samples, the major ions, iron, manganese, dissolved silica, reactive dissolved phosphorus, and pH.

RESULTS AND DISCUSSION

Water Quality Data

All the analytical results have been tabulated (Downes, 1980), but only those central to the discussion of water quality changes are presented here (Figure 7.5a, b, c, d). The parameters are plotted against the bore numbers (Figure 7.1), with the numbering beginning in the recharge zone and progressing "downstream" in the aquifer though the relation between bore number and length of flow path becomes less distinct due to the spread of wells across the lower Valley. Closely similar chemical results were obtained for the two major sampling rounds and for the extended study of selected bores, indicating steady state conditions. The only exceptions to this situation were minor variations in the nitrate levels of observation bores in the unconfined zone. Average values obtained from the year-long survey of the main source of recharge water, the Hutt River, are plotted on the graphs as "R," and ranges are also given where they are significant in relation to ground water levels. No values are reported for pH, carbon dioxide, bicarbonate, or calcium for the main abstraction wells (12) because samples could not be obtained before the lime-addition step.

Major Solutes

Aside from suspended material and biological components, the Hutt River is composed of a dilute solution of sodium, calcium, chloride, and bicarbonate ions, together with dissolved silica, and with lesser concentrations of magnesium, potassium, and sulfate ions. Much of the sodium, chloride, magnesium, and to a lesser extent sulfate, will be contributed by cyclic salt from the precipitation falling in the catchment, but the major source of the other solutes will be weathering-reaction products acquired when the water drains through the soil before reaching the river. Chloride levels of the ground waters are not above that of the river except at bores 15 and 22, which possibly contain a sodium chloride component leached from the sediments. Increases in sodium concentrations as the water progresses through the aquifer are more significant, particularly the approximately 50% elevation on the eastern side. There is about a doubling of the calcium (and magnesium, not shown), with again the larger increases being on the eastern side of the Valley. Sulfate values of the ground waters vary somewhat erratically within

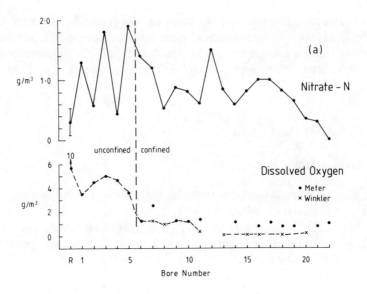

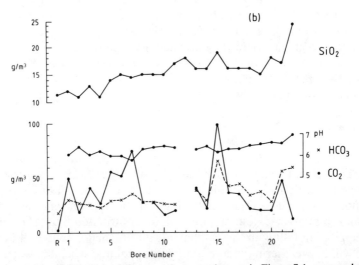

Figure 7.5. Water quality of ground waters, bore numbers as in Figure 7.1, mean values given for R, the river recharge water: (*a*) nitrate and dissolved oxygen; (*b*) silica, pH, bicarbonate, and dissolved carbon dioxide; (*c*) calcium, sodium, and chloride; (*d*) sulfate, iron, phosphate-phosphorus, ammonia-nitrogen, and manganese.

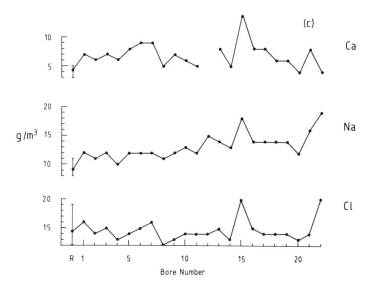

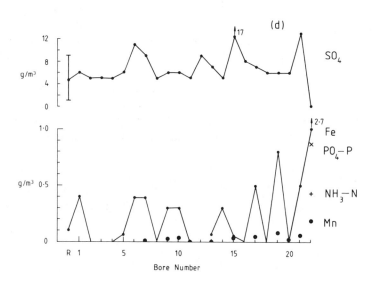

the range observed for the Hutt River, except for the high value of the Feltex bore (15), and the Somes Island bore (22), where the sulfate content is below the detection limit of 1 g/m^3; reasons for both exceptions are discussed in the selection below dealing with oxidation–reduction reactions.

Bicarbonate levels of the aquifer water are elevated above the average river value, with a trend to higher values with increasing distance from the recharge zone. Carbon dioxide content of the ground water is markedly higher than the maximum value of 2 to 3 g/m^3 for the river, with values of 20 to 30 g/m^3 for wells near the recharge area and over 100 g/m^3 on the eastern side of the Valley, but declining to 12 g/m^3 at Somes Island (22). For much of the aquifer, increases in carbon dioxide outstrip that of bicarbonate and result in acidic waters with pH values as low as 5.8 (bore 7).

Dissolved silica values trend distinctly upwards with increasing distance from the recharge area. Although quartz is one of the minerals present in the gravels, it is not the source of the additional dissolved silica because the river water concentration is already above the solubility limit of crystalline silica.

Rock–Water Interaction

These changes in the composition of the water in the aquifer, compared with the river water, are readily explained by carbonic acid attack on minerals in the gravels of the aquifer or in overlying soils in the unconfined area of the aquifer, where there is direct infiltration of rainfall. Oxidation, probably largely bacterial, of organic matter in the water and preexisting organic material in the aquifer is almost certainly the source of this carbon dioxide. Out of direct contact with the diluting effects of the atmosphere, carbon dioxide levels rise and produce acid waters that attack silicate and carbonate minerals in the aquifer gravels. This attack can be written in a general form as

$$\text{Cation-Al-silicate} + H_2CO_3 + H_2O \rightarrow HCO_3^-$$
$$+ H_4SiO_4 + \text{cation} + \text{Al-silicate}$$

and

$$CaCO_3 + H_2CO_3 \rightarrow Ca^{2+} + 2HCO_3^-$$

illustrating the release of cations (Na^+, K^+, Mg^{2+} and Ca^{2+}) together with dissolved silica into the water and also formation of bicarbonate ions. Aluminum is very largely conserved, with only very low levels of "reactive" aluminum appearing in solution (Table 7.1).

Results of calculations of mineral-solution equilibria (Table 7.1a) obtained using the SOLMNEQ program show that of the mineral components of greywacke, the waters are unsaturated with respect to albite, calcite, chlo-

Table 7.1 Reaction States of Selected Bore Waters with Respect to Solution-Mineral Equilibria, ΔG_R (kJ), and Concentrations of Reactive Aluminum (mg/m³)

Bore	ΔG_R (kJ)									Concentration of Reactive Aluminum (mg/m³)
	Albite	Calcite	Mg Chlorite	Anorthite	Quartz	Muscovite	Kaolinite	Boehmite	Gibbsite	
Pomare (1)	−20	−18	−85	−42	+2.6	+29	+13	−1.0	−4.2	9
TV studio (2)	−19	−17	−24	−38	+2.3	+38	+12	−1.1	−4.6	3
Taita School (3)	−20	−18	−84	−41	+2.6	+38	+13	−1.1	−4.2	9
Mabey Road (4)	−19	−18	−80	−41	+2.5	+38	+13	−0.96	−4.1	6
Naenae College (5)	−20	−18	−88	−43	+2.9	+36	+12	−1.8	−5.0	8
Mitchell Park (6)	−18	−18	−80	−41	+3.0	+39	+13	−1.1	−4.2	9
Hutt Grounds (9)	−14	−15	−59	−33	+3.0	+47	+18	+1.1	−2.1	9
HVMT College (10)	−14	−16	−54	−33	+3.0	+46	+17	+0.75	−2.5	6
General Motors (11)	−16	−17	−67	−38	+3.3	+30	+13	−1.6	−5.0	3
Petone Supply (14)	−14	−17	−59	−35	+3.1	+44	+16	−0.04	−3.4	5
Unilever (16)	−19	−15	−71	−44	+3.1	+31	+7.1	−4.2	−7.5	1
McEwan Park (17)	−17	−15	−63	−39	+3.1	+37	+11	−2.5	−5.9	2
Victoria Laundry (18)	−16	−16	−59	−38	+3.1	+39	+12	−1.9	−5.0	2
Petone Memorial (19)	−13	−15	−50	−31	+3.0	+48	+18	+1.0	−2.1	6
Port Road (21)	−14	−13	−50	−35	+3.3	+32	+13	−1.4	−4.6	2
Somes Island (22)	−5.4	−13	−20	−23	+4.2	+59	+23	+2.3	−0.88	4

rite, and anorthite, but are supersaturated with respect to quartz and musco-
vite. The waters are also supersaturated with respect to several aluminosilicates, of which kaolinite is given as an example, suggesting that weathering
reactions are proceeding with the formation of clay minerals. Aluminosilicates formed initially in these reactions may be amorphous precursors of
clay minerals; the deposition of such a phase (allophane) from a cold, dilute
spring water of fairly unexceptional composition apart from a reactive aluminum content of 0.13 g/m^3 has been described previously (Wells et al., 1977).
Several of the bore waters are close ($\sim \pm 1$ kJ) to saturation with respect to
boehmite (AlO(OH)) as is the Somes Island (22) water with respect to gibbsite (Al(OH)$_3$); these saturations are possibly indicative of the phases limiting the concentration of aluminum in solution.

   With current debate on the effects of acid rain on ground water, it is of
interest to calculate from the results of this study the rate of acid attack on
silicate minerals under natural conditions, a measure being provided by the
increase in the bicarbonate concentration of the ground water over that of
the recharge water. Using tritium dating, Grant-Taylor and Taylor (1967)
found the residence time for water from the main wells at Hutt Park (20) to
be about 40 months. As noted previously, neither the bicarbonate nor the
calcium levels of untreated water from these wells can be determined, but
good estimates are possible from trends for the other wells. In their travel
through the aquifer, these ground waters with a pH near 6 and a temperature
of 14°C have lost about 0.1 mole of protons per m^3 per year ($10^{-4}M$ H$^+$ per
year), assuming most of the reaction has been with silicates.

Species Susceptible to Oxidation–Reduction Reactions

Hutt River

Because the Hutt River is shallow and flows moderately fast in the recharge
zone, dissolved gas concentrations would be expected to be close to values
for equilibrium with the atmosphere. At ambient temperatures the concentration of dissolved carbon dioxide in water in equilibrium with air is about
0.5 g/m^3 compared with the values of up to 2 to 3 g/m^3 that can be calculated
from measured values of the pH and alkalinity. As there are no known
discharges contributing appreciable biochemical oxygen demand, dissolved
oxygen values should be close to the equilibrium values of between 13 and 9
g/m^3 at 5 and 20°C, respectively. Under these "oxidizing" conditions nitrate-nitrogen concentrations predominate over those of ammonia-nitrogen,
and much of the small amount of iron present is likely to be associated with
humic acid.

   The average nitrate-nitrogen concentration of weekly samples from the
one year survey was 0.30 g/m^3, while ammonia-nitrogen concentrations
were either below the detection limit of 0.01 g/m^3, or else were in the range
0.01 to 0.05 g/m^3, the only exception being a sample collected after a large
storm.

Unconfined Zone

Dissolved oxygen levels decrease greatly after the water enters the aquifer (Figure 7.5a). Insidious leakage of atmospheric oxygen into pumping and sampling systems cannot be completely discounted. However, the measured values should represent maximum levels; that is, the true value could be lower but is unlikely to be higher than the measured value. On this basis, a maximum of 5 g/m^3 of dissolved oxygen is consumed between Mabey Road (4)–TV (2) and Naenae College (5)–Mitchell Park (6).

Nitrate-nitrogen levels of 0.40 and 0.60 g/m^3 for the Mabey Road (4) and TV (2) bores, respectively, are only slightly above the river value, but the increases for the bores at Pomare (1), Taita (3), Naenae (5), and Mitchell Park (6) are much more significant, the level being near 2.0 g/m^3 NO_3—N for the Naenae samples. That at least a large proportion of the water passing through the aquifer has its nitrate content considerably raised above the average river value of 0.3 g/m^3 NO_3—N is shown by the value of about 1.5 g/m^3 NO_3—N obtained at the main abstraction wells at Hutt Park (12). Mineralization and nitrification of organic nitrogen is one obvious source of the additional nitrate. Although the average value of the soluble organic nitrogen for the Hutt River samples was only 0.20 g/m^3, and accompanied nearly always by negligible amounts of ammonia, there could also be contributions from organic nitrogen bound in larger particles carried into the aquifer by the water, or from preexisting material in the aquifer.

Literature reports indicate 4.35 g of oxygen is used in the nitrification of 1 g of ammonia-nitrogen. If this result is used as an approximation of the oxygen requirements for mineralization and nitrification of organic nitrogen, then consumption of 5 g/m^3 of oxygen would produce 1.15 g/m^3 of NO_3—N, which is sufficient to account for the observed increases. However, this figure represents an unrealistic maximum because an appreciable quantity of the oxygen is used in microbial respiration, as shown by the observed increases in the level of dissolved carbon dioxide. If this is the mechanism for raising the nitrate levels above that of the river, then oxygen must have access in this unconfined part of the aquifer to more than just the uppermost layers of ground water. Samples from a well located between Mitchell Park (6) and the river, which taps shallow, perched ground water, had levels near 0.6 g/m^3 NO_3—N, suggesting soil water in the unconfined zone is probably not of overriding importance as a source of nitrate for the main aquifer.

Significant iron concentrations (up to 0.4 g/m^3) were found in filtered waters from some of the bores in this unconfined area; however, since the iron was not accompanied by detectable manganese (i.e., >0.01 g/m^3), it was possibly a result of corrosion of the casing.

To summarize, in this northern unconfined section of the aquifer, nitrate levels are appreciably above those of the river while dissolved oxygen levels have been depleted to less than half the saturated values of the river water, with a general decline as the water progresses through the aquifer.

Confined Zone

This area is characterized by a continuing trend toward low levels of dissolved oxygen. Reliance had to be placed on values determined using the dissolved oxygen meter for those waters containing sufficient iron to interfere with the Winkler method. Differences in values obtained by the two methods, where both were used, warrant comment. In laboratory tests in which residual oxygen was removed chemically from aqueous solutions, the meter indicated values of no more than a few tenths of 1 g/m^3, a variance insufficient to explain the differences between the two methods for the field measurements. Some of the discrepancy in nearly anaerobic natural waters could be due to the presence of a species interfering in the electrochemical method. In another study of ground waters, with dissolved oxygen levels about 50% of air-saturation values, the meter and Winkler methods were in excellent agreement. Nevertheless, it is clear that conditions become more reducing in the confined area of the aquifer; moreover, the unambiguous, low dissolved oxygen values obtained by the Winkler method for wells near the foreshore—for instance 0.05 and <0.05 g/m^3 for Unilever (16) and Victoria Laundry (18) respectively—indicate the water is close to being anaerobic by the time it has reached this part of the aquifer.

There is a somewhat erratic decline in nitrate levels from the highest values of the unconfined zone (Figure 7.5a). If the nitrate values for wells 10 to 21 are replotted as a transect across the lower valley a clearer picture emerges—there are low nitrate values for bores on either side of the valley with a peak at the main abstraction bores at Hutt Park (12). It could be argued that the low nitrate values on the west result from the fact that water passing down this side of the valley has never had its nitrate content raised to the levels of the observation bores in the unconfined zone or even as high as the Hutt Park bores, or that there is simply a continuous increase in nitrate levels between the river and Hutt Park (if results from the observation bores in the unconfined zone are excluded); but there appears to be no other explanation for the low nitrate values of bores 15 and 21 on the eastern side and Somes Island (22) other than that denitrification has occurred. Whether the extent to which denitrification takes place is more strongly influenced by the residence time than by the availability of reducing substances, for instance peat, in different parts of the aquifer is not clear because these two factors probably vary in unison.

Sulfide Oxidation

The sulfate content of the Feltex bore (15) is high in relation to the other bores (Figure 7.5b). Oxidation of iron sulfides would appear to be the only source of this additional sulfate:

$$FeS + \tfrac{9}{4}O_2 + \tfrac{5}{2}H_2O \rightarrow Fe(OH)_3 + SO_4^{2-} + 2H^+$$

$$\tfrac{1}{2}FeS_2 + \tfrac{15}{8}O_2 + \tfrac{7}{4}H_2O \rightarrow \tfrac{1}{2}Fe(OH)_3 + SO_4^{2-} + 2H^+$$

These equations include the oxidation of iron to the ferric state, and thus its removal as insoluble hydrated oxides, because the dissolved iron concentration for this bore is only 0.05 g/m^3—and not the 0.88 or 0.50 g/m^3—per 1 g/m^3 dissolved oxygen consumed, to be expected from FeS and FeS$_2$, respectively, in the production of sulfate without oxidation of the iron. Because only about 1 g/m^3 of dissolved oxygen will be available, resulting in no more than 1.3 and 1.6 g/m^3 sulphate from FeS and FeS$_2$, respectively, these reactions supply insufficient sulfate to explain the approximately 8 g/m^3 elevation of the Feltex bore.

The only other oxidizing agent available is nitrate acting as shown by the following equations:

$$FeS + \tfrac{9}{5}NO_3^- + \tfrac{8}{5}H_2O \rightarrow Fe(OH)_3 + SO_4^{2-} + \tfrac{9}{10}N_2 + \tfrac{1}{5}H^+$$

$$\tfrac{1}{2}FeS_2 + \tfrac{3}{2}NO_3^- + H_2O \rightarrow \tfrac{1}{2}Fe(OH)_3 + SO_4^{2-} + \tfrac{3}{4}N_2 + \tfrac{1}{2}H^+$$

Relative to the main abstraction wells at Hutt Park (12) and the bores on the eastern side in the unconfined zone, the Feltex bore (15) has been depleted in NO$_3$–N by about 1 g/m^3. This would produce 3.8 and 4.6 g/m^3 sulfate from FeS and FeS$_2$, respectively, which combined with that from the oxygen-consuming reactions nearly accounts for the raised sulfate level of the Feltex bore. It is therefore suggested that in the vicinity of this bore there is oxidation of iron sulfides coupled with the reduction of oxygen and denitrification.

Somes Island

In comparison with water quality parameters of the other bores, this water has several distinguishing characteristics: the absence of nitrate and sulfate, moderate concentrations of ammonia and manganese (ammonia was not detected in any other water from the main aquifer), and relatively high concentrations of iron and reactive dissolved phosphate. (In other bores reactive dissolved phosphate did not exceed a few mg/m^3 of PO$_3$–P.) These features result in a water of low quality, one manifestation of which is excessive biological growth in the reservoir, no doubt stimulated by the high concentrations of nutrients.

Controls on the concentrations of iron, phosphate, and aluminum by solution–mineral equilibria were investigated using the SOLMNEQ program (Table 7.2), which was also used to assess the sensitivity of the calculated values of ΔG_R to variations in pH and concentration of solutes. This ground water is approaching saturation with respect to siderite, and it is possible the level of dissolved iron is controlled by the precipitation of an impure carbonate phase. Because the water is well above saturation with respect to vivianite and hydroxylapatite (and also chlor- and fluorapatite), these phases are ineffective in limiting the concentration of dissolved phosphate. The dissolved aluminum concentration is close to that predicted by the solubility of gibbsite (Al(OH)$_3$) for the observed pH and, in turn, probably limits

Table 7.2   Reaction States, ΔG_R (kJ), of Somes Island (22) Bore Water for Mineral-Solution Equilibria Possibly Controlling the Concentrations of Iron, Phosphate, and Aluminum, and Their Sensitivity to Changes in Water Quality Parameters

Mineral	ΔG_R (kJ)	Change in ΔG_R (kJ)	Parameter	Change (g/m^3)
Siderite	−2.2	0.3	pH	0.05
FeCO$_3$		0.1	Fe	0.1
		0.1	HCO$_3$	3.0
Vivianite	+8.5	0.2	pH	0.05
Fe$_3$(PO$_4$)$_2$ · 8H$_2$O		0.25	Fe	0.1
		0.1	PO$_4$-P	0.04
Hydroxylapatite	+6.3	1.6	pH	0.05
Ca$_5$(PO$_4$)$_3$OH		1.4	Ca	0.5
		0.4	PO$_4$-P	0.04
Gibbsite	−0.88	0.25	pH	0.05
Al(OH)$_3$		0.6	Al	0.001
Boehmite	+2.3	0.25	pH	0.05
AlOOH		0.6	Al	0.001
Variscite	−1.1	0.2	pH	0.05
AlPO$_4$ · 2H$_2$O		0.6	Al	0.001
		0.1	PO$_4$-P	0.04

the level of dissolved phosphate through the formation of variscite (AlPO$_4$ · 2H$_2$O). While the various solid phases identified above may control the level of dissolved species, they may be unimportant products in the longer term because of the slower formation of more stable (less soluble) minerals. A possible limitation of this approach is that no account has been taken of complex formation involving organic ligands.

Oxidation–Reduction Sequence

When dissolved oxygen has been largely consumed, soluble iron and manganese appear in easily detectable concentrations, and under more reducing conditions there is clear evidence of denitrification. Hydrogen sulfide was not detected in any samples, but sulfate reduction is obviously occurring before the water reaches Somes Island (22) and the resultant sulfide will have been precipitated mainly as iron sulfides. Water reaching this position in the aquifer is expected to be significantly older than that withdrawn from the main bores at Hutt Park (12); unfortunately, an age for this water was not obtained at a time when the tritium method was still feasible.

Nitrogen reduction as a source of ammonia in ground water is less well established and accepted than, for example, sulfate reduction. An alternative source is the mineralization of organic nitrogen under anaerobic conditions, but since soluble organic nitrogen did not generally exceed 0.2 g/m^3 throughout the aquifer, there would need to be contributions from particulate organic nitrogen sources for this alternative to be viable. Certainly, the results do not support nitrate reduction as being the source of ammonia because denitrification occurs before the appearance of ammonia. Although methane was not determined, it would be expected, on the basis of the reaction sequence, to be present in the Somes Island (22) water. Copious amounts of gas rich in methane were encountered during exploratory drilling of the deeper saline aquifers.

The development of the range of oxidation–reduction conditions over the length of the main aquifer, extending to the very reducing state at Somes Island (22) after a comparatively short flow path and residence time of the water, will be enhanced by the low levels of nitrate and sulfate in the recharge water.

Most of the increase in nitrate levels of the Hutt Park (12) bores over the last twenty years is almost certainly due to increased abstraction rates decreasing the residence time and hence the extent to which denitrification occurs before the water reaches this part of the aquifer. This drawing of slightly oxygenated water toward the main abstraction wells also explains the elevated sulfate level of the Feltex bore (15), which arises from the oxidation of iron sulfides in what would have previously been a highly reducing environment.

Applications

The position that water from a particular bore has reached in this sequence will depend, assuming there is no lack of reactants, on the rates of reaction and the residence time. These factors need to be assessed when considering the effect of new wells or changing the pattern of withdrawal from an aquifer. Transfer of the main abstraction point from Hutt Park (12) to Knights Road (7) will increase the residence time of water in the lower Valley and is expected to move it further down the reaction sequence, resulting in lower nitrate levels and higher concentrations of iron and manganese. Possibly the sulfate reduction zone, which at present must be between the shore and Somes Island, will migrate toward the recharge area and the effects become apparent in the on-shore bores; for reasons discussed previously sulfate reduction must have occurred in the vicinity of the Feltex bore (15) in the past.

In many other aquifers the range of oxidation–reduction conditions encountered will be smaller, or will be extended over much greater distances, so changes in the abstraction pattern will have a lesser or more localized effect. For instance, a greywacke gravel aquifer on the Canterbury Plains of New Zealand is superficially similar to the Hutt aquifer, yet wells most distant from the recharge zone still contain several g/m^3 of dissolved oxygen

(Bathurst et al., 1979), even though the water has traveled further underground than the length of the Hutt aquifer (and much of it under confined conditions), presumably because the gravels are cleaner with respect to the content of oxidizable organic matter. The presence of this level of dissolved oxygen, together with the absence of dissolved iron and manganese, indicates that the low nitrate concentration (0.2 g/m^3 NO$_3$-N) is not the result of denitrification. At least at the time when this water was passing through the unconfined part of the aquifer, significant contamination of the predominantly river recharge water by nitrate from soil waters in the unconfined zone was not occurring. In the absence of a determined age for waters from these wells, this conclusion need not apply to the present. However, low dissolved oxygen concentrations (<0.2 g/m^3) and the presence of dissolved iron and manganese indicate that the low nitrate levels (<0.03 g/m^3 NO$_3$-N) for water in the same area from an overlying aquifer are probably due to denitrification.

CONCLUSIONS

Observed changes in the concentrations of oxidizable and reducible species as water passes through the Hutt aquifer closely follow the sequence predicted from thermodynamic data. Rock–water interactions influence this sequence by their effects on the pH of the water and by providing species which, through solubility controls, limit the concentrations of some constituents.

Characterization of ground waters in terms of this oxidation–reduction sequence should be useful in predicting spatial variations of water quality throughout an aquifer and assessing effects of changes in the recharge–discharge pattern or in the quality of the recharge water.

ACKNOWLEDGMENTS

This study was made possible through the cooperation of the Wellington Regional Water Board in providing field assistance. Dr. C. D. Stevenson is thanked for helpful discussions.

SEA WATER INTRUSION OF A KARST AQUIFER

The Waikoropupu Springs, a major karstic resurgence located 4 km inland from the shores of Golden Bay on the north coast of the South Island of New Zealand, have an average discharge of 14 m^3/sec and rank as springs of the first order. The springs issue from the floor of a small subsidiary valley in the northwest corner of Takaka Valley, which is a wedge shaped depression 9 km across at its widest point on the coast, 26 km long, and oriented north–south. This off-center siting of the resurgence is unusual for karst terrain. Due to the eastward tilt of the blocks on either side, the catchment of Takaka Valley is strongly asymmetric, most of it lying in the Tasman Mountains on the western side, which are composed of Palaeozoic schists. A fault scarp rising to a karstic plateau of Arthur Marble at 850 m forms the steep eastern wall of the Valley, and relatively little precipitation drains to the Valley from the plateau. Much of the Valley is covered with Quaternary gravels and sands over Tertiary sediments, including the Takaka Limestone and Motu-pipi Coal Measures exposed in several places. These Tertiary sediments lie unconformably on Arthur Marble, which extends to the coast and is the main karst aquifer.

Aquifer recharge is by the Takaka River, which enters the apex of the Valley from a narrow gorge, and since about 9 m^3/sec can be absorbed (Williams, 1977), the river bed is dry 7 km from the gorge except under flood conditions. Other recharge is from precipitation in the mid-Valley region. Williams (1977) gives details of the hydrology, and the application of isotopes in identifying sources of recharge and determining flow-through times is discussed by Stewart and Downes (1981).

In the next section water quality data are presented which show that there is intrusion of sea water into the Waikoropupu Springs and that the extent of intrusion depends on the discharge rate.

RESULTS AND DISCUSSION

There are significant differences in chemical composition between the discharge from the Main Spring vent of the Waikoropupu Springs at 10 m above sea level and water from Low's bore in marble in the recharge zone (Table 7.3), the most obvious being the higher sodium chloride content of the Spring water. Apart from calcium, and to a lesser extent potassium, there is good agreement of the "observed" with the "expected" changes in composition between Low's bore and the Main Spring (Table 7.4), the "expected" changes being calculated on the basis of the increased chloride content arising from intrusion of sea water. Infiltrating soil waters in the mid-Valley region will have greater solvency than the river water, which is the major source of Low's bore water, resulting in higher calcium levels for the freshwater contribution to the Springs than that indicated by the analysis of Low's bore. Also, the water at Low's bore is not saturated with respect to calcite (ion activity product is 1.29×10^{-9} mole^2/L^2 compared with the solubility product of 5.24×10^{-9} mole^2/L^2), so calcium levels can rise through continued contact with the marble.

Table 7.3   Chemical Composition of Takaka Ground Waters (in g/m³)

	Main Spring 15 March 1979	Low's Bore 14 March 1979
pH	7.55	7.30
T (°C)	11.4	12.3
Na	47	3
K	3.9	0.5
Mg	6.6	2.4
Ca	57	43
Sr	0.16	0.09
Cl	75	3.3
HCO₃	194	139
SO₄	15	4
SiO₂	6	7

Table 7.4   Comparison of Observed Increases in Ion Concentrations (in g/m³) Between Low's Bore and the Main Spring with Values Calculated Assuming the Increased Chloride Level is Due to Sea Water Intrusion

	Observed	Expected
Na	44	40
K	3.4	1.4
Mg	4.2	4.8
Ca	14	1.5
SO₄	11	10

[a] Reproduced by permission of Northern Illinois University Press, DeKalb, IL from M. K. Stewart and C. J. Downes, "Isotope Hydrology of the Waikoropupu Springs, New Zealand," in *Isotope Studies of Hydrologic Processes,* ed. Eugene C. Perry, Jr., and Carla W. Montgomery, NIU Press, pp. 15–23, 1982.

Evaporites are not known in the geological sequence, nor is there any convincing reason to suspect contributions from saline metamorphic waters. Though the $^{18}O/^{16}O$ and D/H ratios of sea water are very different from those of fresh ground waters in the Takaka district, the proportion of sea water required (~0.5%) to account for the observed salinity is too small for the existence of a detectable effect on the stable isotope ratios of the water in the Springs' discharge. However, the $^{18}O/^{16}O$ and $^{32}S/^{34}S$ ratios of sulfate in the discharge are in accord with the existence of a sea water contribution (Stewart and Downes, 1981).

A second water quality feature of the Springs is that the chloride content increases with increased discharge (Figure 7.6). This is contrary to the expectation that increased recharge, leading to higher discharge, would shift the fresh water–salt water interface seaward and result in lower salinity. Because the discharge is gauged at a position that includes runoff from the small valley in which the Springs are located, the actual dependence of salinity on discharge from the Springs will be slightly stronger than that indicated by the analytical data. This effect of increased salinity with increased discharge rate could be due to unique hydrological conditions, but should be noted when considering the exploitation of ground waters from aquifers in comparable geological settings.

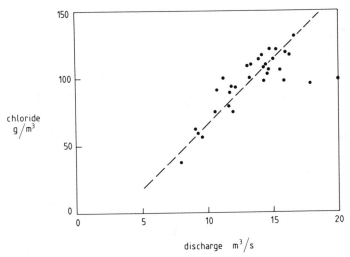

Figure 7.6. Variation of the chloride concentration with discharge from the Main Spring vent of the Waikoropupu Springs. (Reproduced by permission of Northern Illinois University Press, DeKalb, IL from M. K. Stewart and C. J. Downes, "Isotope Hydrology of the Waikoropupu Springs, New Zealand," in *Isotope Studies of Hydrologic Processes*, ed. Eugene C. Perry, Jr. and Carla W. Montgomery, NIU Press, pp. 15–23, 1982.)

CONCLUSIONS

The salinity of the Waikoropupu Springs, the only on-shore discharge point for a coastal karst aquifer, increases with the flow, an effect which on chemical and isotopic evidence is due to sea water intrusion.

REFERENCES

APHA (1976). *Standard Methods for the Examination of Water and Wastewater*, 14th ed. American Public Health Association, Washington, D.C.

Barnes, R. B. (1975). The determination of specific forms of aluminum in natural water. *Chem. Geol.* **15**:177–191.

Bathurst, E. T. J., Downes, C. J., Williamson, R. B., and Ayrey, R. B. (1978). Water quality of Christchurch aquifers. In: Technical Publication No. 2, Dept. of Agricultural Microbiology, Lincoln College, N.Z., pp. 169–181.

Brown, E., Skougstad, M. W., and Fishman, M. J. (1970). Methods for collection and analysis of water samples for dissolved minerals and gases. In: *Techniques of Water-Resources Investigations of the United States Geological Survey*. U.S. Government Printing Office, Washington, D.C.

Champ, D. R., Gulens, J., and Jackson, R. E. (1979). Oxidation-reduction sequences in ground water flow systems. *Can. J. Earth Sci.* **16**:12–23.

Chao, T-t and Kroontje, W. (1966). Inorganic nitrogen transformations through the oxidation and reduction of iron. *Soil Sci. Soc. Am. Proc.* **30**:193–196.

Donaldson, I. G. and Campbell, D. G. (1977). Groundwaters of the Hutt Valley–Port Nicholson alluvial basin. New Zealand Department of Scientific and Industrial Research, Information Series No. 124.

Downes, C. J. (1980). Chemistry of the Hutt Valley underground waters. Chemistry Division, New Zealand DSIR, unpublished report.

Edmunds, W. M. (1973). Trace element variations across an oxidation-reduction barrier in a limestone aquifer. In: E. Ingerson (Ed.), Proceedings of the Symposium on Hydrogeochemistry and Biogeochemistry, Tokyo, 1970, Clarke Co., Washington, D.C., pp. 500–527.

Grant-Taylor, T. L. and Taylor, C. B. (1967). Tritium hydrology in New Zealand. In: *Isotopes in Hydrology*. International Atomic Energy Agency, Vienna, pp. 381–400.

Hutton, P. (1965). Report on the hydrology of the lower Hutt Valley. Hutt Valley Underground Water Authority (unpublished).

Kharaka, Y. K. and Barnes, I. (1973). SOLMNEQ: Solution–mineral equilibrium computations: Menlo Park, Calif., U.S. Geological Survey Computer Contr., available from U.S. Dept. Commerce, National Technical Information Service, Springfield, VA 22151, as report PB-215 899.

McNabb, J. F. and Dunlap, W. J. (1975). Subsurface biological activity in relation to ground-water pollution. *Ground Water* **13**:33–44.

Stevens, G. R. (1956). Stratigraphy of the Hutt Valley, New Zealand. *N.Z. J. Sci. Tech. B* **38**:201–235.

Stevenson, C. D. (1975). Some thoughts on the chemistry of the Lower Hutt aquifer system. Chemistry Division, New Zealand DSIR, unpublished report.

Stewart, M. K. and Downes C. J. (1981). Isotope hydrology of the Waikoropupu Springs, New Zealand. In: Eugene C. Perry, Jr., and Carla W. Montgomery (Editors), *Isotope Studies of Hydrological Processes*. Northern Illinois University Press, De Kalb, pp. 15–23.

Stumm, W. and Morgan, J. J. (1970). *Aquatic Chemistry,* Wiley-Interscience, New York.

Wells, N., Childs, C. W., and Downes, C. J. (1977). Silica Springs, Tongariro National Park, New Zealand: Analyses of the spring water and characterisation of the alumino-silicate deposit. *Geochim. Cosmochim. Acta* **41**:1497–1506.

Williams, P. W. (1977). Hydrology of the Waikoropupu Springs: a major tidal karst resurgence in northwest Nelson (New Zealand). *J. Hydrol.* **35**:73–92.

8

BEHAVIOR OF ORGANIC POLLUTANTS IN PRETREATED RHINE WATER DURING DUNE INFILTRATION

G. J. Piet
J. G. M. M. Smeenk

National Institute for Public Health and Environmental Hygiene
Leidschendam, The Netherlands

As a part of the national drinking-water supply system, pretreated Rhine water has been infiltrated for almost 25 years in the dunes of The Netherlands. A large part of the province of North-Holland, including the city of Amsterdam, is still dependent on infiltrated Rhine water, which enters the dunes after transport through a 60–100 km pipeline. This water is pretreated by coagulation, sedimentation, and rapid sand filtration before the transport. Chlorination before transport is applied only when the water temperature is above 10°C (see Figures 8.1, 8.2).

A study of 20 different water supplies in The Netherlands (Zoeteman, 1980) showed that dune infiltration combined with additional polishing procedures appeared to be a water treatment method leading to an acceptable drinking-water quality.

Another study of the National Institute for Water Supply had the object of measuring organic substances in the aquifer after induced recharge. During the latter study in 1978, it was observed that a variety of industrial chemicals such as trichloroethene, tetrachloroethene, tetrachloromethane, bis(2-chloroisopropyl)ether, chlorophosphates, and small amounts of chloroanilines

DUNE INFILTRATION
STUDIES 1981

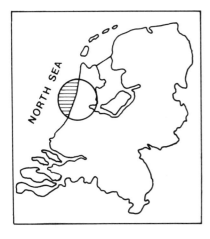

Figure 8.1. Dune infiltration of Rhine water, areas in The Netherlands.

and chlorotoluidines are present in water after bank infiltration, even after residence times in the soil of several years. In later studies other volatile organic chlorine compounds such as 1,2-dichloroethane have been detected. It had been observed already that ongoing bank infiltration of Rhine water caused a slowly moving zone of odor-intensive compounds in the aquifer (Sybrandi, 1980; Figures 8.3 and 8.4).

Industrial organic compounds present in the river Rhine have been detected in drinking-water of The Netherlands when this water has been used as a raw water source. For these reasons more attention has been paid to industrial organic compounds in the aquifers of sites used for artificial recharge (Piet and Zoeteman, 1980).

In the meantime other scientists have studied the effects of the organic content of the soil and the specific properties of organic compounds based on their water–octanol partition coefficients (Schwarzenbach and Westall, 1981; Roberts and Valocchi, 1981). Some studies have indicated the effects of microbiological decay of organic compounds in the soil (van Genuchten, 1981; McCarthy et al., 1981). It is evident that many processes in the soil affect organic compounds. These processes can vary with the seasons. The infiltrated water and the soil play a role, but the effect of a variety of microorganisms under aerobic and anaerobic conditions in the soil should not be underestimated. The field study in the dunes of the province of North-Holland has been set up as an orientation to get more information on the behavior of chemicals that might endanger soil and drinking-water quality and to identify the dominant processes taking place during infiltration. A major aspect presented is the evaluation of differences in water quality

LOCATION OF INFILTRATION AREAS AND THEIR SUPPLIES (1976)

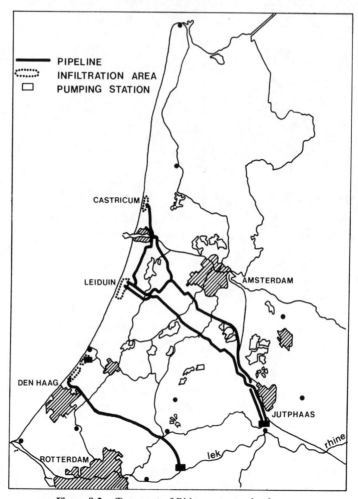

Figure 8.2.   Transport of Rhine water to the dunes.

under aerobic and anaerobic conditions as a function of the residence time. In this paper the results of this orientation are reported.

DESCRIPTION OF THE SITES

Two sites have been selected for investigation:

1. The dune infiltration area of the Provincial Water Works (North-Holland) at Wijk aan Zee (see Figure 8.2, Castricum).

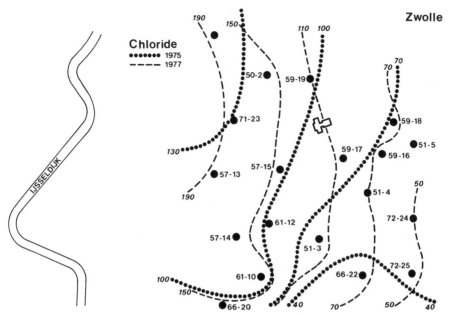

Figure 8.3. Bank infiltration of Rhine water, Zwolle, Provincial Water Supply—increase of chloride in the subsurface aquifer (movement from left to right: 1 cm = 20 m).

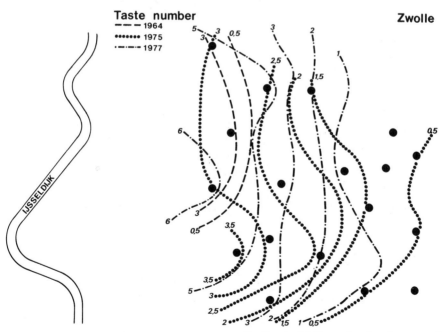

Figure 8.4. Bank infiltration of Rhine water, Zwolle, Provincial Water Supply—increase of odor intensive compounds in the subsurface aquifer (movement from left to right: 1 cm = 20 m).

2. The dune infiltration area of the Municipal Water Works of Amsterdam (Leiduin).

The study of Wijk aan Zee was carried out in winter (November 1980 to February 1981). The temperature of the infiltrating water was 2–3°C. The study in Leiduin was carried out from April 1981 to July 1981. During this time the temperature of the infiltrating water ranged from 4 to 17°C. At both sites pretreated Rhine water was introduced. Both areas have a system of open infiltration with artificial canals fed from a central introduction pond located at the end of the transport line (Figures 8.5 and 8.6). The soil of both sites consists of fine-grained dune sand mixed with clay. At the Leiduin site aggregates of peat are nonhomogeneously distributed in the soil. The infiltration conditions of both sites are comparable.

The Wijk aan Zee Site

An outline of the situation at Wijk aan Zee is shown in Figure 8.7. An oversight of the sampling of infiltrating water at the wells and at basin 9 is given in Tables 8.1 and 8.2.

DUNE INFILTRATION NORTH-HOLLAND PROVINCIAL WATER SUPPLY

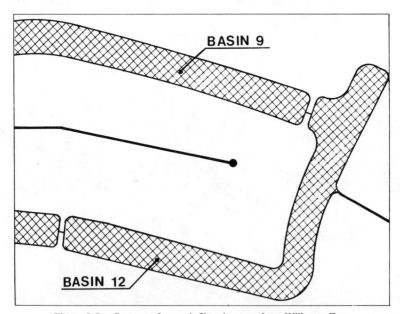

Figure 8.5.   System of open infiltration canals at Wijk aan Zee.

Percolation area GWA

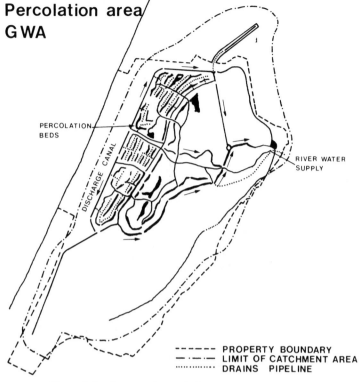

PERCOLATION BEDS

DISCHARGE CANAL

RIVER WATER SUPPLY

```
------  PROPERTY BOUNDARY
-·-·-·  LIMIT OF CATCHMENT AREA
·······  DRAINS  PIPELINE
```

Figure 8.6. System of open infiltration canals at Leiduin.

DUNE INFILTRATION NORTH-HOLLAND PROVINCIAL WATER SUPPLY

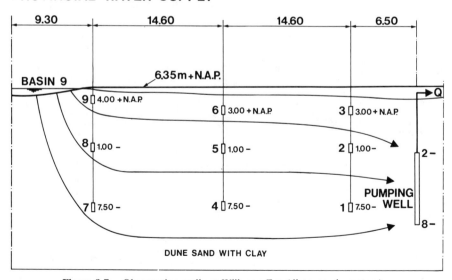

9.30	14.60	14.60	6.50

BASIN 9

6.35m+N.A.P.

Q

9 ☐ 4.00 +N.A.P.

6 ☐ 3.00 +N.A.P.

3 ☐ 3.00+N.A.P.

8 ☐ 1.00 −

5 ☐ 1.00 −

2 ☐ 1.00−

2 −

7 ☐ 7.50 −

4 ☐ 7.50 −

1 ☐ 7.50 −

PUMPING WELL

8 −

DUNE SAND WITH CLAY

Figure 8.7. Observation wells at Wijk aan Zee (distances in meters).

127

Table 8.1   Sampling Points at Wijk aan Zee

Location[a]	Horizontal Distance from Basin (m)	Location below the Surface (m)	Average Linear Velocity of Water (m/day)	Chloride Residence Time (days)
Basin 9	0	0	—	0
Well 8	0.5	7.4	1.5	3
Well 5	14.6	7.4	1	18

[a] Wells 8 and 5 have been selected for study on the basis of the taste of the water. The best taste was observed at well 5, the worst taste at well 8.

Table 8.2   Nitrate and Oxygen Levels of Sampling Wells at Wijk aan Zee (average from 10 observations from November 1980 to February 1981)

Location[a]	Nitrate (mg/L)	Oxygen (mg/L)
Basin 9	4.3	11.1
Well 8	3.8	5.4
Well 5	3.9	5.4

[a] Wells 8 and 5 have been selected for study on the basis of the taste of the water. The best taste was observed at well 5, the worst taste at well 8.

DUNE INFILTRATION LEYDUIN 1981

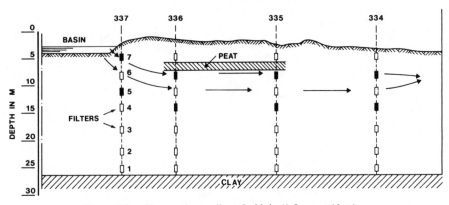

Figure 8.8.   Observation wells at Leiduin (1.5 cm = 10 m).

Table 8.3   Sampling Points at Leiduin

Location	Horizontal Distance from Basin (m)	Location below the Surface (m)	Average Linear Velocity of the Water (m/day)	Chloride Residence Time (days)
337/6	2	5	1	7
337/5	2	8	0.5	20
336/6	12.5	5	0.7	20
335/6	32	5	0.7	50

The Leiduin Site

The infiltration site at Leiduin is pictured in Figure 8.8. In the aquifer at this site, 28 filters are placed at several depths and at several distances from the infiltration basin. The wells which have been sampled during the period from April 1981 to July 1981 are mentioned in Table 8.3.

Sampling Procedures

In the aquifer of the Leiduin site, small filters are connected to PVC tubes (i.d. 2 cm) leading to the surface (Figure 8.9). A polyethylene tube (i.d. 0.4

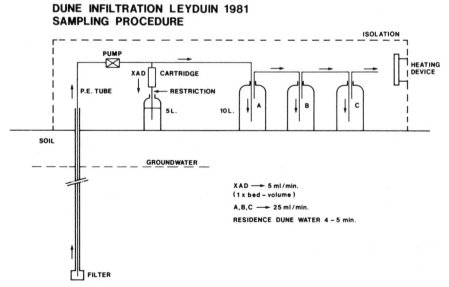

Figure 8.9. *In situ* sampling system for field studies at Leiduin.

cm) connected to a stainless steel membrane pump placed in an insulated chest on top of the soil ends in a corresponding filter of an observation well. The temperature in this chest is regulated by a heating device.

The membrane pump continuously draws water from the filter and guides it through an XAD cartridge (5 mL/min) filled with 5 mL of a mixture of XAD-4 and XAD-8, and simultaneously through a series of three glass sampling vessels (25 mL/min). The XAD procedure was chosen because it could be applied *in situ* and because a wide spectrum of compounds, including those with mutagenic activity, are concentrated by XAD (Van Kreijl et al., 1980; Kool et al., 1981, 1982).

The residence time of the water in the polyethylene tube is short (4–5 min) in order to avoid as much as possible migration of organic compounds through the wall of the tube.

Every 24 hours the XAD cartridges and the sampling vessels are replaced. The XAD cartridges are transported to a nearby field laboratory for further treatment, and the glass vessel next to the membrane pump is taken to the analyzing laboratory of the Municipal Water Works of Amsterdam. The second vessel is connected to the membrane pump, and at the end of the line a new vessel is installed. In this way almost no loss of volatile organic compounds in the glass vessels is observed since two of the three vessels are always filled with water. Biological activity in the XAD cartridges is inhibited by small pieces of silver wire on top of the XAD. After replacement, the XAD cartridges are immediately eluted with diethyl ether. Because the concentration of the organic compounds from the ground water takes place *in situ,* artifacts introduced during transport and storage of the water and contamination in the laboratory during the concentration is greatly reduced. Sampling of the ground water for analytical purposes started after the whole sampling system had been operating continuously for 2 weeks. The eluates of the cartridges were stored at −20°C.

The sampling periods of each well were chosen in such a way that within a period of about 8 weeks the same body of water in the aquifer is analyzed at different locations after different residence times of the water in the underground, as shown in Figures 8.10 and 8.11. By this procedure the change of water quality during residence in the soil could be studied.

The infiltrating water is sampled about 1 cm above the sediment layer in the basin just before it enters the soil (these samples are called "basin"). The sampling procedure at the Wijk aan Zee site is similar to the procedure at the Leiduin site with the only exception that no membrane pumps had to be used at Wijk aan Zee. The water could be introduced into the cartridges and glass sampling vessels by the hydrostatic pressure of the aquifer.

Analytical Procedures

XAD Cleaning

Analysts generally purify XAD by flushing it with several organic solvents (Webb, 1975). However, when water is passed through this supposed "clean

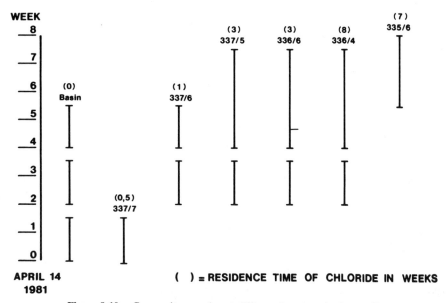

Figure 8.10.   Composite samples at different locations in the aquifer.

DUNE INFILTRATION LEYDUIN 1981
TIME SCALE OF SAMPLES
(CORRESPONDING TO BASIN WATER)

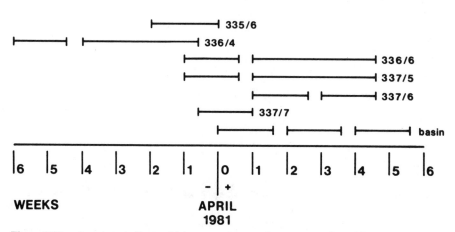

Figure 8.11.   A scale to indicate which composite samples correspond to which body of water.

XAD,'' about 50 organic compounds not present in the solvents are eluted. The precleaning procedure of XAD had to be changed, and washing with purified water had to be introduced as a final step in the cleaning procedure.

A diethyl ether eluate of the treated XAD after the passage of several liters of pure water contains only very small traces of toluene, meta/para xylene, 2 alkanes, and 3 unidentified compounds if the solvent itself is pure. This known background is subtracted.

XAD Eluate Treatment

Prior to analysis the diethyl ether eluates must be concentrated. To this end excessive water is removed by freezing the extract at $-20°C$. Butylcyclohexane is added as an internal standard, and the eluate is concentrated a hundredfold without loss of rather volatile and high-boiling compounds according to the method of Fritz (1975).

Representative organic compounds have been used as external standards to calculate recoveries of classes of organic compounds, so that a quantitative analysis can be carried out.

Mass Spectrometry

One microliter (μL) of the final concentrate is introduced by splitless injection (Giger et al., 1976). The capillary column is connected to a Finnigan 4000 mass spectrometer equipped with an Incos datasystem. The scan rate is 1 scan/second at 70 eV ionization potential. One run of 75 min consists of 4500 scans from 20–500 a.m.u. An example of a total ion current chromatogram is presented in Figure 8.12.

To confirm the identity of the isomers found in the samples, the Incos search system is supplemented with additional information concerning the gas chromotographic indices of these compounds.

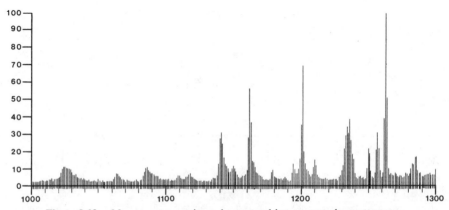

Figure 8.12. Mass spectrometric analyses, total ion current chromatograms.

Gas Chromatography

Quantitative analysis of organic chlorine compounds is carried out with a fused silica OV-1 capillary column connected to a double detector system— Electron Capture Detector and Flame Ionization Detector in series (Piet et al., 1980). Volatile halogenated organic compounds are analyzed by a static headspace technique (Piet et al., 1978) applying a whisker-coated OV-225 capillary column. All procedures are calibrated for quantitative analysis using standard-addition methods.

Miscellaneous Analytical Procedures

For the determination of the following parameters, the water of the vessels is used.

Nitrate, nitrite, ammonia and sulfate are measured with an autoanalyzer (Technicon, type 2). Total dissolved organic carbon (DOC) is measured as CO_2 after decomposition of the organic compounds in a filtered sample with persulfate and UV destruction. For the "total" adsorbed organic chlorine (AOCl), determined organic compounds are adsorbed on activated carbon and analyzed with a micro-coulometer (Glaze, 1977). A part of the AOCl concentrate is used to determine "total" organic bromine (AOBr) by means of ion chromatography.

For the determination of organo-phosphorus and organo-nitrogen, concentrated XAD 4/8 eluates are used. These eluates are analyzed by The Netherlands Water Works Testing and Research Institute according to the procedure described by Puijker et al. (1981).

Trihalomethanes are determined by a headspace method.

RESULTS

The results of two different investigations are reported. The first investigation concerns the measurements during winter (December 1980 to February 1981) at the Wijk aan Zee site and second the measurements from April 1981 to July 1981 at the Leiduin site.

This organization has been chosen because in winter the behavior of several classes of organic compounds seems to be different from the behavior of these classes during summer, a difference probably due to increased microbiological activity in the soil when the temperature of the water rises.

Wijk aan Zee

More than 400 organic compounds have been registered by gas chromatography and mass spectrometry. In this article a selection is made of some compounds at concentrations greater than 0.05 μg/L.

Table 8.4 Organic Compounds in the Water of the Wijk aan Zee Site

Compound	Infiltrated Water	Well 8 (3 days)	Well 5 (18 days)	Estimated Half-life (days)
Aromatic				
Benzene	1	5	12	Long
Toluene	1	6	2.3	Long
Ethylbenzene	1	3.5	0.8	Long
m/p-Xylene	1	1.2	1.2	Long
o-Xylene	1	34	14	Long
Naphthalene	1	6	5	Long
Organic Chlorine				
Tetrachloroethene	1	1.9	0.7	Long
bis(2-Chloroisopropyl)ether	1	1.7	0.3	50
bis(2-Chloropropyl)ether	1	1.3	0.3	50
Chlorocresol	1	0	0	1
A chloronitrobenzene isomer	1	0.1	0.1	2
A chloronitrobenzene isomer	1	1	0.8	Long
A dichloroaniline isomer	1	4	6	Long
A dichlorocresol isomer	1	0.4	0	3
tri(2-Chloroethyl) phosphate	1	3.6	2.9	Long
Unknown chloro compound	1	4.1	4.0	Long
Organic Nitrogen				
Unknown heterocyclic	1	0.7	0.1	5
Nitrobenzene	1	0.4	0.1	3
Ethylmethyl/pyridine	1	0.2	0.2	1
A nitrotoluene isomer	1	0.3	0.1	2
A nitrotoluene isomer	1	1.1	0.8	Long
Dinitrotoluene	1	0	0	1
Alkanes				
C-11	1	1,9	3.5	Long
Phthalates				
Diethyl	1	3,3	2.9	Long
Dibutyl	1	60	45	Long
Others				
2-Methylthiobenzthiazole	1	0.5	0.4	3

More than 60% of the registered constituents consists of unidentifiable substances. They either are present in the infiltrated water or appear after infiltration in the soil.

In Table 8.4 an oversight is given of compounds in the infiltrating water, in the water of well 8 (3 days chloride residence time), and in the water of well 5 (18 days chloride residence time).

The concentration of a compound at different locations of the site is expressed in numerical units. In the infiltrated water it is expressed as 1 unit, so that the concentration at other locations in the aquifer can be related to the original value in the infiltrated water. The estimated half-life (in days) is site-specific and depends on processes in the aquifer at the moment of study. As the compounds are determined in the same body of water, which is moving underground, the change of the water quality as a function of the residence time is indicated by the concentration of the organic compounds. When the estimated half-life of a compound is denominated "long," it means that no evident decrease of a compound at that moment is observed, so that almost no degradation under the given conditions in the aquifer has taken place.

Some compounds are enhanced after a given residence time. This may indicate, though collective samples have been analyzed, that memory effects of the soil can affect the results. It is clear that in these cases the half-life may be affected and not be regarded as a general constant. Table 8.5 demonstrates the different behaviors of isomeric compounds.

The tables seem to indicate that aromatic compounds and some organo-chlorine compounds have a tendency to break through in winter and pass into the drinking-water supply. Benzene is of particular concern in this respect. The same is true for several volatile organo-chlorine compounds produced by industry. Only a few compounds are eliminated after 18 days.

Table 8.5   Some Unknown Isomers in the Water of the Wijk aan Zee Site

Major Ions of the Mass Spectrum	Infiltrated Water	Well 8 (3 days)	Well 5 (18 days)
45, 41, 79, 75, 155, 157	1	1.3	0.4
45, 41, 75, 77, 155, 157	1	0.9	0
43, 87, 59, 75, 101, 161	1	6	4
43, 87, 59, 75, 101, 161	1	8.5	9.5
75, 79, 49, 39, 111, 113, 189	1	4.5	0.3
75, 79, 49, 39, 111, 113, 189	1	2.5	0.6
41, 55, 127, 69, 170	1	1.6	0.3
41, 55, 69, 127, 170	1	0.7	0

Isomers can show a mutually different behavior, as confirmed by the results presented in Table 8.5. A chloronitrobenzene isomer is almost unaffected during the transport in the aquifer, as is tri(2-chloroethyl) phosphate. These above-mentioned compounds with a tendency to break through have all been detected in drinking-water derived from infiltrated Rhine water.

The bad taste of the water of well 8 is—according to perfume experts of Naarden International, who examined the odor of XAD extracts of well 8 and well 5—probably due to microbiological metabolites present at well 8.

The observations at Leidiun, made during spring and summer, will be presented below.

Leiduin

During the measurements at Leiduin, the average temperature of the infiltrating water (basin) was about 8°C in the middle of April. A sudden increase of the temperature took place in the first week of May, and during the second week of May the temperature of the water in the shallow basin rose to about 14°C. When this warmer water reached the observation points, a rather sudden decrease of the nitrate level occurred (Figure 8.13). At an observation well (337-5) with anaerobic conditions (nitrate < 1 mg/L, sulfate 65–70 mg/L), the sulfate level decreased to less than 25 mg/L in the second week of June. These sudden changes are probably caused by increased microbiological activity because all the measurements have been made in the same moving body of water.

In the following presentation of the results, the term "aerobic" refers to conditions with unchanging nitrate levels in well 337-6 (1 week residence time) and well 336-6 (3 weeks residence time), as this was the case during the first 3 weeks of the study from April to May (Figure 8.13).

The indication "anaerobic" refers to the condition with suddenly decreasing nitrate levels in 337-6 and 336-6. "Deep anaerobic" refers to the condition in 337-5 when the sulfate level decreased. These notations are chosen because the redox potentials have not been measured.

The behavior of the organic compounds during the changing conditions in the aquifer is of interest. It should be kept in mind that the field observations at Leiduin as well as at Wijk aan Zee are site-specific.

Results of Sum Parameters

To study classes of compounds, sum parameters such as total adsorbable organic halogen (AOX), XAD-adsorbable organic phosphorus (XOP), and XAD-adsorbable organic nitrogen (XON) are measured. The results are presented in Figure 8.14 (AOX and AOBr) and Figure 8.15 (XOP and XON). From Figure 8.14 it is evident that a 35% decrease of organic chlorine occurred within 3 weeks residence time during aerobic conditions. The organo-bromine compounds decreased about 55 and 65% during aerobic and anaerobic conditions, respectively.

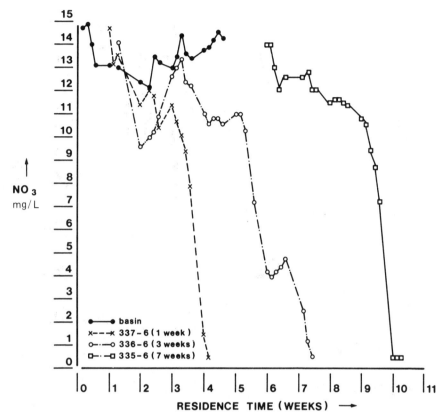

Figure 8.13. Representation to demonstrate the rather sudden decrease of the nitrate level in the same body of water at the moment of the rise of the temperature of the water at different locations in the aquifer.

A 70% decrease for XOP was observed after 1 week residence time (Figure 8.15). After 3 weeks residence time a reduction of about 50% is reached during aerobic and anaerobic conditions.

The reduction of organo-nitrogen compounds is more drastic, particularly during aerobic conditions (90%). Under anaerobic conditions the reduction is somewhat less (80%).

Results of Specific Compounds

In the aquifer of the infiltration area, an abundance of organic compounds below the μg/L level is present; about 350 compounds have been registered. When the conditions in the aquifer changed rather suddenly, several previously unobserved compounds appeared in the chromatograms of wells 337-6

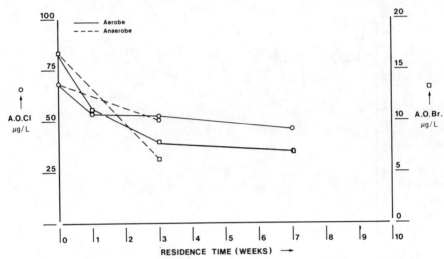

Figure 8.14. Adsorbable organic chlorine and adsorbable organic bromine in the same body of water (in $\mu g/L$).

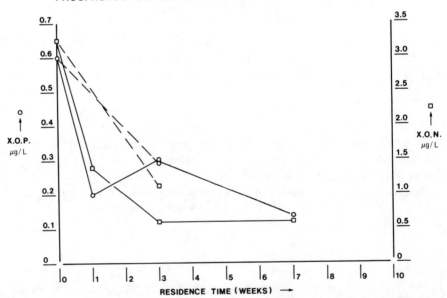

Figure 8.15. Adsorbable organic phosphorus and adsorbable organic nitrogen in the same body of water (in $\mu g/L$).

Table 8.6   Reduction of Low-Boiling Aromatic Compounds (percent decrease with regard to the average value in the corresponding infiltrating water)

Compound	Aerobic		Anaerobic		Deep Anaerobic
	1 Week	3 Weeks	1 Week	3 Weeks	3 Weeks
Benzene	70	80	95	85	95
Toluene	50	60	95	85	90
Ethylbenzene	50	60	80	40	95
m/p-Xylene	40	60	95	20	60
o-Oxylene	60	60	90	50	90

and 336-6. More than 70% of all compounds could not be identified with existing databanks. This could indicate that a large number of metabolites was present. A selection of compounds present in concentrations ranging from 0.05 to 5 μg/L has been made and is presented below.

Aromatic Compounds

The reduction of low-boiling aromatic compounds is given in Table 8.6. For most aromatic compounds the half-life is less than 1 week.

Haloforms

The results presented in Table 8.7 have been obtained by measuring the concentration of the haloforms. Haloforms are strongly reduced; the half-life is less than 1 week. Chloroform was found, however, at low levels after residence times of more than 10 weeks at 334-6. No bromoform was detected at that location.

Table 8.7   Reduction of Haloforms (percent decrease with regard to the average value in the corresponding infiltrating water)

Compound	Aerobic		Anaerobic		Deep Anaerobic
	1 Week	3 Weeks	1 Week	3 Weeks	3 Weeks
Chloroform	80	90	95	95	95
Bromoform	70	70	85	90	95

Organo-Chlorine Compounds

Some rather persistent organo-chlorine compounds are bis(2-chloroisopro-pyl)ether and tri(2-chloroethyl) phosphate, compounds with a more hydro-philic character in the river Rhine; both are frequently detected in drinking-water derived from infiltrated Rhine water. In Table 8.8 measurements of these compounds are listed. Both substances show a similar behavior. The anaerobic conditions with 3 weeks residence time refer to the observations in well 337-5 before the sulfate level decreased at that location and the condition became "deep anaerobic." The rather sudden decrease of the sulfate level after water with a higher temperature arrived resulted in in-creased reduction of the compounds, probably due to increased microbiolog-ical activity under anaerobic conditions.

A variety of organo-nitrogen compounds, including nitro-aromates, are present in the infiltrated water; the nitro-aromates are almost completely eliminated within a few days. In anaerobic conditions of the aquifer at low concentrations (<0.1 $\mu g/L$), substances such as chloroanilines, N-phenyl-aniline and methylaniline have been detected. There are indications that anilines can be formed from corresponding nitro-compounds. Oxygen con-taining compounds such as phthalates (particularly dibutyl phthalate) and esters such as ethyl acetate are detected at all observation points. Ethyl acetate, which can be formed by microbiological processes, increases as nitrate levels decrease. Several unknown ethers, partly already present in the basin water, are not significantly reduced and are present at all observa-tion wells.

Triethyl phosphate is almost completely reduced within 1 week.

When the nitrate level decreases, previously unobserved compounds ap-pear, including several alkanes. Among the unidentified compounds formed are some organo-chlorine compounds with a persistent character. One of the compounds present at several locations in the aquifer is an ester of phos-phorothioacid, 0,0,0-trimethyl, which is most probably a herbicide metab-olite.

Table 8.8   Reduction of Two Rather Persistent Hydrophilic Organo-Chlorine Compounds (percent decrease with regard to the average value in corresponding infiltrating water)

Compound	Aerobic		Anaerobic		Deep Anaerobic
	1 Week	3 Weeks	1 Week	3 Weeks	3 Weeks
Bis(2-chloroiso-propyl)ether	80	85	80	10	60
Tri(2-chloro-ethyl) phosphate	50	55	45	10	65

Chlorobenzenes and traces of pesticides containing chlorine are found at low levels (0.01 μg/L) in almost all samples at different locations. Chlorobenzenes in the aquifer include 1,2,3,4-tetrachlorobenzene and hexachlorobenzene.

DISCUSSION

The field studies in the dunes demonstrate that a great variety of organic compounds move in the aquifer if a reversible exchange between the soil and the water phase exists. Among these compounds are lower aromates, low-boiling organo-chlorine compounds and chloroaromates. Compounds with a hydrophilic character such as chloroethers, chlorophosphates, esters, and some chloronitro-aromates seem to travel unhindered through the soil, if biodegradation does not take place. Depending on the organic content of the soil, the progress of organic compounds is more or less retarded, although they may break through in course of time if the sorption capacity of the soil is limited. During winter the quality of the infiltrated water as a function of the residence time is probably governed to a great extent by physical and chemical processes in the aquifer.

When the activity of microorganisms in the soil increases, several organic compounds are affected, and their pathways can change. A number of metabolites of unknown nature are thus formed, some of which seem to have a rather persistent character. The measurement of "total" adsorbable organic phosphorus and nitrogen demonstrates that a complete mineralization of organic compounds of this nature does not occur.

A class of compounds that travels more or less unhindered in the aquifer is the adsorbable organic chlorine containing fraction. The nature of these compounds is not yet well known, but it is evident that they can easily break through even during increased activity of the soil ecosystem. When the nature of organic compounds prevents a reversible exchange between soil and the water phase, lasting soil contamination may result, with adverse effects on the ecosystem of the soil. Substances such as chloroanilines have a very limited mobility in the soil (TNO, 1980). These compounds are present in the river Rhine (Wegman, 1981b), could be formed during anaerobic conditions in the aquifer (Piet, 1980a), and are products of the breakdown of herbicides such as phenylcarbamates, phenylurea and acylanilides.

Volatile organic chlorine compounds such as 1,2-dichloroethane, vinylchloride, tetrachloroethene, trichloroethene, tetrachloromethane, 1,1,1-trichloroethane, dichloromethane, methyl chloride and chloroform compose a class of compounds produced by industry in large amounts. Once present in the soil, most of these substances are retarded only by organic material in the soil; the nature of most of these compounds, however, does not make them susceptible to microbiological degradation under aerobic conditions. Several of these substances are adverse to human health. For this reason an open infiltration system with shallow canals is favorable because these compounds are volatilized to a great deal before they can enter the soil.

Odor-intensive substances have an adverse effect when infiltrated water is used for the drinking-water supply. Though these substances can have an industrial origin (Zoeteman, 1980), there are indications that they are produced by microorganisms. The identity of these substances is unknown at this moment.

CONCLUSIONS AND RECOMMENDATIONS

The organic content of the soil of an infiltration site probably determines to a great extent the behavior of nonpolar and slightly polar compounds, and at this moment this is not well known of many infiltration sites. Therefore it is frequently not possible to predict when several types of compounds will break through when physicochemical processes are the primary regulators of the mobility of organic compounds.

Microbiological activity plays an important role in the behavior and fate of the organic chemicals. Though several chemicals are of a persistent nature and must be watched, two aspects should be kept in mind: not only the effects of microorganisms on chemical substances, but also the effects of chemicals on the ecosystem. Methods to evaluate the activity of the ecosystem under field conditions must be developed, and possible contaminants of the soil or their precursors must then be removed from the water before it is infiltrated.

For several hydrophilic compounds of a persistent nature, which seem to travel rather unhindered in the subsurface aquifer, no adequate analytical procedures are developed at this moment. Their influence on water quality and effect on human health is not well understood.

To get more insight into the complicated processes in the field, studies at an infiltration site must be continued during different seasons in combination with adequate model experiments which give information on breakdown of compounds under field conditions. This could lead to better prognoses on the behavior and fate of chemicals in the environment.

Pretreatment of infiltrated water should remove organic compounds that break through during winter and can induce lasting soil contamination. In this respect aeration, or a system with open infiltration, combined with improved coagulation and sedimentation could be applied. Chemical treatment such as prechlorination, which leads to the formation of many persistent organo-chlorine compounds, should be avoided as much as possible.

To maintain a hygienic reliability of the water, prechlorination is not recommended if the sorption capacity of the soil is adequate and no shortcuts in the infiltration area are available for the breakthrough of microorganisms.

The analysis of the soil of an infiltration site should be extended to substances which are almost irreversibly bound to the soil at pH 7. Research is underway to determine these, but it is still uncertain whether quantitative analyses of these compounds are sufficiently reliable.

A final conclusion might be that under certain conditions the soil acts as a good purification system; the pitfalls of breakthrough of compounds, possible soil pollution, and adverse effects on the soil ecosystem should, however, be kept in mind.

ACKNOWLEDGMENT

The dedication and assistance of the staff and technical assistants of the Municipal Water Works of Amsterdam and the National Institute for Water Supply supported the investigations. Particularly, the technical assistance of Mr. R. Westerveld and Mr. T. P. Spierenburg in the field was of great value. The organizing committee of the "First International Conference on Ground Water Quality Research" enabled the presentation of this paper.

REFERENCES

Fritz, J. S. (1975). Concentrating organic impurities from trace levels. *Ind. Eng. Chem., Prod. Res. Dev.* **14**(2):94–96.

Genuchten, van, M. T. (1981). Analytical solution for chemical transport with simultaneous adsorption, zero-order product and first-order decay. *J. Hydrol.* **49**:213–233.

Giger, W., Reinhard, M., Schaffner, C., and Zürcher, F. (1976). Analyses of organic constituents in water by high-resolution gas chromatography in combination with specific detection and computer-assisted mass spectrometry. In: L. H. Keith (Ed.), *Identification and Analysis of Organic Pollutants in Water.* Ann Arbor Science, Ann Arbor, MI, Chapter 26.

Glaze, W. H., Peyton, G. R., and Rawley, R. (1977). Total organic chlorine as water quality parameter. Adsorption/Microcoulometric method. *Environ. Sci. Technol.* **11**:685–690.

Grob, K. and Grob, K., Jr., (1978). On-column injection on to glass capillary columns. *J. Chromatogr.* **151**:311–320.

Kool, H. J., Van Kreijl, C. F., Van Kranen, H. J., and De Greef, E. (1981). The use of XAD-resins for the detection of mutagenic activity in water. II. Studies with drinking water. *Chemosphere* **10**:99–108.

Kool, H. J., Van Kreijl, C. F., De Greef, E., and Van Kranen, H. J. (1982). Presence, Introduction and Removal of Mutagenic Activity during the Preparation of Drinking Water in The Netherlands. *Environm. Health Persp.* **46**:207–214.

Kreijl, C. F. van, Kool, H. J., De Vries, M., Van Kranen, H. J., and De Greef, E. (1980). Mutagenic activity in the rivers Rhine and Meuse in The Netherlands. *Sci. Total Envir.* **15**:137–147.

McCarty, P. L., Reinhard, M., and Rittmann, B. E. (1981). Trace organics in groundwater. *Environ. Sci. Technol.* **15**(1):40–51.

Morra, C. F. H., Linders, J. B. H. J., den Boer, A. C., Zoeteman, B. C. J. (1979). Organic chemicals measured during 1978 in the river Rhine in The Netherlands. RID-report 79-3. National Institute for Water Supply, Leidschendam, The Netherlands.

Piet, G. J. and Zoeteman, B. C. J. (1980a). Organic water quality changes during sand-, bank-, and dune infiltration of surface waters in The Netherlands. *J. Am. Water Works Assoc.* **72**:400–404.

Piet G. J., Morra, C. H. F., de Kruijf, H. A. M. Schultink, L. J., Smeenk, J. G. M. M. (1981). The behaviour of organic micropollutants during passage through the soil. In: *Quality of Groundwater,* Elsevier, Amsterdam, pp. 557–565.

Piet, G. J., Slingerland, P. B., Bijlsma, G. H. and Morra, C. F. H. (1980b). A fast quantitative analysis of a wide variety of halogenated compounds in surface-, drinking-, and groundwater. *Environ. Sci. Res.* **16**:69–81.

Piet, G. J., Slingerland, P., de Grunt, F. E., V.d. Heuvel, M. P. M., Zoeteman, B. C. J. (1978). Determination of very volatile halogenated organic compounds by means of direct head-space analysis. *Anal. Lett.* **A11**(5):437–448.

Puijker, L. M., Veenendaal, G., Janssen, H. M. J., and Grieping, B. (1981). Determition of organophosphorus and organosulphur at the sub-ng-level for use in water analysis. *Fresenius Zeit. Anal. Chem.* **306**:1–6.

Roberts, P. V. and Valocchi, A. J. (1981). Principles of organic contaminant behavior during artificial recharge. In: *Quality of Groundwater,* Elsevier, Amsterdam, pp. 439–451.

Schwarzenbach, R. P. and Westall, J. (1981). Transport of non-polar organic pollutants in a river water–groundwater infiltration system: A systematical approach. In: *Quality of Groundwater,* Elsevier, Amsterdam, pp. 569–575.

Sybrandi, (1980). Personal communication.

Vonk, J. W. (1980). Behaviour of chloroanilines in the soil, the decomposition products and xenobiotic substances (in Dutch). *Netherlands Organization for Applied Research.* TNO, Institute of Organic Chemistry, Utrecht.

Webb, R. G. (1975). Isolating organic water pollutants, XAD resins, urethane foams, solvent extraction. EPA 660/4-75-003. U.S. Environmental Protection Agency, Corvallis, Oregon.

Wegman, R. C. C. and De Korte, G. A. L. (1981a). Aromatic amines in surface waters of The Netherlands. *Water Res.* **15**:391–394.

Wegman, R. C. C. and De Korte, G. A. L. (1981b). The gas chromatographic determination of aromatic amines after bromination in surface water. *Intern. J. Environ. Anal. Chem.* **9**:1–61.

Zoeteman, B. C. J. (1980). *Sensory Assessment of Water Quality* Pergamon Press, London.

9

THE EFFECTS OF DISCHARGING A PRIMARY SEWAGE EFFLUENT ON THE TRIASSIC SANDSTONE AQUIFER AT A SITE IN THE ENGLISH WEST MIDLANDS

K. M. Baxter

Groundwater, Environmental Protection
Water Research Centre, Medmenham Laboratory
United Kingdom

The discharge of sewage and sewage effluents directly to the ground as a means of disposal is widespread within the United Kingdom. The environmental impact of such disposal has become increasingly troublesome in recent years, and as a result of this concern the Water Research Centre (WRC) has been conducting a research program since 1976 to investigate the effects of recharging sewage effluents to the Cretaceous Chalk and Triassic Sandstone aquifers at six sites in the United Kingdom (Baxter and Edworthy, 1979). It has been estimated that about 150,000 m³/day of effluent is discharged into the Chalk, 42,000 m³/day into the Triassic Sandstones, and 22,000 m³/day into minor aquifers such as glacial-fluvial gravels (Baxter and Clark, 1984).

The aim of the research is to formulate "aquifer management" guidelines so that new sites can be chosen—and older ones operated—more efficiently.

The research results from a number of Chalk sites have been reported elsewhere (Edworthy et al., 1978; Baxter et al., 1981; Clark and Baxter, 1981; Baxter and Clark, 1984). This chapter reports on the investigation into the effects of effluent recharge at one site on the Triassic Sandstone in the West Midlands.

SITE DESCRIPTION

The area of effluent disposal is located close to the river Stour between the towns of Kinver and Stourbridge. The land surface is gently undulating between 67 and 130 m above Ordnance Datum (AOD) and is used predominantly for agriculture. There are a number of public supply boreholes (let-

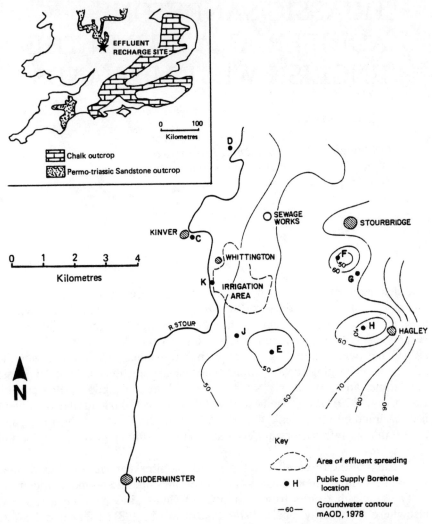

Figure 9.1. Study area showing ground water contours and extent of spreading area.

tered C to J, Figure 9.1) and one observation borehole (K, Figure 9.1) nearby.

Primary sewage effluent has been spread over the site by surface irrigation for about 100 years. The source of the effluent comprises part of the Wolverhampton/Birmingham conurbation, and the sewage effluent is therefore mostly of domestic origin with a small proportion (15 to 20%) of industrial effluent. One-third of the total flow from the source area is sent to the site after primary treatment, the remaining two-thirds diverted to two other sewage treatment works. However, these two other works have no sludge treatment capacity, and so since 1928 all the primary sludge has also been pumped to the site for spreading.

At present the area available for irrigation is about 170 ha, although only about 100 ha are used in any one year. Before 1967, when a sewage works to the north of the site was constructed, over 260 ha were used for irrigation. Since that date the works has discharged its effluent straight in the river Stour, and consequently the area necessary for spreading has been reduced. Irrigation occurs from some 120–150 discharge chambers situated at topographically high points over the undulating land surface. The effluent is distributed over the site by four gravity and one pumped carrier that connect all the chambers across the site. These can be opened and closed by manual

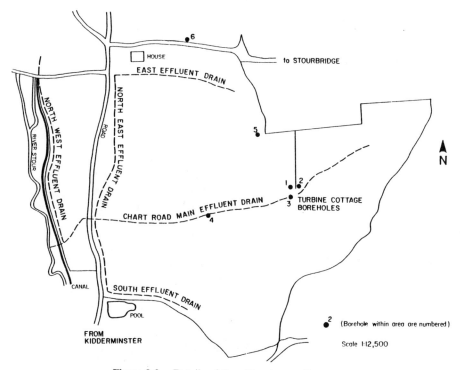

Figure 9.2. Details of the effluent spreading area.

sluice gates to modify the pattern of irrigation to suit the requirements of the tenant farmer. The average dry weather flow (DWF) is 12 megaliters/day although the maximum permitted rate is 27 megaliters/day. Any flow in excess of this latter figure is diverted into storm tanks at one of the source sewage works. Irrigation continues until clogging of the surface by suspended solids prevents further infiltration.

Two main collecting drains taking any natural and artificial surface runoff into the river Stour run through the center of the area and along its western boundary (Figure 9.2).

GEOLOGY AND HYDROGEOLOGY

The geological succession in the Stourbridge area is summarized in Table 9.1 and shown in Figure 9.3; it is composed chiefly of arenaceous and rudaceous deposits of Triassic age.

As the area lies at the northern end of the Worcester Graben, the Triassic sediments are considerably faulted. The stratigraphic dip is generally to the

Table 9.1   Geological Succession in the Stourbridge Area

Period	Epoch	Age	Formation/Zone Lithology	Thickness (m)
Quaternary	Recent and Pleistocene		Alluvium	0–3
			Glacial sands and gravels	0–4
		Unconformity		
Mesozoic	Triassic	Keuper	Lower Keuper Sandstone (red and brown sandstone with pebbly bands)	46–152
		Bunter	Upper Mottled Sandstone (soft red sandstone)	61–152
			Bunter Pebble Beds (pebbly red conglomeritic sandstone)	113–122
			Lower Mottled Sandstone (soft red sandstone)	0–259
		Unconformity		
Paleozoic	Carboniferous	Upper Coal Measures	Keele Beds	38
			Highley Beds (grey clays, shales, sandstone, coals and limestones)	91–183

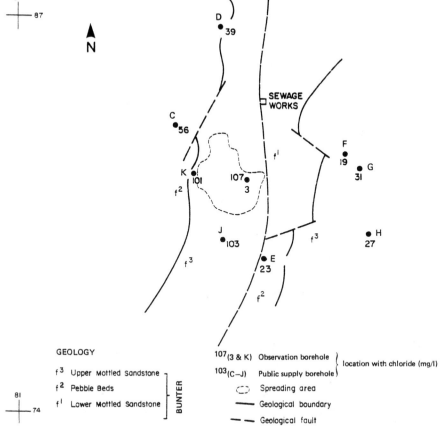

GEOLOGY

f³ Upper Mottled Sandstone

f² Pebble Beds

f¹ Lower Mottled Sandstone

BUNTER

107(3 & K) Observation borehole

103(C-J) Public supply borehole

⟨⟩ Spreading area

—— Geological boundary

— — Geological fault

location with chloride (mg/l)

Figure 9.3.   Regional chloride ground water concentration and geology.

east at between 3 and 6°, but locally may be as high as 9° (Whitehead and Pocock, 1947).

The water table is between 25 and 30 m below ground level in the vicinity of the irrigation area and the ground water gradient is between 0.3 and 1.0° to the west, toward the river Stour and across the stratigraphic dip. The ground water gradient has locally been increased by pumping at the nearby public supply borehole C.

Test pumping of boreholes in the study area has given results for transmissivity (T) as high as 1500–3500 m²/day and a specific yield (S) of between 1.1 and 1.5 × 10⁻³. Transmissivities of 500–1000 m²/day and a specific yield of 0.1 are believed to be more typical for the area as a whole, by analogy with other Permo-Triassic sandstone areas.

Average annual rainfall and effective rainfall at Edgbaston, Birmingham, Meteorological Station are 688 mm (1951–1975) and 308 mm (1972–1978), respectively.

SITE INVESTIGATION

The site investigation has centered on the drilling of seven boreholes and on a soil survey.

Five shallow boreholes (Nos. 1–2 and 4–6, Figure 9.2) have been drilled to 30 m depth or less to sample the unsaturated zone beneath the irrigation area. Two deeper boreholes (122 m and 84 m) were also drilled to sample the saturated zone beneath the area of infiltration and down the ground water gradient from this area (No. 3, Figure 9.2 and J, Figure 9.1, respectively). Core samples were taken from all the investigatory boreholes and porewater extracted. Samples of ground water were also taken for chemical analysis as drilling progressed.

A series of downhole geophysical logs were run in the two deep boreholes to assist the interpretation of the porewater quality data and to select suitable depths for the installation of five WRC *in situ* ground water samplers (see next section).

The soil survey was undertaken to determine the adsorption capacity of the soil for heavy metals to 1 m depth. Hand-auger samples were taken at six depth intervals at the top, middle, and base of a slope over which effluent and sludge had been irrigated since 1884 and 1967, respectively. In order to obtain sufficient soil for analysis, and also to achieve statistically significant sampling, ten such depth profiles were taken at 10 m intervals at each site along the contour of the slope, and bagged together as one depth profile. Samples from one profile on an unirrigated site were also taken to act as control.

GROUND WATER SAMPLER INSTALLATION, SPECIAL AND ROUTINE MONITORING

The WRC *in situ* sampler (Joseph, 1980) is essentially a 0.5 L chamber buried at a specific depth in a borehole but connected to the surface by two tubes with an OD of 6 mm. Gas, usually nitrogen, is applied under pressure to one tube, and the chamber containing water is evacuated via the other. This is usually done two or three times before a representative sample is taken. In installation, the sampler is surrounded by clean inert gravel and is sealed above and below the gravel-packed interval with a thick layer of bentonite. Thus, any water drawn into the sampler must come from the adjacent section of the aquifer (Figure 9.4). The standard sampler has a uPVC body and is connected to the surface by nylon tubes. Five such samplers were installed in borehole 3 within the effluent infiltration area at 57.5, 70.1, 80.4, 90.5, and 96.0 m below surface.

Sampling began in May 1979 and continued at approximately monthly intervals until December 1980. In addition to these regular samples, which

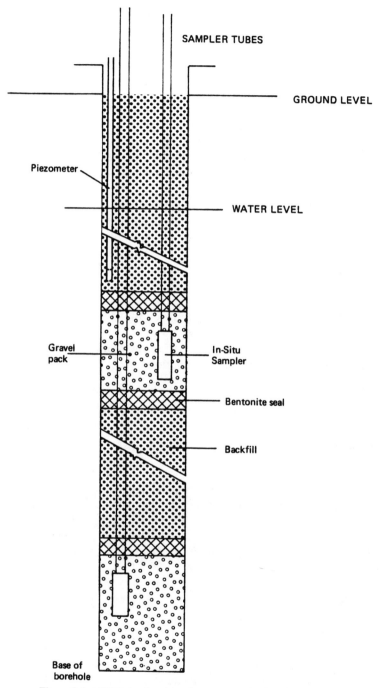

Figure 9.4. WRC *in situ* ground water sampler installation.

were analyzed for the major ions, samples for microbiological analyses were taken from each of the five samplers in June 1980 and again in June 1981.

LABORATORY STUDIES

Porewater samples were extracted from unsaturated and saturated zone core samples by high-speed centrifuging and have been analyzed for the major ions and for the heavy metals lead, zinc, and copper. Additional centrifuging and analysis was carried out separately by the University of Sheffield on samples from boreholes 5 and J to determine the effects on porewater composition of different extraction methods.

Ground water samples taken during drilling from the observation boreholes 3 and J have been analyzed for the major ions only. Samples from a number of public supply boreholes were, however, analyzed for lead and copper.

Whole-rock analysis of a number of core samples has been carried out by X-ray fluorescence (XRF) and X-ray diffraction (XRD).

Grain-size analysis, porosity, and permeability determinations have also been made on 36 samples. Fourteen thin-sections were prepared and examined under a polarizing microscope.

Soil survey samples were subjected to sequential leaching to remove the following progressively less soluble fractions (Ross, 1978):

1. Water soluble.
2. Carbonate-associated.
3. Cation-exchangeable.
4. Sulphide and organic.
5. Moderately reduced oxide.
6. More resistant oxide.

Each fraction was then analyzed for cadmium, chromium, lead, and copper. In addition, the soil survey samples were subjected to XRF trace element analysis for nickel, cobalt, manganese, vanadium, chromium, zinc, copper, rubidium, strontium, yttrium, zirconium, lead, and barium.

RESULTS

Effluent Quality

The average long term quality of irrigated effluent has varied only slightly since 1975, although there has been considerable monthly variation, especially in the BOD, suspended solids and heavy metal concentrations. The mean quality from 1975–77 is given in Table 9.2. There is little information

Table 9.2   Mean Values of Concentrations of Constituent for Final Effluent (1975–1977) (all results in mg/L unless otherwise stated)

Constituent	n	$\bar{x}$	σn	C^n of V^n
pH units	27	7.03	0.20	0.03
BOD	27	63.5	85.1	1.34
COD	27	194	79.5	0.41
Suspended Solids	28	104	63.9	0.61
Nitrate-N	27	16.5	5.5	0.33
Ammonia-N	27	13.3	5.1	0.38
Chloride	15	111	25	0.23
Total phosphorus-P	5	6.9	5.8	0.84
Phenols	8	2.5	0.94	0.38
Metals (µg/L)				
Chromium	26	35	23	0.66
Copper	26	95	34	0.36
Nickel	26	48	29	0.60
Zinc	26	578	275	0.48
Cadmium	16	11	9	0.82
Lead	17	104	61	0.59

on the composition of the effluent previous to 1975, but it is believed that the dilution would have increased over the last 100 years and that the phosphorus levels would also have risen.

As well as a change in the composition of the effluent over the life of the site, the *per capita* flow has probably trebled. By analogy with records from other areas, the average 1880 flow would have been around 4 megaliters/day, increasing rapidly between 1880 and 1900 to 8 megaliters/day and again between 1930 and 1980 to 12 megaliters/day.

Regional Ground Water Quality

Public supply boreholes D, E, F, G, and H (Figure 9.3) are considered on the basis of their historical quality data to be unpolluted and their average composition representative of natural regional ground water (Table 9.3).

By comparison it is evident from the chloride concentrations of boreholes C, J and K (Figure 9.3) that there is some contamination to the west and south of the irrigation area. The chloride concentration (Table 9.4) has been used to estimate the percentage of effluent reaching these boreholes. It can be seen that the ground water in boreholes K and J contains over 70%

Table 9.3 Average Regional Natural Ground Water Quality (mg/L unless shown otherwise)

Ammonia-N	<0.04	Calcium	64.7
Nitrate-N	9.1	Magnesium	4.97
Nitrite-N	<0.01	Sodium	12.3
Chloride	25.6	Potassium	2.54
Sulfate	43.1	Lead	<0.001
Alkalinity (CaCO$_3$)	102	Copper	0.004
O-phosphate	<0.02		

effluent, while at public supply borehole C the degree of contamination approaches 30%.

Historical quality data from borehole C shows a threefold increase in the nitrate, chloride and sulfate concentration over the last 40 years, while at boreholes E and H there has been only a 20% increase in chloride over the last 20 years. However, at all three locations over the past 11 years, the ground water quality has remained fairly constant, suggesting steady state conditions have been reached.

Geochemistry of the Unsaturated Zone

Porewater Quality

Porewater quality in the unsaturated zone within the irrigation area has been investigated at boreholes 1, 2, and 5 (Figure 9.2). Boreholes 1 and 2 were

Table 9.4 Variation in Chloride and Nitrate Concentration in Ground Water Used to Estimate Percentage of Effluent

Sample Location	Chloride		Nitrate-N			
	$S_C{}^a$	$\%E_C{}^b$	$S_N{}^a$	$\%E_N{}^b$	$E_N/E_C{}^c$	Total-N
Effluent	129	100	16.5	100	1.0	29.8
BH 3	112	83	33.0	323	3.9	33.9
BH K	101	72	40.5	420	5.8	41.1
BH C	56	29	9.1	0	—	9.1
BH J^d	103	75	29.6	277	3.7	29.6
Background	26	0	9.1	0	—	9.1

^{a} Concentration of chloride (S_C) and nitrate (S_N) in ground water.

^{b} % effluent $= \dfrac{\text{C sample} - \text{C background}}{\text{C effluent} - \text{C background}} \times 100$ for chloride (E_C) and nitrate (E_N).

^{c} Ratio of E_N to E_C.

^{d} Depth samples.

drilled to 10 m depth and located 30 m apart so that borehole 2 was on land irrigated directly from the surface while borehole 1 was on unused land.

The difference between the porewater quality in the two boreholes is a result of the direct surface recharge of effluent and is restricted to the upper 2.5 m of the profile for the major ions (Figure 9.5). Here chloride and o-phosphate are at greater concentrations in borehole 2 while nitrate and ammonia are at lower levels. Analyses for the heavy metals (Figure 9.6) differ only in the top 0.5 m of the profile, where both copper and zinc are at higher levels in samples from borehole 2. Higher zinc concentration in samples from borehole 2 are also observed at 2.75 and 5.25 m depth.

Borehole 5 was drilled some 400 m to the northwest of boreholes 1 and 2 on another part of the irrigated area, as close as possible to the site of the soil survey (Figure 9.2). The upper portion of the quality profile from borehole 5 resembles that of borehole 1 rather than that of borehole 2, as might be expected (Figure 9.7). This suggests that very little direct surface recharge of effluent occurs at the site of borehole 5.

Comparison of the results of borehole 5, on the irrigated area, with those from the unsaturated zone of borehole J, 1 km down ground water gradient beneath unirrigated land, is given in Table 9.5. It is apparent that potassium, cadmium, chromium, nickel, and iron are depleted in the porewater beneath the irrigated area, while sodium, magnesium, and calcium are enriched.

Whole-Rock Analyses

Mineralogical analysis by X-ray diffraction shows the Upper Mottled Sandstone to contain quartz, feldspar, calcite, kaolinite, illite, and some chlorite.

Table 9.5 Comparison of Unsaturated Zone Porewater Quality Between Irrigated and Unirrigated Sites (all results in mg/L)

	$\Delta \bar{x}$	A—Irrigated (BH 5)				B—Unirrigated (BH J)			
	(A − B)	n	$\bar{x}$	σn	C of V	n	$\bar{x}$	σn	C of V
Sodium	33.1	25	96.5	28.0	0.29	15	63.4	15.7	0.248
Potassium	−12.9	25	30.7	6.93	0.23	15	43.6	18.3	0.420
Magnesium	13.2	22	32.3	13.5	0.42	15	19.1	6.0	0.314
Calcium	77.5	22	175	46.0	0.26	15	97.5	19.8	0.203
Cadmium	−0.172	10	0.0085	0.003	0.35	15	0.18	0.24	1.33
Chromium	−0.47	10	<0.002	—	—	15	0.47	0.79	1.68
Nickel	−0.152	10	0.038	0.026	0.68	15	0.19	0.28	1.47
Copper	0.279	10	0.309	0.230	0.74	15	0.03	0.07	2.33
Zinc	0.873	10	0.889	0.528	0.59	15	0.016	0.039	2.44
Lead	0.011	10	0.058	0.096	1.7	15	0.047	0.09	1.91
Iron	−0.838	10	0.052	0.040	0.77	14	0.89	0.81	0.910

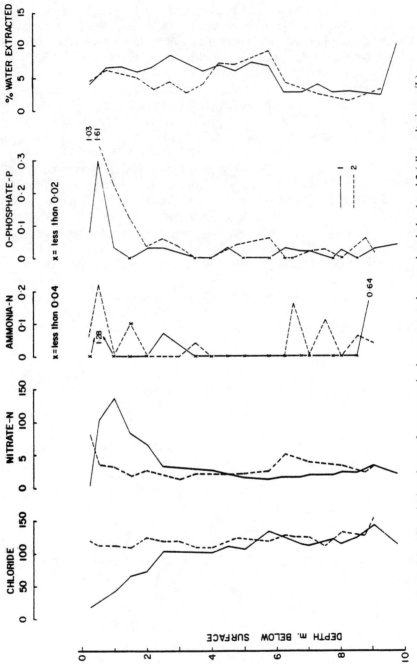

Figure 9.5. Chemical analyses of porewater in the unsaturated zone—boreholes 1 and 2 (all results in mg/L).

156

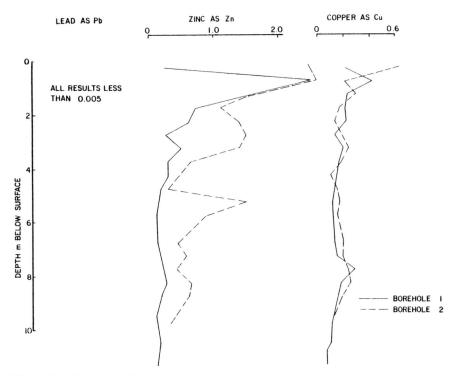

Figure 9.6. Heavy metal analyses of porewater in saturated zone—boreholes 1 and 2 (all results in mg/L).

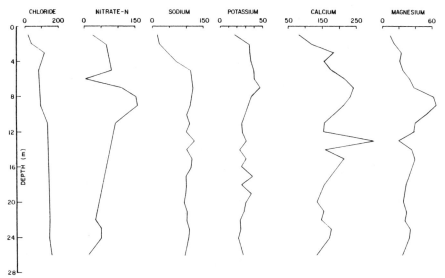

Figure 9.7. Chemical analyses of porewater in the unsaturated zone—borehole 5 (all results in mg/L).

157

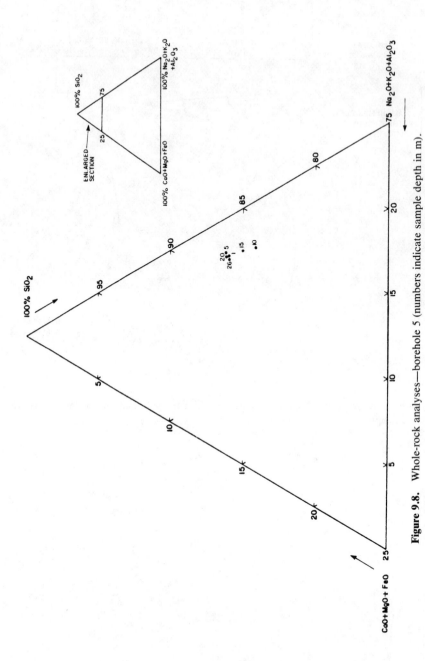

Figure 9.8. Whole-rock analyses—borehole 5 (numbers indicate sample depth in m).

Elemental analysis by X-ray fluorescence of samples from borehole 5, beneath the irrigated area, indicates a decreasing percentage of silica with depth to 10 m. Below this depth the silica percentage increases to the base of the borehole (Figure 9.8). By comparison, samples from borehole J, beneath unirrigated land, show the percentage of calcite increasing with depth and reaching a maximum value midway down the unsaturated zone (Figure 9.9). Samples also exhibit a decreasing clay and feldspar mineral content with depth.

X-ray fluorescence trace-element analyses from the unsaturated zones of boreholes 3, J and 6 are compared in Table 9.6. Two profiles from beneath unirrigated land (J and 6) are included to establish any natural variability. Beneath the irrigated area there are relatively higher concentrations of most of the trace elements except manganese and chromium, which are depleted. Comparison of the two unirrigated sites does show some natural variation, especially for manganese (Table 9.6).

Table 9.6 Comparison of Trace Element Concentration in Sandstone in the Unsaturated Zone Between Irrigated and Unirrigated Sites (all results in ppm)

	Irrigated		Unirrigated					
	Borehole 3		Borehole J		Borehole 6		Comparison	
Element	Mean $(A\bar{x})$	Depth Trend[a]	Mean $(B\bar{x})$	Depth Trend[a]	Mean $(C\bar{x})$	Depth Trend[a]	$A\bar{x}-B\bar{x}$	$B\bar{x}/C\bar{x}$
Ni	10.6	C	7.3	D	6.4	I	4.2	1.14
Mn	491	I	634	D	158	I	−143	4.01
V	25.2	C	21.6	D	21.0	I	3.6	1.02
Ba	443	C	393	D	390	I	5.0	1.01
Sr	69.3	C	56.9	D	60	I	12.4	0.99
Zr	293	D	151	D	203	I	142	0.74
Rb	49.7	C	24.2	D	34	I	25.5	0.71
Zn	38.7	D	28.9	D	21	I	9.8	1.38
Cr	36.7	D	37.5	C	31	I	−0.8	1.21
Y	26.5	C	14.3	C	13.3	I	12.2	1.08
Cu	9.0	D	7.5	C	8.9	C	1.5	0.84
Pb	93.8	C	82.5	D	85	C	11.3	0.97
Co	65.0	I	50.0	I	78	D	15.0	0.64

[a] Depth trend: I = increasing concentration with depth; D = decreasing concentration with depth; C = constant concentration with depth.

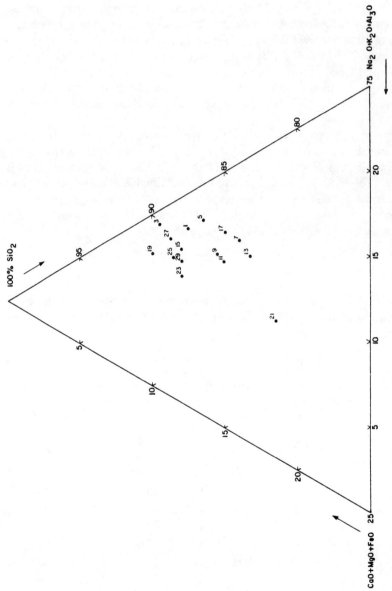

Figure 9.9. Whole-rock analyses—borehole J (numbers indicate sample depth in m).

Geochemistry of the Saturated Zone

Porewater Quality

Porewater quality depth profiles in the saturated zone beneath the effluent irrigation area have been determined in borehole 3 (Figure 9.2) from 25 to 122 m depth (Figures 9.10, 9.11, 9.12). From these profiles four intervals of different porewater quality can be recognized (Table 9.7). Interval 2 is characterized by a greater variation in the concentrations of most constituents than intervals 1 and 3. Interval 4 has minimal variation and an overall lower level of all determinants, indicating that this interval is only slightly contaminated.

Samples of ground water taken during drilling represented the average quality over an increasing thickness of aquifer as drilling progressed. The chloride concentration of about 115 mg/L in these samples was generally lower than in the porewater. The levels of nitrate, orthophosphate, ammonia, sodium, and potassium, however, were all quite similar.

The second borehole cored through the saturated zone was borehole J (Figure 9.1), located 1 km down ground water gradient from the irrigation area. The porewater quality profile shows a gradual increase in most of the major ions at a depth between 35 and 67 m, followed by a decrease below this depth and a return to their initial concentrations at 80–85 m. Magnesium increases only slightly in concentration below 67 m (Figure 9.13). It would therefore appear that borehole J has penetrated a body of contaminated interstitial ground water some 50 m thick lying between 35 and 85 m below surface.

Examination of the heavy metal porewater profile (Figure 9.14), although indicating zones of more contaminated ground water between 45 and 62 m

Table 9.7 Summary of Porewater Composition in Saturated Zone (mean range of concentrations shown in mg/L)[a]

	Interval 1 (25–55 m)	Interval 2 (55–75 m)	Interval 3 (75–91 m)	Interval 4 (91–122 m)
Chloride	125–150	60–150	140–165	125–140
Nitrate-N	29–38	3–34	29–43	20–33
Ammonia-N	0.1–0.8	<0.04–0.2	<0.04–0.5	<0.04–0.1
Sodium	65–105	18–80	65–100	55–78
Potassium	10–20	6–11	11–15	8–13
O-phosphate-P	<0.02–0.25	<0.02–0.05	0.03–0.25	<0.02–0.1
Lead	<0.005–0.15	<0.005–0.75	0.04–0.15	0.04–0.08
Zinc	0.14–0.32	0.15–0.42	0.1–0.35	0.05–0.15
Copper	0.07–0.26	0.08–0.3	0.1–0.3	0.05–0.14

[a] See Figures 9.10–9.12.

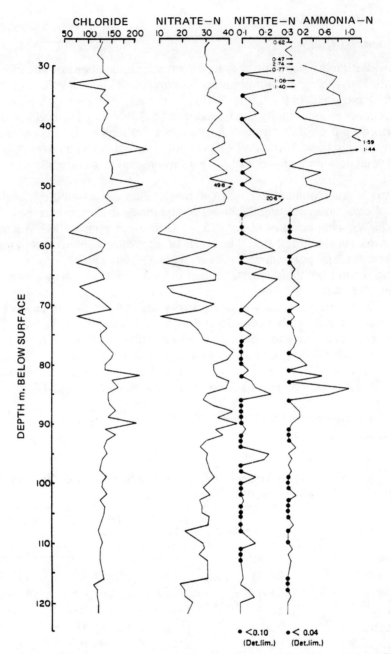

CHLORIDE · NITRATE—N · NITRITE—N · AMMONIA—N

Figure 9.10. Porewater analyses—borehole 3—zone of saturation (all results in mg/L).

162

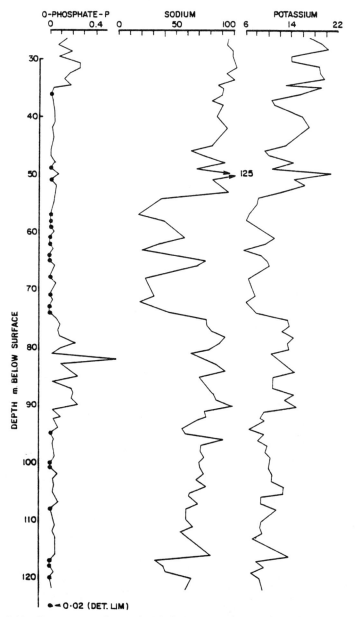

Figure 9.11. Porewater analyses—borehole 3—zone of saturation (all results in mg/L).

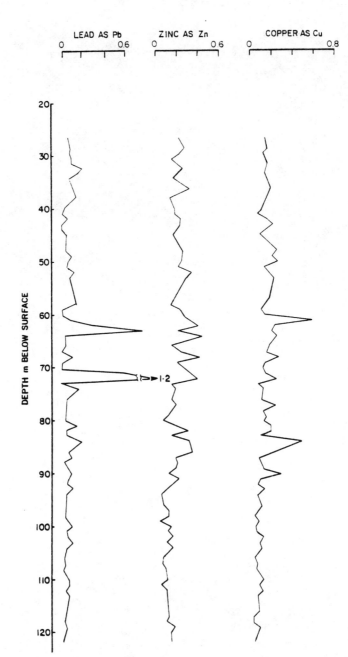

Figure 9.12. Porewater analyses—borehole 3—heavy metals in zone of saturation (all results in mg/L).

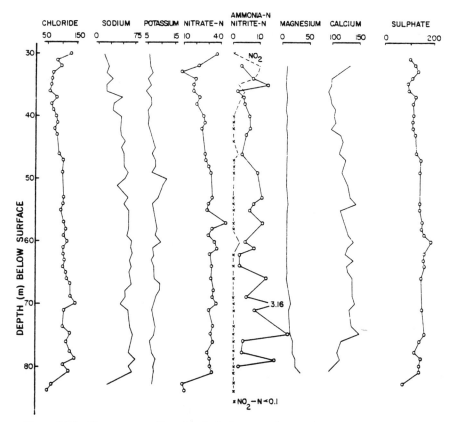

Figure 9.13. Porewater quality—borehole J—zone of saturation (all results in mg/L).

and again between 67 and 77 m, does not confirm the presence of the zone of contamination between 35 and 85 m observed from the major ion chemistry.

Whole-Rock Analyses

Major element analyses of core from the saturated zone at borehole 3 was undertaken for CaO only (Table 9.8a). A more complete analysis was carried out on samples from the saturated zone of borehole J outside the area of effluent irrigation (Table 9.8b).

The absence of CaO in the upper samples from borehole 3 and its overall lower percentage there, as compared with borehole J, is most probably the result of enhanced leaching of calcite beneath the irrigation area. This is confirmed by the fact that this difference in abundance decreases with depth. The concentrations of the other oxides have a positive correlation with the Al_2O_3 abundance on the clay-mineral fraction.

The trace-element composition of the saturated zone samples from borehole J (Table 9.9) show no distinct trends with depth. Values are approxi-

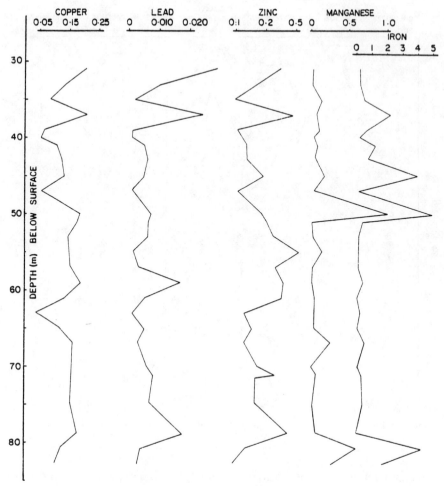

Figure 9.14. Heavy metal composition of porewater—Borehole J—zone of saturation (all results in mg/L).

mately equal to the unsaturated zone analyses (Table 9.6) apart from cobalt and lead, which are, respectively, one-half and twice the concentration in the saturated zone samples.

A comparison of XRF trace-element analyses between borehole 3 and borehole J (Table 9.10) shows an appreciable depletion of cobalt and manganese and some loss of lead beneath the irrigated area.

Soil Survey

The results of the analyses of the soil samples for the heavy metals copper, lead, chromium, and cadmium from the upper, middle, lower, and control

Table 9.8 Major Element Analyses of Saturated Zone Samples (all figures are percentages, standard deviations in brackets)[a]

(a) Beneath an Irrigated Area

Depth (m)	CaO
26.6	nd
39.1	nd
46.3	1.2
55.1	1.8
55.8	3.0
62.0	1.2
68.1	3.9
71.0	0.9
74.5	1.2
86.1	nd
96.1	2.8
Mean	2.0(1.0)

(b) Beneath an Unirrigated Area

Depth (m)	SiO_2	CaO	Al_2O_3	TiO_2	Fe_2O_3	MgO	Na_2O	K_2O
31.1	90.5	0.12	3.9	0.13	1.2	0.49	0.01	2.2
40.0	79.8	4.3	5.8	0.22	2.1	0.79	0.01	3.0
50.1	76.8	3.7	7.3	0.33	2.0	0.90	0.01	3.6
60.9	83.0	3.7	4.9	0.17	1.6	0.59	0.01	2.7
69.9	73.5	9.4	4.0	0.20	2.1	0.88	0.01	1.6
80.7	80.7	3.1	3.6	0.22	4.0	1.7	0.01	1.2
Mean	80.7(5.3)	4.1(2.7)	4.9(1.3)	0.21(0.06)	2.2(0.88)	0.89(0.89)	0.01	2.4(0.82)

[a] n.d. = not detected.

Table 9.9 XRF Trace Element Whole-Rock Analysis for Samples from Borehole J (all results in ppm)

Depth (m)	Ni	Co	Mn	V	Cr	Zn	Cu	Pb	Y
31.1	3	228	98	14	19	35	7	22	13
40.0	5	91	233	26	18	55	9	32	20
50.1	9	100	1500	26	32	29	6	110	24
60.9	6	114	323	18	22	19	6	21	16
69.9	6	75	980	23	25	16	5	21	18
80.7	8	104	1459	35	27	19	9	25	25
Average	6.2	119	766	24	24	29	7	39	19

Table 9.10   Comparison of XRF Trace Element Whole-Rock Analysis of Saturated Zone Samples from Boreholes 3 and J (results are mean and standard deviation)

	Borehole 3 (A)	Borehole J (B)	Mean A − Mean B
Ni	6.8 ± 2.5	6.2 ± 2.0	+0.6
Co	18.2 ± 7.5	119 ± 50.4	−100.8
Mn	430 ± 450	766 ± 576	−336
V	19.9 ± 5.3	23.7 ± 6.6	−3.8
Cr	31.7 ± 11.6	23.8 ± 4.8	+7.9
Zn	29.9 ± 11.8	28.8 ± 13.4	+1.1
Cu	7.5 ± 1.7	7.0 ± 1.5	+0.5
Pb	27.6 ± 7.7	38.5 ± 32.2	−10.9
Y	21.6 ± 5.9	19.3 ± 4.2	+2.3
Number of analyses	10	6	

profiles are given in Figure 9.15 as a series of depth profiles (A to F). The six profiles (A to F) represent the cumulative results from the analyses of the progressively less soluble phases 1 to 6 mentioned above (i.e., profile A results from the analysis of phase 1 only, while profile B is the sum of the analyses of phases 1 and 2, etc).

It can be seen from Figure 9.15 that apart from chromium, which was found to be at above background levels in all three traverses, the concentrations of the other three metals exceed the control concentration only in the lower traverse. The concentrations of all four heavy metals in each traverse decrease with depth, background levels being reached by 1 m depth. It is also worthy of note that the heavy metals were generally found to be held in the less soluble phases 4 to 6.

When these results are compared with those obtained by XRF (Table 9.11), it is clear that leaching did not successfully extract all of the heavy metals present in the soil samples, particularly lead and chromium. The leaching also appeared less successful for the shallower samples in every case. The difference between the analytical results for copper, lead, and chromium in the lower traverse is shown in Figure 9.16. The principal difference between the XRF and leaching results is that the XRF results indicate lead to be retained in the top 0.5 m of profile E.

Lithological and Geophysical Logs

Lithological logging of the recovered core shows that the Upper Mottled Sandstone and the underlying Bunter Pebble Beds can be recognized. Their junction is placed at 95 m below surface (Figure 9.17).

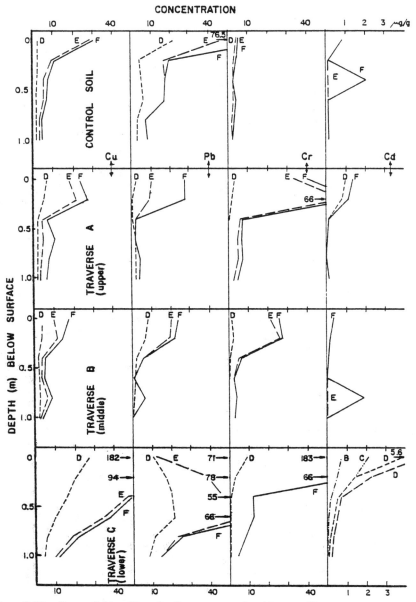

Figure 9.15. Sequential leaching of soil survey samples. Note: profiles A–F are cumulative sums of results from phases 1–6.

Table 9.11 Comparison of Analytical Results for Total Heavy Metal Concentrations in Soil Samples by XRF (X) and Leaching (L)[a] (all results in μg/g)

Traverse and Sample No.[b]	Copper (X)	Copper (L)	Lead (X)	Lead (L)	Chromium (X)	Chromium (L)	(X) – (L) Cu	Pb	Cr
A1 Upper	86	24	117	27	134	41	62	90	93
2	42	28	73	28	105	66	14	45	39
3	15	8	46	2	49	7	7	44	42
4	13	10	55	2	53	7	3	53	46
5	12	7	51	4	50	7	5	47	43
6	11	6	54	4	50	5	5	50	45
B1 Middle	31	18	72	24	74	26	13	48	48
2	29	15	64	22	66	27	14	42	39
3	34	5	56	5	46	6	29	51	40
4	13	5	39	1	43	3	8	38	40
5	10	8	44	6	38	4	2	38	34
6	11	4	41	0	44	3	7	41	41
C1 Lower	274	182	339	71	351	183	92	268	168
2	208	94	288	78	187	66	114	210	121
3	88	50	134	55	57	12	38	79	45
4	39	39	83	68	53	13	0	15	40
5	23	22	65	25	43	8	1	40	35
6	24	11	52	15	41	5	13	37	36
X1 Control	73	31	188	77	58	5	42	111	53
2	36	11	87	19	58	5	25	68	53
3	23	7	53	18	49	3	16	35	46
4	20	6	52	17	62	4	14	35	58
5	19	5	46	8	50	3	14	38	47
6	18	4	42	8	51	2	14	34	49

[a] XRF analysis by University of Sheffield. Leaching by Water Research Centre, Medmenham.
[b] Sample 1 = 0 − 0.2 m, 2 = 0.2 − 0.3 m, 3 = 0.3 − 0.45 m, 4 = 0.45 − 0.6 m, 5 = 0.6 − 0.75 m, 6 = 0.75 − 1.0 m.

The Upper Mottled Sandstone is predominantly a greyish-red, fine-grained, well-sorted sandstone with minor shale and marl partings. The color is fairly uniform apart from some slight mottling within the top 40 m. Opaque minerals are visible in hand specimens below 55 m depth, and mica is observed in samples between 35 and 83 m.

The Bunter Pebble Beds are darker red, medium to fine-grained, more fractured, less well sorted sandstones becoming conglomeratic below 108 m

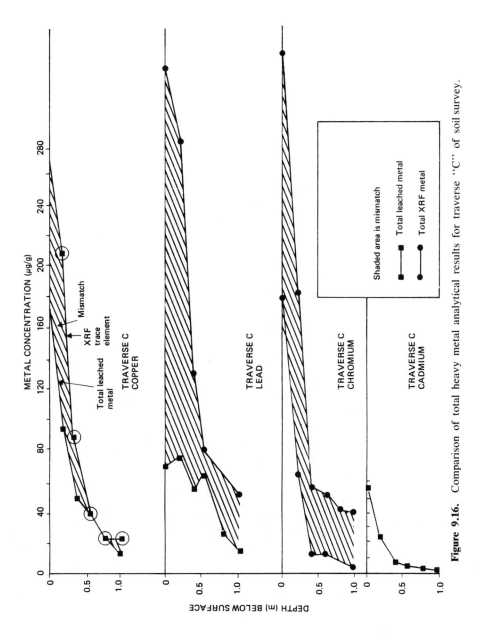

METAL CONCENTRATION (μg/g)

DEPTH (m) BELOW SURFACE

Mismatch

XRF trace element

Total leached metal

TRAVERSE C
COPPER

TRAVERSE C
LEAD

TRAVERSE C
CHROMIUM

TRAVERSE C
CADMIUM

Shaded area is mismatch

Total leached metal

Total XRF metal

Figure 9.16. Comparison of total heavy metal analytical results for traverse ''C'' of soil survey.

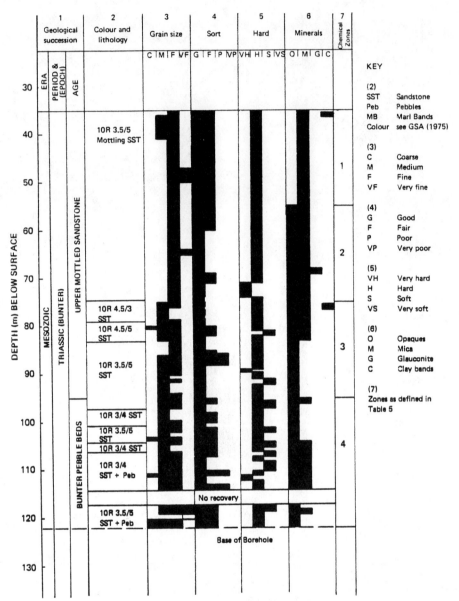

Figure 9.17.   Geological log—borehole 3.

depth. Opaques are observed over the full extent of the Pebble Beds with occasional micaceous partings (especially between 104–113 m depth).

Features similar to borehole 3 are observed in the geological log for borehole J (Figure 9.18), except that the Upper Mottled Sandstone is less well-sorted and contains a few pebbly intervals in borehole J. The Upper Mottled Sandstone/Bunter Pebble Bed junction has been identified at 82 m depth.

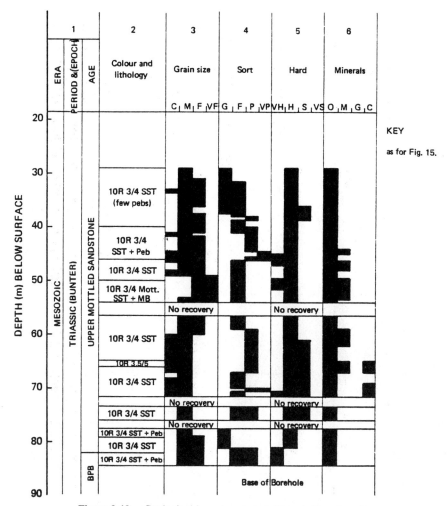

Figure 9.18.   Geological log—borehole J (Breech Tree Lane).

The downhole geophysical logs from borehole 3 are shown in Figure 9.19. The fluid temperature and fluid conductivity logs show that colder, more polluted water is entering the borehole at a depth of 33.5 m from behind the casing, then flowing downward and passing out into the formation at between 70 and 85 m depth. The formation resistivity and natural gamma logs both show a change at 95 m (Figure 9.19), which confirms the top of the Pebble Beds to be at this depth.

Six distinct zones can be recognized in the geophysical logs from borehole 3 (Figure 9.19), the lower five corresponding quite closely with those identified with the porewater quality of the saturated zone (Table 9.7).

The corresponding logs for borehole J indicate the Upper Mottled Sandstone to be more highly fractured than in borehole 3. Additionally, the 50 m

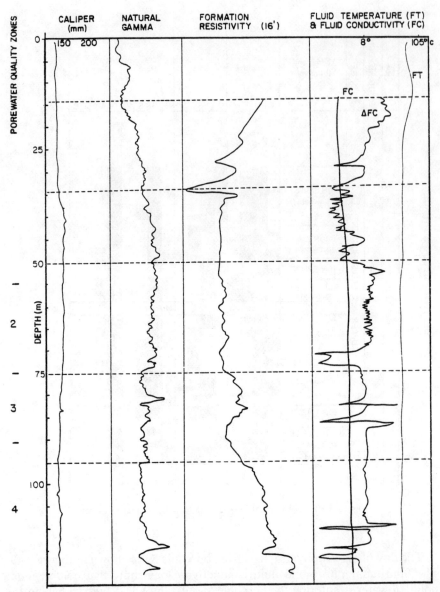

Figure 9.19. Downhole geophysical logs—borehole 3—showing possible zonation for comparison with porewater quality zones.

Table 9.12   Summary of Thin-Section Examination BH 3

Depth (m)	Grain Size[a]			Angularity[b]				Sphericity[c]		Type[d]	
	VF	F	M	A	SA	SR	R	L	H	LG	FG
-20											
-30											
-40											
-50											
-60											
-70											
-80											
-90											
-100											
-110											
-120											

[a] VF = very fine; F = fine; M = medium.

[b] A = angular; SA = subangular; SR = subrounded; R = rounded.

[c] L = low; H = high. LG = lithic greywacke; FG = felsic greywacke.

[d] Determined by 100 point counts.

thick contaminated zone observed from the porewater profile in borehole J (Figure 9.13) is detected by the fluid logs. The Upper Mottled Sandstone/ Bunter Pebble Bed junction is seen at 82 m depth.

Laboratory Examination of Core Material

Grain-Size Analysis

Grain-size analyses of samples of the Upper Mottled Sandstone from borehole 3 show that the sandstones were generally fine to medium-grained, moderately well-sorted, positively skewed, and leptokurtic; they were found to contain between 2 and 17% silt/clay. Samples from below 95 m depth, however, are generally coarser-grained, very positively skewed, and very leptokurtic. This confirms the base of the Upper Mottled Sandstone at 95 m depth in borehole 3.

Porosity and Permeability Determinations

Values of porosity determined on core samples range between 22 and 31%, the lowest value being at 95 m and the highest at 106 m depth. Porosity and permeability are strongly related, as might be expected in such an aquifer, K_H varying between 2 and 0.01 m/day. The vertical permeability is generally 20 to 60% of the horizontal permeability.

Table 9.13   Comparison of Clay Fractions, <2 μm, from Borehole 5 (irrigated) and Borehole J (unirrigated)

	Percentage (based on peak height)		
Sample Depth (m)	Kaolinite	Illite	Smectite
BH 5			
1	13	20	67
5	12	24	64
10	12	34	54
15	11	31	58
20	9	24	67
26	12	25	63
BH J			
1	28	26	46
15	18	15	67
29	23	25	52

Thin-Section Examination

Microscopic examination of 14 thin sections from within the saturated zone of the aquifer from borehole 3 show the rock type to be a felsic greywacke becoming more lithic in character with depth (Pettijohn, 1954). There is also an increase in grain-size; grains becoming less angular and more rounded with depth. The individual grains of feldspar and quartz are coated with an iron oxide film, which becomes less developed below 98 m depth. The results of thin-section examination are summarized in Table 9.12.

Clay Mineralogy

The main clay minerals found in samples from borehole 5 were smectite as the dominant mineral, illite, and secondary kaolinite produced as a weathering product of feldspar. Some mixed-layer clays were also present. No systematic variation with depth was observed. These three clay minerals are also present in samples from borehole J. The percentage of kaolinite and illite is higher in the top and bottom of the borehole while smectite is more abundant in the middle sample (Table 9.13).

WRC *IN SITU* SAMPLERS AND RESULTS OF SPECIAL MONITORING

Figures 9.20(*a*)–(*c*) shows the temporal variations in ground water quality as determined by the five *in situ* samplers in borehole 3. Initially there was

Table 9.14 Results of Microbiological Examination

Date		Virus (P.F.U.)	Colony Count @ 37°C (per mL)	E. coli (per 100 mL)	Streptococcus fecalis (per 100 mL)
Effluent	6/80	2924/L[a]	160,000	1,300,000	12,000
Sampler 1	6/80	0/5[b]	11	0	0
	6/81	0/5	34	0	0
Sampler 2	6/80	0/5	700	0	0
	6/81	0/5	120	0	0
Sampler 3	6/80	0/5	40	0	0
	6/81	0/5	130	0	0
Sampler 4	6/80	0/5	48	0	0
	6/81	0/5	20	0	0
Sampler 5	6/80	0/5	600	0	0
	6/81	0/5	60	0	0

[a] Poliovirus types 1, 2, and 3.

[b] 0/5 = none in 5L.

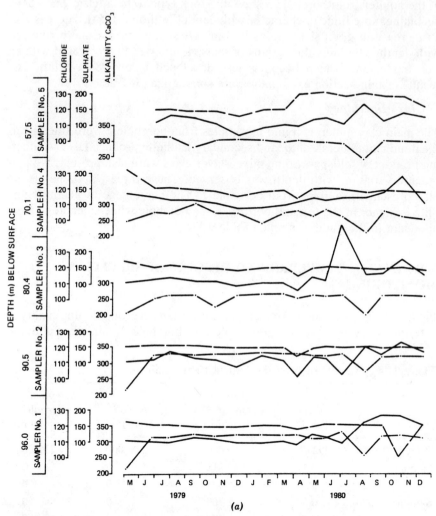

Figure 9.20a. Chemical quality of ground water with time at different depths as determined from the five *in situ* samplers in borehole 3 (all results in mg/L).

178

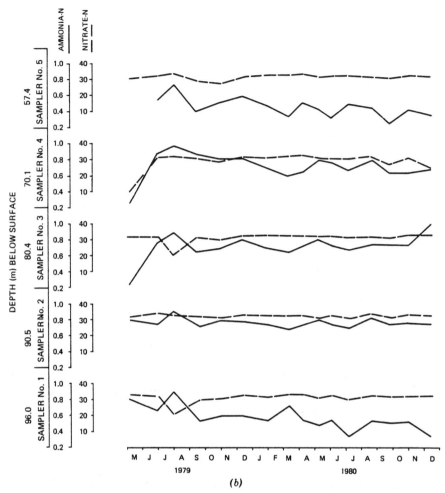

Figure 9.20*b***.** Chemical quality of ground water with time at different depths as determined from the five *in situ* samplers in borehole 3 (all results in mg/L).

considerable variation in quality between samplers, but with time these differences were less obvious and a more uniform composition was reached. Slight perturbations in the chloride, nitrate, and magnesium concentrations are seen in the samples taken in July 1979 and July 1980. Additionally, chloride concentrations increased and ammonia decreased slightly between June and December 1980.

Two special sets of samples were taken in June 1980 and June 1981 for microbiological analysis by the Thames Water Authority, the results of which are given in Table 9.14. Both sets of results are consistent in showing that no viruses, *E. coli,* or *streptococcus fecalis* are detected within the saturated zone despite their high concentration in the effluent.

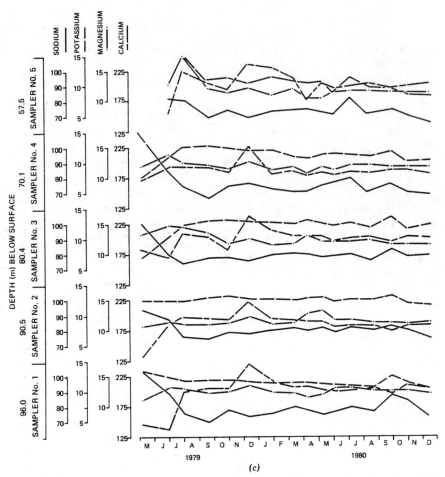

Figure 9.20c. Chemical quality of ground water with time at different depths as determined from the five *in situ* samplers in borehole 3 (all results in mg/L).

DISCUSSION

1. There is a good agreement between lithological and geophysical logs in defining the Upper Mottled Sandstone/Bunter Pebble Bed junction. The percentage of silt/clay is much reduced in the Bunter Pebble Beds with a corresponding increase in permeability. There is also an increase in formative resistivity in the Bunter Pebble Beds. The results of the geological investigation agree with those of the Geological Survey Memoir (Whitehead and Pocock, 1947).

Additionally, the geophysical downhole logs have agreed with porewater quality data in identifying five hydrogeologically significant intervals within the saturated zone.

2.   Ground water samples collected by Edmunds and Morgan-Jones (1976) from the Upper Mottled Sandstone at Albrighton, to the north of the area of investigation, possessed lower Ca/Mg ratios than were found in samples from borehole 3. These differences are most probably related to variations in the clay mineralogy of the sandstone.

3.   The composition of the effluent has been fairly constant since 1975, yet the quality over the last 100 years is considered to have improved due to increased dilution. At the same time, however, there has probably been an increase in total phosphorus. The high chloride and ammonia concentrations with low NO_3/NH_4 ratio of 1.2/1 is typical of primary domestic effluent. Assuming the chloride ion to be geochemically conservative and knowing the average chloride concentration of the effluent and natural ground water, it has been possible to estimate the percentage of effluent reaching the observation and public supply boreholes. On this basis, the ground water beneath the irrigation area is over 80% effluent, making the ratio of artificial/natural recharge 4.9/1. The catchment area to borehole 3 has been estimated at 2.60 km^2, over which the average effective natural infiltration is 308 mm/year. The natural recharge to the water table is therefore 0.8×10^6 m^3/year. As the annual volume of effluent artificially recharged is around 4.4×10^6 m^3, the expected dilution available to the effluent is 4.7/1, very close to that predicted from the chloride concentration alone.

4.   Analysis of the sequential leachates from the soil survey samples shows that in the upper and middle traverses only chromium is retained while in the lower traverses lead, copper, and chromium are retained within the first 1.0 m of infiltration. The whole-rock XRF analyses of the same samples show much the same pattern but overall greater concentrations. Since the lower traverse is situated at the base of the slope over which the effluent flows and is close by a drainage ditch, there was probably greater infiltration at this point and therefore a greater retention of metals. The differences in analytical results between the two methods (Table 9.11 and Figure 9.16) suggest that the sequential leaching technique is unable to extract all metal held by the soil, especially lead and chromium. Porewater from the unsaturated zone beneath the infiltration area has a lead concentration equal to only one half of that of the applied effluent, whereas for copper the levels are higher than the effluent. Copper, therefore, is retained deeper in the unsaturated zone while lead is carried on down to the water table. Cadmium concentrations in the soil samples have been determined only by the sequential leaching experiment, and are unusual in that a peak at 0.5 m depth in the control traverse and one at 0.8 m in the middle traverse suggests elution of the metal by the infiltrating effluent. This hypothesis is supported by the porewater analyses from the unsaturated zone, where the ratio between the irrigated and unirrigated values is 0.05/1. Chromium, too, may be leached from the unsaturated zone at depth, but no figures for its concentration in the saturated zone are available to confirm this. Soil samples were not analyzed for zinc, but porewater results from the unsaturated zone suggest retention within this zone.

The principal removal mechanism for chromium, copper, and to a lesser extent lead from the effluent within the soil and unsaturated zones is adsorption by clay minerals.

5. The results of the soil survey suggest the following order of mobility: Cr < Cu ≤ Pb < Cd. Experiments carried out at the Uffington Lysimeter site by workers of the Institute of Geological Sciences (Bromley and Boreham, 1978; Ross et al., 1980) on the adsorption of heavy metals by the Lower Greensand report the following order—Pb < Cr < Cu < Cd—much in agreement with the work of Griffen and Shimp (1976). The reason for the greater mobility of lead as compared with these other investigations is uncertain, but it may be related to the differing clay mineralogy.

6. The porewater composition within the unsaturated zone beneath the irrigated area is similar to that of the effluent, apart from lower concentrations of nitrate, ammonia, and phosphate. The changing concentration of nitrate with depth appears to be related to the oxygen level, as there is a twofold increase in nitrate concentration at 6 m depth and a subsequent decrease below this more aerobic zone. Physicochemical processes are known to remove phosphorus by adsorption onto clay minerals and nitrogen by denitrification under anaerobic conditions. These processes are operating within the unsaturated zone, although nitrification may also occur if oxygen is present in the zone. Research carried out at a number of sites on the Cretaceous Chalk aquifer of the United Kingdom indicates that the Chalk has a much greater ability than the Triassic sandstone to remove nitrogen from infiltrating sewage effluent (Baxter and Clark, 1984). The reason for this is uncertain but is thought to be due to the higher storativity of the sandstone, which allows more oxygen into the unsaturated zone thus eliminating the shallow anoxic zone necessary for denitrification (Lance, 1975).

7. The full thickness of the zone of saturation sampled at borehole 3 has been contaminated by effluent as shown by the porewater chemistry profile. Four zones can be distinguished on the basis of porewater quality below the water table, and there is reasonable correlation between these zones and those identified from the geophysical logs. In general terms the porewater concentrations are similar from the water table to the base of the borehole. The full thickness of the contaminated zone at this location was not, therefore, penetrated in drilling. The variation in porewater quality is determined not just by the vertical movement of contamination down through the saturated zone but also by the lithological variation within the aquifer. This explains the good agreement between geophysics, lithology, and porewater chemistry. "Interval 2" in borehole 3, for example, coincides with a more "flaggy" lithology responsible for preferential flow along the fractures, which results in a rapidly changing quality profile. Correlation coefficients between the sodium, potassium, chloride, nitrate, and heavy metal concentrations in each interval show ion exchange to be occurring in interval 4 and the preferential adsorption of heavy metal in all four zones. The concentra-

tion of ammonia in the top 30 m of the saturated zone beneath the irrigation area is high, although substantially less than in the effluent. This confirms the ground water flow pattern as determined by geophysics in that this is an interval of dominantly downward flow resulting in anaerobic conditions. There is some nitrification in this zone as shown by the high nitrate levels. At borehole J, about 500 m down ground water gradient from the edge of the infiltration area, a zone of contaminated ground water is observed from 35–85 m below surface; the top of this zone lies 15 m below the water table. This has been confirmed by geophysical logs. An increase in ammonia, nitrate, and chloride porewater concentrations with depth is also observed in borehole J beneath the unirrigated area, from the water table to 75 m below surface. This deeper zone of contamination can be correlated with interval 2 (about 65 m depth) in borehole 3. Interval 2 of borehole 3 is believed to be a zone of rapid horizontal to near-horizontal flow, as indicated by the large variations in porewater quality (Figure 9.10). This flow is most probably responsible for the relatively high degree of contamination observed near the base of borehole J, both zones occurring within the same stratigraphic interval.

8.   The strong quality stratification observed in borehole 3 clearly illustrates the need for special care in the sampling and subsequent monitoring of ground water quality. An open borehole, even if depth-sampled, will not exhibit the same zonation as observed from the porewater profile. Convective circulation and atmospheric intrusion must be prevented, yet it must still be possible to sample at a number of depths without affecting ground water quality. For this reason WRC *in situ* samplers were installed in borehole 3 during May 1979 to monitor temporal changes in quality at five depths within the saturated zone. Although there has been little variation with time since installation, there is appreciable variation between samplers, particularly in ammonia concentration and alkalinity and to some extent in the chloride concentration. The reason for the complete absence of viruses, *E. coli* or fecal *streptococci* within the saturated zone of borehole 3 is uncertain; a more intensive sampling program would be necessary before any definite conclusions could be made. It is possible that they are removed by filtration and/or adsorption by clay minerals within the unsaturated zone.

9.   Whole-rock analysis has shown that the leaching of calcite is enhanced by the infiltration of effluent through the unsaturated zone although there is a considerable natural variation due to lithological differences. Silica concentration is also significantly lower in the sandstone below the irrigation area than elsewhere probably as a result of the dissolution of the cementing matrix. The leaching of calcite from the unsaturated zone is reflected in the high calcium concentration in the porewater of that zone. The trace elements rubidium, barium, and zirconium appear to be precipitated in the unsaturated zone as a result of effluent infiltration. No such enrichment is observed away from the irrigated area in borehole J. No depth trends are observed for

the other trace elements analysed, and their distribution is therefore considered to be original. Within the saturated zone there is a significant depletion of cobalt and manganese as a result of effluent irrigation. Any changes in the abundances of the other trace elements are hidden by the natural variation. Edmunds and Morgan-Jones (1976) also report that clay minerals and iron oxide cement naturally present in Triassic sandstones exert the major control over the trace-element geochemistry of the ground water. This would also appear to be the case in the study area.

10.   The drilling investigation has shown the aquifer immediately beneath the irrigation area to be contaminated to at least 120 m depth. The plume of contaminated ground water is known to be 50 m thick at 1.5 km radius from the center of the area. Assuming radial flow from the rectangular recharge area and a specific yield of 0.15, the minimum volume of contaminated ground water is about 1.2×10^8 m^3. There is good agreement, considering the limitations of the data, with the estimated total effluent volume recharged over the last 100 years—2.2×10^8 m^3—assuming an average 30% dilution. The extent to which the flow of contaminated ground water is radial can be determined by estimating effluent dilution within a series of two-dimensional concentric zones of 1, 2, 3, and 4 km radii about the irrigation area. Chloride concentrations within these zones should be 91, 59, 38, and 25 mg/L respectively. Public supply boreholes C and D (Figure 9.1) at 2.5 and 3.7 km radius have average chloride concentrations of 56 and 39 mg/L, which, considering the simplistic nature of the model, are in reasonable agreement. Actual concentrations at public supply boreholes F and G (Figure 9.1) are lower than predicted as these boreholes are to the east of the irrigation area, up the ground water gradient. These calculations suggest that steady-state conditions have been reached, a picture confirmed by the historical quality data.

11.   The first appearance of contaminated ground water in public supply borehole C, at a 2 km radius from the edge of the irrigation area occurred after 60 years. It is estimated, therefore, that if irrigation were to stop immediately it would take at least 60 years for all traces of contamination to disappear. This would be a minimum time because without any artificial recharge there would be reduced ground water flow due to a smaller ground water gradient between the irrigation area and the public supply borehole.

CONCLUSIONS AND RECOMMENDATIONS

1.   There is good agreement between lithological and geophysical logging, both in determining geological boundaries and in identifying hydrogeologically significant zones.

2.   Physicochemical processes operating in the unsaturated zone affect the removal of nitrogen, phosphorus, heavy metals, and microbes. The clay minerals exercise a major control in this area.

3.   Flow along fractures and bedding planes has been found to be of importance in determining the pattern of contamination above and below the water table.

4.   Beneath the irrigated area, ground water contains a very high proportion of effluent down to over 120 m below surface. The ground water is contaminated by effluent to a radius of several kilometers, this contaminated zone being 50 m thick at 1.5 km radius.

5.   Strong vertical stratification in porewater quality, due mainly to bedding and lithological contrasts, have been observed in the saturated zone beneath the irrigation area. Because of this, five WRC *in situ* samplers were installed to monitor temporal variation in quality.

6.   As the effluent moves further from the site of disposal, mixing with native ground water is the main mechanism of quality improvement.

7.   Whole-rock analysis has shown that effluent infiltration through the unsaturated zone has substantially leached calcite while enriching the rubidium, barium, and zirconium concentrations. No depth trends are observed for other minor or trace constituents as there is considerable natural variation due to litho-stratigraphic differences.

8.   Using the geochemically conservative chloride ion as a tracer for sewage effluent, it has been determined that the first arrival of contamination appeared in the nearest public supply well some 60 years after the start of irrigation in 1880. Some 30 years later equilibrium conditions were established that have persisted for the last 10 years. Assuming such steady-state conditions, it is estimated that even if recharge were to stop immediately, the aquifer within a 2 km radius of the irrigation area would remain contaminated for at least 60 years.

9.   The results of the research at this site indicate that if properly managed such sewage disposal methods considerably improve the quality of the recharged effluent and also conserve ground water resources.

10.   Complementary studies at effluent-discharge sites on the Chalk confirm these results and point to the following factors as being important in the management of aquifers artificially recharged with sewage effluent:

(a)   The hydraulic loading must not exceed the empirical treatment capacity of the unsaturated zone.

(b)   The minimum thickness of unsaturated zone in a fissured aquifer required is about 20 m although this would depend on the quality of the recharge effluent and the hydrogeology.

(c)   Prior to recharge the effluent should be pretreated only to "primary" standards.

(d)   Public supply wells must be at least 1–2 km from the recharge area, especially if they are directly down ground water gradient from the recharge site.

(e)   Public supply and observation boreholes should be monitored at monthly intervals for chemical and microbiological contamination. If necessary, additional observation boreholes should be drilled in the direction of ground water movement and depth samplers installed.

Each site must, however, be treated individually, and it is recommended that extensive engineering and hydrogeological testing be carried out before any new effluent recharge sites are commissioned.

ACKNOWLEDGMENTS

The work described in this paper was partially financed by the Department of the Environment. The author would like to thank Mr. K. J. Edworthy, who was responsible for the research program, and Dr. L. Clark for helpful discussion.
    Thanks must also be expressed to the Department of the Environment and the Director, W. R. C. Medmenham for permission to publish this chapter.

REFERENCES

Baxter, K. M. and Clark, L. (1984). Effluent recharge. The effects of effluent recharge on groundwater quality. Technical Report TR 199. Water Research Centre, Medmenham.

Baxter, K. M. and Edworthy, K. J. (1979). The impact of sewage effluent recharge on groundwater for an area in S.E. England. In: *Artificial Groundwater Recharge,* Volume III—Bulletin 13 of the German Association for Water Resources and Land Improvement, Bonn. Parry, Hamburg/Berlin, 1982.

Baxter, K. M. and Edworthy, K. J. (1980). Virology of wastewater recharge to the Chalk aquifer. In: M. Goddard and M. Butler (Eds.), *Viruses and Wastewater Treatment.* Pergamon, Oxford.

Baxter, K. M., Edworthy, K. J., Beard, M. J., and Montgomery, H. A. C. (1981). Effects of discharge sewage to the Chalk. *Sci. Total Environ.* 21:77–83.

Bromley, J. and Boreham, D. (1978). A comparison of data on the movement of metal ions through sands and clays from the Lower Greensand formation. WLR Technical Report No. 21, Dept. of the Environment, London.

Clark, L. and Baxter, K. M. (1981). Organic micropollutants in effluent recharged to groundwater. *Water Sci. Technol.* 14(12):15–30.

Edmunds, W. M. and Morgan-Jones, M. (1976). Geochemistry of groundwater in British Triassic sandstones: The Wolverhampton–E. Shropshire Area. *Q. J. Eng. Geol.* 9(2):73–102.

Edworthy, K. J., Wilkinson, W. B., and Young, C. P. (1978). The effect of the disposal of effluents and sewage sludge on groundwater quality in the Chalk of the United Kingdom. *Prog. Water Technol.* 10(5/6):479–93.

Griffen, R. A. and Shimp, N. F. (1976). Effect of pH on exchange-adsorption or precipitation of lead from landfill leachates by clay minerals. *Environ. Sci. Technol.* **10**(13):1256–1261.

Joseph, J. B. (1980). Installation and operating manual for the WRC *in situ* sampler. WRC Report 27-M, Water Research Centre, Medmenham, U.K. (unpublished).

Lance, J. C. (1975). Fate of nitrogen in sewage effluent applied to soil. *J. Irrig. Drain. Div. ASCE,* **101 No IR3:** 131–144.

Pettijohn, F. J. (1954). Classification of sandstones. *J. Geol.* **62:**360–365.

Ross, C. A. M. (1978). Extractive characterisation of heavy metal distribution in contaminated Lower Greensand. WLR Technical Report No. 61, Dept. of the Environment, London.

Ross, C. A. M., Rees, J. F., and Lewis, G. N. J. (1980). Uffington lysimeters—operations and results (Part 5). WRL Technical Report, Dept. of the Environment, London (unpublished).

Whitehead, T. H. and Pocock, R. W. (1947). Dudley and Bridgenorth; explanation of one-inch geological sheet 167. *Mem. geol. Surv. Gt. Br.*

10

LAND APPLICATION OF MUNICIPAL WASTE WATER

M. B. Tomson
Carol Curran
S. R. Hutchins
M. D. Lee
Gordon Waggett
C. C. West
C. H. Ward

National Center for Ground Water Research
Rice University
Houston, Texas

Land application of municipal waste water promises to be one of the more efficient methods of waste water treatment, with costs of 4–15¢ per 1000 gal compared with 7–40¢ per 1000 gal for secondary activated sludge treatment (Rice and Gilbert, 1978). There are three types of land application: (1) overland flow, (2) slow rate infiltration or irrigation, (3) rapid infiltration (Loehr et al., 1979). Of the three, rapid infiltration is the least land intensive with application rates of 30–120 m/yr and will be the primary topic of the present chapter.

The objective of rapid infiltration is not only disposal of waste water, but treatment and purification by the soil and soil microorganisms followed by ground water recharge. Typically, a series of basins of about 1 ha are inundated with waste water for 1 week and then allowed to dry and aerate for 2 or 3 weeks. A "living filter" mat of microorganisms is built up in the top few centimeters of the basin. As water percolates through this active mat, microorganisms normally reduce the biochemical oxygen demand to less than 1 mg/L.

Rapid infiltration is capable of producing tertiary treatment quality water with a few notable exceptions: nitrate, cadmium, viruses, and chromatographable trace level organics (C-TLOs). The nitrate level can be reduced by application-regime management. Bouwer et al. (1978) reported 65% denitrification by reducing the loading rate in a rapid infiltration system. Cadmium levels in the effluent are only occasionally above the United States Public Health Service limit of 0.01 mg/L and can probably be eliminated by controlling the input sources. Virus removal may be a more pernicious problem and is discussed in Chapter 5 by C. P. Gerba (see also Keswick and Gerba, 1980). Little field information existed for rapid infiltration sites on the breakthrough of C-TLOs into ground water prior to this study. The objective of the present research on rapid infiltration systems was to determine if and to what extent C-TLOs in the waste water were passing through the topsoil filter mat of microorganisms and contaminating the associated ground water.

TOXICITY OF GROUND WATER CONTAMINANTS

One of the major concerns with rapid infiltration sites is contamination of the ground water by C-TLOs and the effects these compounds have on man and the environment. Hutchins et al. (1983) identified several compounds that have been found in ground waters at four land application sites. They were bis(2-ethylhexyl) phthalate, dibutyl phthalate, diethyl phthalate, dimethyl phthalate, naphthalene, toluene, o-, p-, and m-dichlorobenzenes, tetrachloroethylene 2,6-di-t-butyl-p-benzoquinone, butyl benzenesulfonamide, benzophenone, p(1,1,3,3)-tetramethyl butylphenol, xylene, and 2-(methylthio)-benzothiazole. Leach et al. (1980) found that lindane was the only pesticide whose levels were significantly elevated over the background concentrations in the ground waters at four rapid infiltration sites. These compounds are probably representative of the types of compounds which persist in the ground water after rapid infiltration application. The toxicity of these compounds to representative aquatic organisms including algae, invertebrates, and fish, to man, and to other mammals determines the extent for concern over their presence in ground waters. The following discussion presents the toxicity of these compounds.

Several of the compounds identified by Hutchins et al. (1983) are on the U.S. Environmental Protection Agency's (EPA) Priority Pollutant List. These compounds can be grouped into phthalate esters, aromatic compounds, and tetrachloroethylene. The phthalate esters include bis(2-ethylhexyl) phthalate, dibutyl phthalate, diethyl phthalate, and dimethyl phthalate. These compounds are used in the production of plastics and as carriers for pesticides, oils, and insect repellants (USEPA, 1980c). They are found throughout the environment; part per billion (ppb) levels are present in many surface waters, with higher levels near industrial centers. Bis(2-ethylhexyl) phthalate is the most heavily used phthalate plasticizer. It is highly toxic to aquatic life and slightly toxic to mammals with some chronic effects. It has been shown to be teratogenic and cytotoxic but generally not mutagenic or carcinogenic. The level in ambient water proposed by the EPA to protect human health was 10 mg/L (Majeti and Clark, 1980). Dibutyl phthalate is extremely toxic to aquatic organisms and moderately toxic to mammals (USEPA, 1980c). It is teratogenic, but not typically considered to be carcinogenic or mutagenic. Dibutyl phthalate is considered one of the more hazardous phthalates with the recommended human health criterion set at 5 mg/L (Majeti and Clark, 1980). Diethyl and dimethyl phthalate have similar toxicological properties; both are less toxic to aquatic organisms than are the other phthalates (USEPA, 1980c). Diethyl phthalate is generally more toxic to mammals than the dimethyl ester. Both have been shown to be teratogenic, mutagenic, and carcinogenic in some test systems. The proposed human health criteria in ambient water are 160 mg/L for dimethyl phthalate and 60 mg/L for diethyl phthalate.

The aromatic compounds identified by Hutchins et al. (1983) on the priority pollutant list are naphthalene, toluene, and the dichlorobenzenes. The dichlorobenzenes are used as chemical intermediates, insecticides, and air deodorants (USEPA, 1980a). They have been found in surface and ground waters, in municipal and industrial waste waters, in air, and in soil, usually at the ppb level. Aquatic toxicity is generally in the 1 to 100 mg/L range. Mammalian toxicity can occur at relatively low levels. Chronic ingestion of low doses (0.01–0.1 mg/kg) has been shown to affect the central nervous system adversely, cause blood disorders, and disrupt normal enzyme activity in the livers and kidneys of rats. In addition to these target organ systems, the skin and respiratory tract of humans have been affected. Teratogenic and mutagenic activity have not been demonstrated by animal testing. Although strong evidence for the carcinogenic activity of dichlorobenzenes has not appeared, it seems prudent to consider them to be potential carcinogens. The recommended human health criterion for all isomers in ambient water is 0.230 mg/L (Majeti and Clark, 1980).

Naphthalene, a naturally occurring chemical, is found in petroleum and coal tar and is used in the production of dyes, solvents, lubricants, motor fuels, and phthalic anhydrides, and as a pesticide (USEPA, 1980b). It has been detected in ambient water (up to 2 μg/L), sewage effluents (up to 22 μg/L), and drinking-water supplies (up to 1.4 μg/L). The toxicity of naphthalene to aquatic organisms occurs at the 1–5 mg/L level. The recommended 96-hr threshold limit to protect aquatic organisms is 1–10 ppm (USHEW, 1977). Naphthalene is moderately toxic to mammals and can cause chronic problems such as the development of cataracts (USEPA, 1980b). Metabolites of naphthalene may be more toxic. Naphthalene may be teratogenic, but is probably not mutagenic, and most studies show that it is not carcinogenic. The human health criterion for ambient water is 0.46 mg/L (Majeti and Clark, 1980).

Toluene is used principally for the production of benzene and other chemicals, as a solvent for paint, and as a gasoline additive (USEPA, 1980e). It is derived from petroleum products. Levels of toluene in water supplies may reach 0.1–11 μg/L. In general, the aquatic toxicity of toluene is relatively low, with the recommended 96-hr threshold level set at 10–100 ppm (USHEW, 1977). Acute exposure appears to have only a small potential for toxicity to mammals beyond the inhibition of the central nervous system. Chronic exposure to toluene does not generally result in tissue damage or toxicity. It has not been shown to be mutagenic or carcinogenic, but it may be teratogenic. The recommended human health criterion in ambient water is 17.4 mg/L (Majeti and Clark, 1980).

Tetrachloroethylene, a man-made compound, has been used primarily as a solvent in dry cleaning and as a degreasing agent in metal industries (USEPA, 1980d). It has been detected in trace amounts in water, municipal and industrial waste water, air, foodstuffs, and aquatic organisms. Toxic effects on aquatic organisms have been found as low as 450 μg/L, but most of its lethal effects are from the 10 mg/L range up. Tetrachloroethylene affects the central nervous system primarily; inhalation by rats of levels as low as 100 mg/m^3 for 4 hr/day for 15–30 days affected their nervous systems. Exposure to high concentrations can cause damage to the kidney and liver. Some evidence suggests that tetrachloroethylene is teratogenic, mutagenic, and carcinogenic. The recommended criterion for tetrachloroethylene, based primarily on a human lifetime carcinogenic risk of 10^{-5}, is 8.0 μg/L.

Two compounds identified in the ground water near land application sites (Hutchins et al., 1983) have not been investigated to determine their toxicological properties. 2,6-Di-t-butyl-p-benzoquinone is similar to 2,5-dimethyl benzoquinone and benzoquinone, which have some background toxicological information. These compounds are highly toxic; a man's death has resulted from ingestion of 40 mg/kg, and carcinogenic activity has been reported (USHEW, 1977). Butyl benzenesulfonamide is the other compound that has not been evaluated. Ethyl benzenesulfonamide, which has a similar structure, is moderately toxic. N,N-Diethyl benzenesulfonamide is teratogenic, which suggests that butyl benzenesulfonamide may also be.

Not many data exist for three of the other compounds, but the available evidence suggests that they are not likely to be highly toxic. Benzophenone and p(1,1,3,3)-tetramethyl butylphenol would probably be ranked as only slightly toxic based on the available mammalian data. Benzophenone is used as a fixative for perfumes and insecticides (Merck, 1976). p(1,1,3,3)Tetramethyl butylphenol is a nonionic detergent with surface-tension-reducing properties. More data is available for xylene, but based on its toxicity data, it should not be greatly toxic. Xylene is obtained from coal tar and is used as a solvent and feedstock for various compounds. Xylene is toxic to algae. When inhaled, it can produce toxic effects or irritation to man. The mammalian toxicity is moderate.

Lindane and 2-(methylthio)benzothiozole have higher toxicity and are greater concerns. The 2-(methylthio)benzothiozole has a high toxicity to mice (USHEW, 1977). Lindane, a man-made pesticide, can be lethal at relatively low doses (Merck, 1976). It has been shown to be highly toxic in animal studies (USHEW, 1977), especially to cattle where oral LD50s as low as 5–25 mg/kg are reported (Sax, 1979). Chronic exposure may lead to liver damage (Merck, 1976). Because lindane is considered carcinogenic, the recommended human health criterion for ambient water that allows one case of cancer per 100,000 people is 0.02 μg/L (Majeti and Clark, 1980). Many of the compounds present in the ground waters associated with rapid infiltration sites are hazardous at relatively high concentrations. However, most of the compounds are present at only trace levels in ground water and may be a problem only if they are carcinogenic, bioaccumulate in the body, act synergistically, or have chronic sublethal effects. There may be no safe level for carcinogens (Doull et al., 1980). The compounds that have been implicated as potential carcinogens are diethyl and dimethyl phthalate, the dichlorobenzenes, tetrachloroethylene, and lindane; further testing may show that these compounds are not carcinogenic or that some of the remaining compounds are carcinogenic. Synergistic action, environmental factors, or extreme susceptibility may increase the toxicity of the trace organics from rapid infiltration systems.

METHODS

In Figure 10.1 an overall diagrammatic summary of the procedures used to collect and analyze field samples is presented. Most of the analytical details have been presented elsewhere and will only be summarized herein (Junk et al., 1974b; Tomson et al., 1979, 1980, 1981; Hutchins et al., 1983; Dunlap et al., 1976). Generally, a water sample of about 25 L was pumped at 30 to 45 mL/min through a 10 mL Teflon column containing XAD-2 macroreticular resin (Rohm & Haas, Dallas, TX). The ends of the resin column were then sealed in the field with Swagelock (Houston, TX) stainless steel end caps for air transport to the laboratory. Two or three 15 mL volumes of ether were used to strip C-TLOs from the resin in the columns into a 100 mL round-bottomed flask, to which a graduated test tube had been attached and sealed (Figure 10.1). The ether volume was reduced to 1 mL by using a Kuderna Danish evaporator over steam. The 1 mL concentrate was dried with K_2SO_4 and used for all subsequent analyses. The final gas chromatographic analyses varied considerably for each sample, but the following general procedure was used. A temperature-programmed gas chromatogram was run [60°C (4 min) to 250°C @ 8°C/min] on a Tracor Model 560 gas chromatograph (Austin, TX) using either a 50 m WCOT SP2100 (Hewlett-Packard, Palo Alto, CA) capillary column in the splitless mode or a 2-mm I.D. 3% OV-17 on 100/120 packed column. The flame ionization detector output was recorded on a Spectra Physics Model 4100 integrating computer (Palo Alto, CA). A gas chromatograph/mass spectrogram (GC/MS) was produced with matched columns and conditions on a Finnigan Model 4400 GC/MS (Palo Alto, CA). The GC/MS data system (GC/MS/DS) produced suggested identities of each chromatographic peak. In some cases internal standards of either D_8-naph-

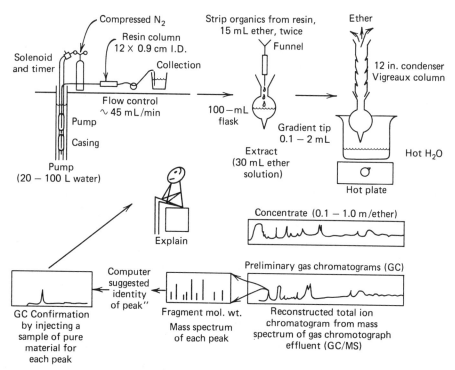

Figure 10.1.   Pictorial summary of laboratory procedures used to sample and analyze C-TLOs.

thalene or D_{10}-anthracene were used. Finally, when pure standards were available, confirmation of suggested compound identity was made by GC retention time matching.

If the water table was less than about 8 m, ground water samples were obtained by sucking the water through the resin column with a peristaltic pump. In every field study a "resin column blank" was produced by taking a resin column to the field site and handling it in a manner similar to that which was used to handle the sample resin columns, except that no water was pumped through it. This "blank" column was returned to the laboratory for analysis along with the sample resin columns. Recovery efficiencies for our apparatus, pumping rates, analyses, and so on, for many of the compounds listed below were determined and generally ranged from 50 to 100% for the concentrations encountered in these studies; similar results were reported by Junk et al. (1974b). Many of the wells were cased with PVC pipe, and there was concern that the PVC might either release C-TLOs into or sorb C-TLOs from the ground water samples. To test this concern, laboratory leaching studies were performed using PVC, Teflon, and several other plastics (Curran and Tomson, 1983). This study suggested that there was probably little reason for concern, but the long term exposure studies by Miller (1982) and the flow through leaching study using flexible tubings by Junk et al. (1974a) should also be consulted.

FIELD SITES

C-TLO results from five land application sites will be presented. In Table 10.1 a summary is presented of the site characteristics, history, and appropriate literature references for each site. Unfortunately, at the time of this writing, the Ft. Polk, Louisiana, site was still not operational and, hence, only laboratory column simulations are presented.

Ft. Devens, Massachusetts

A schematic plan view and a typical cross-section are presented in Figures 10.2a and 10.2b. Numerous ground water, sewage, and soil samples have

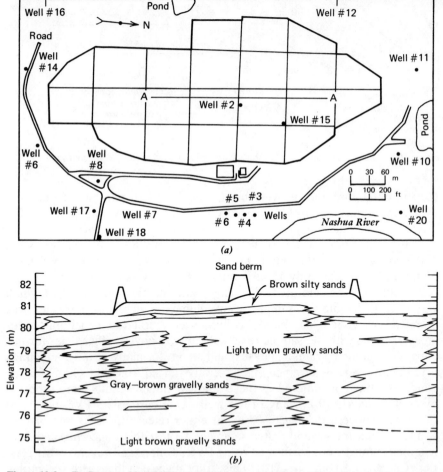

(a)

(b)

Figure 10.2.   Ft. Devens, Massachusetts rapid infiltration site. (a) Plan view showing location of sampling wells and ground water flow. (b) Cross-sectional view of site (see Satterwhite, 1976 for further discussion of soil characteristics). (c,d,e) GC traces from samples obtained from Ft. Devens at corresponding attenuation for direct visual comparison.

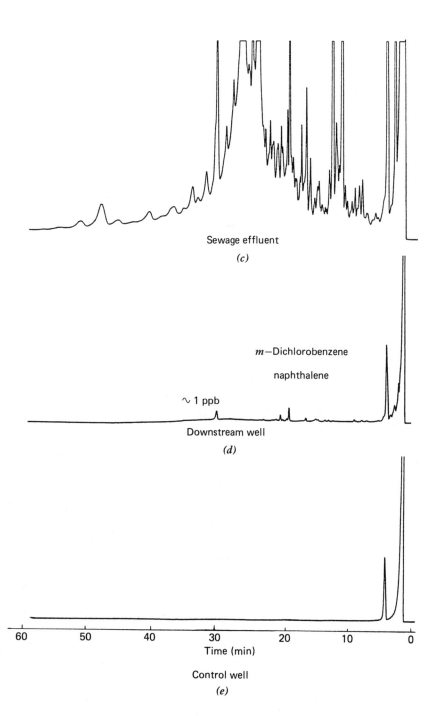

Sewage effluent

(c)

m–Dichlorobenzene

naphthalene

∿ 1 ppb

Downstream well

(d)

Control well

(e)

Time (min)

60   50   40   30   20   10   0

195

Table 10.1 Characteristics of Land Application Field Sites Studied

Site	Fort Devens, Massachusetts	Lubbock, Texas Frank Gray Farm	Phoenix, Arizona 23rd Ave. Project	Boulder, Colorado	Fort Polk, Louisiana Laboratory Column Study
Location	U.S. Army fort 52 km NW of Boston	Lubbock, Texas, 100 mi NW of Dallas	Near Salt River, west of Phoenix	75th Street Waste Water Treatment Plant	Central Louisiana, 300 mi north of New Orleans
Years in Operation	42	46	6	2	0
Basin number and size	22 @ 0.8 acres each	3000–5000 acres of irrigation	4 @ 10 acres each	3 @ ~0.7 acres each	14 @ 31.6 ha total
Soil type	Sandy gravel with clay lenses	Loamy soil to a caliche layer @ ~5–15 ft	Sandy	Coarse sand and gravel	Fine sand with broken clay lenses
Average influent	1.3 MGD	11,000 acre-ft/yr	13 MGD	Experimental	5.2 MGD, design
Influent type	Unchlorinated primary after Imhoff tank (6 hr)	Mixed, mostly chlorinated secondary	Chlorinated secondary	Chlorinated secondary and primary	Chlorinated secondary
Application schedule	3 basins inundated for 2 days followed by 14 days rest	Large holding ponds (2–3 mos. of effluent) and irrigation of	2 basins, 21-day inundation followed by 21-day drying	Experimental	2 day flooding/12 day drying, planned

Application rate	94 ft/yr	Up to 15 ft/yr in some areas for disposal above (grain sorghum, wheat, cotton, and grasses)	364 ft/yr	20–500 ft/yr	Slow, at present
Dates sampled	Oct. 1978; Mar. 1981; May 1983	June–Aug. 1978	Feb. 26–Mar. 1, 1980; Jan. 1980	Aug. 1978	Only background water samples have been taken
References	Satterwhite et al. 1976; Bedient et al. 1982; Loehr, 1977; Loehr et al. 1979	Wells et al. in Loehr et al. 1979; Jeffrey, 1978	Tomson et al. 1981; USEPA, 1977; Loehr et al. 1979; Bouwer et al. 1981	Carlson et al. 1982; Pendleton and Miller, 1980	Hutchins et al. 1983
Comments	This is often presented as a "model" rapid infiltration site which produces "tertiary" quality drinking water in the associated ground water	Major concern has been high NO_3 levels in the ground water. Ground water is now pumped from Gray's farm to recreational lakes in the area	Site was designed as a result of extensive studies on a smaller site nearby (Loehr et al., 1979)	Carlson et al. 1982 suggested that primary effluent may be better than secondary when applied to the land. Occasional NO_3 problems occurred	Full-scale operation has not yet begun owing to low infiltration problems

been obtained from Ft. Devens. There are 22 monitoring wells around and in the site (Satterwhite et al., 1976). Many of these wells have been sampled for C-TLO analysis. The results in Figures 10.2c–e are typical and are reproduced here from gas chromatograms run at the same effective attenuation for direct visual comparison. For analysis, the chromatogram in Figure 10.2d was run so that the ~1 μg/L peak, for example, was at full scale; still most of the peaks were base-line resolved. Chromatograms for ground water from well Nos. 4 and 5 were very similar. The peak at about 4 min is due to an ether impurity, probably ethylacetate. When the chromatogram for the control well, No. 13, was run at maximum sensitivity, the only additional peak that occurred was a small peak representing about 0.03 μg/L in the original sample corresponding to bis(2-ethylhexyl) phthalate. This indicated that well 13 was a good control well and that our overall analytical procedure was probably working satisfactorily at a sensitivity of a few ng/L level and above.

Suggested identities of about 100 compounds were obtained from the Finnigan 4400 GC/MS/DS runs on the sewage and ground water samples. The identities of about 40% of these compounds were confirmed by GC retention time matching with pure samples. An abstracted list of the concentration of compounds found at Ft. Devens is presented in Table 10.2. As can be seen from Table 10.2, most compounds of the C-TLO-type were removed almost completely, with an average of 96% removal. A few notable exceptions are the chlorinated hydrocarbons, tetrachloroethylene and dichlorobenzene, and the phthalate esters—Note: dibutyl phthalate, No. 42, @ 449 μg/L in the sewage, may be an analytical error because this high concentra-

Table 10.2   An Abstracted List of Concentrations (μg/L) of Compounds Found at the Fort Devens, Massachusetts Rapid Infiltration Site

Contaminant	Basin (μg/L)	No. 4 + No. 5 / 2 (μg/L)	Removal (%)
Toluene	89.0	0.01	100
Tetrachloroethylene	2.57	0.63	76
o-Xylene	1.23	0.15	88
Unidentified (BL) Styrene (5)	1.26	0.03	98
o-Ethyltoluene	6.52	0.16	98
sec-Butylbenzene	10.7	ND[a]	100
m-Dichlorobenzene	3.16	0.56	83
β-Methylstyrene	ND	0.04	+[b]
3-(Methylbutyl)benzene	ND	0.04	+
3,7-Dimethyl-1,b-octadien-3-ol (Linalool)	2.4	ND	100

Table 10.2 *(Continued)*

Contaminant	Basin (μg/L)	No. 4 + No. 5 / 2 (μg/L)	Removal (%)
m-Cresol	66.0	0.01	100
2,5-Dimethyl-2,5-hexanediol	28.0	0.06	100
5-Methyl-2-(1-methylethenyl) cyclohexanol	10.6	0.04	100
Menthol	7.01	0.10	99
Borneol or isoborneol	20.7	0.05	100
Triethylphosphate	ND	0.36	+
α-Terpineol	85.2	0.05	100
2,3-Dichlorophenol	ND	0.20	+
p-Allylanisole	4.14	ND	100
Naphthalene	4.94	0.10	98
p-t-Butyl-cyclohexanone	ND	0.41	+
2-Methylnaphthalene	3.89	ND	100
t-Butylphenol	ND	NQ[c]	+
1-Methylnaphthalene	2.93	ND	100
Indole	10.8	ND	100
Styrene glycol	4.61	ND	100
di-t-Butyl-p-benzoquinone (#4)	6.11	0.02	100
3,4-Dichloroaniline	6.11	0.20	97
Dimethyl phthalate	13.7	0.10	99
o-Hydroxybiphenyl	48.8	1.35	97
1,1,3,3-Tetramethylbutylphenol	ND	1.57	+
Diethyl phthalate	9.41	0.87	91
Benzophenone	5.17	0.86	83
Hexadecanol	6.52	ND	100
Phensuximide	ND	0.03	+
Simazine	ND	0.21	+
Tris-(2-chloroethyl)phosphate	ND	0.57	+
N-Butylbenzenesulfonamide	ND	0.09	+
Dibutyl phthalate	(449.0)[d]	ND	100
Bis-(2-ethylhexyl) phthalate	5.60	1.40	75
			Average 96

[a] ND—denotes compound was not detected.

[b] +—well concentrations > basin concentrations.

[c] NQ—denotes compound could not be quantitated.

[d] Compound concentrations in parenthesis are probably in error.

199

tion for this compound has not occurred in any subsequent samplings. Unfortunately, due to the sampling and workup scheme used, the very volatile compounds such as chloroform were not detected. Finally, it should be added that probably none of the compounds in wells 4 and 5 are of serious health risk to humans at the concentration levels observed.

Phoenix, Arizona—23rd Avenue Project

A plan view of the 23rd Avenue Project site is presented in Figure 10.3a along with a cross-sectional view showing depth to water table, and so on, in Figure 10.3b. The site characteristics and brief history are listed in Table 10.1, column (3). A thorough discussion of our C-TLO analysis and results for this site is presented in the paper by Tomson et al. (1979). Several sampling trips were made to this site to establish background information and develop field procedures, but only analytical results of the February/March 1980 sampling trip are summarized herein. An objective of the sampling on this trip was to determine the short term statistical variation of the ground water contaminant concentrations. To do this, identical samples were taken from the CW60 well on five successive days. Fifty-eight (58) compounds were identified and measured in the sewage influent on the first and last day of the sampling, and 20 of these compounds were detected and measured in the ground water sample each day. In addition, nine compounds not in the secondary sewage influent were found in the ground water each day. Several of these additional nine compounds, such as 2-propyl-1-heptanol or 1-chloro-2, 3-dihydro-1H-indene, might be accounted for as products of either chlorination, oxidation, or nature, but generally their concentrations were only a few ng/L. A two-factor analysis of variance without replication was performed on the data for the five days of sampling, and it was concluded that there was no difference in ground water concentrations over this short time. For illustration purposes, the concentrations of eight compounds in the basin influent waste water and in the ground water are listed in Table 10.3. The median removal efficiency was 91% with tetrachloroethylene (70%), 2,6-di(t-butyl)-p-cresol (71%), and dibutyl phthalate (13%) as notable exceptions. Removal efficiencies may be inversely related to water solubility or directly related to octanol–water partition coefficients and biodegradability. Removal efficiencies will be discussed below. None of the compounds occurred in the ground water at the 23rd Avenue project at concentrations greater than 1 μg/L and, consequently, there appears to be no immediate health risk from C-TLO contamination of the ground water.

Boulder, Colorado

A small experimental rapid infiltration site was built on the grounds of the 75th Street Waste Water Treatment Plant in Boulder, Colorado. A plan view is shown in Figure 10.4a, and a cross-sectional view showing the clay dikes,

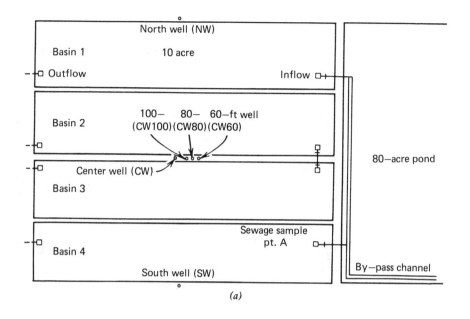

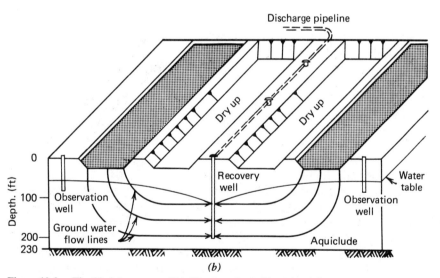

Figure 10.3. The 23rd Avenue rapid infiltration site in Phoenix, Arizona. (*a*) Plan view showing location of sampling wells. (*b*) Cross-sectional view of site showing ground water flow.

Table 10.3   Breakthrough Concentration of C-TLOs on Five Successive Days at Phoenix, Arizona, 23rd Avenue Project

Compound	Purity Index[a]	GC	Basin Input 1	Basin Input 5	Ground Water from Well CW 60 1	2	3	4	5	Average Removal (%)
Toluene	958	X[b]	964	1310	17	9	8	19	24	99
Tetrachloroethylene	979	X	251	50	29	21	34	65	74	70
p-Xylene	968	X	2140	4050	49	10	40	11	19	99
p-Dichlorobenzene	943	X	617	355	42	22	43	72	69	90
2,6-Di(t-butyl)-p-cresol	887		267	250	56	(200)[c]	71	96	79	71
Benzophenone	894	X	193	202	28	4	23	16	11	92
p-(1,1,3,3-Tetra-methylbutyl)phenol	923	X	757	809	17	10	14	13	12	98
Dibutylphthalate	957		248	260	316	(729)[c]	274	314	239	13

[a] These "purity indexes" are measures of goodness of fit between the GC/MS/DS library spectra and the sample spectra. A "1000" indicates perfect match. By studying numerous suggested identities confirmed by CG retention time matching, a cut-off of about 700 has been established for "reasonable" certainty of identity.

[b] Indicates that the compound identity was confirmed by GC retention time matching.

[c] Compound concentrations in parenthesis are probably in error.

and so on, is illustrated in Figure 10.4b. Samples were taken before, during, and after an inundation cycle. Table 10.4 lists the concentrations of compounds found in respective samples after the inundation (see Table 10.1) was complete, along with an analysis of a sample of the waste water being applied to the basins.

Over 80 organic compounds were identified either in the waste water or in the ground water at this site. The extent of C-TLOs in the off-site samples may indicate general contamination of the ground water in the area, but samples from the nearby creek and from wells not on the waste water treatment plant grounds would be needed before any general concern is expressed. The presence of compound No. 13 in Table 10.4, ferrocene—cyclopentadienyl iron (II)—is strong evidence of ground water contamination from products produced by a local chemical manufacturing plant because this local plant is of one of the major producers of ferrocene, a zenobiotic compound, in the United States.

A statistical analysis of the Boulder C-TLO data has been performed by one of the authors (S. Hutchins), and the details will be published elsewhere. He concluded that (1) the concentrations of the different organics conformed to a log-normal distribution, (2) there was no significant variation in concen-

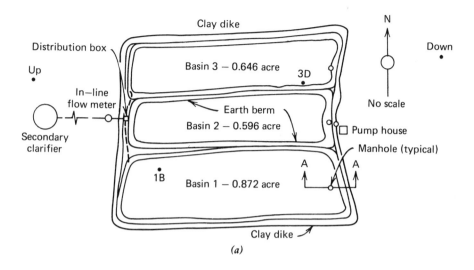

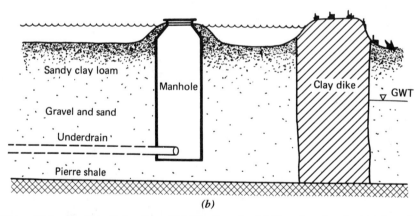

Figure 10.4. Boulder, Colorado rapid infiltration site. (*a*) Plan view showing location of sampling wells. (*b*) Cross-sectional view of site.

Table 10.4 Boulder, Colorado Ground Water Analysis for C-TLOs (ng/L)[a]

Compound	Basin Load	Upstream After	On-Site 3D, After	On-Site 1B, After	Downstream After
o-Xylene	17				
Styrene		230		20	10
Isopropylbenzene	980				
m-Dichlorobenzene		20	30	30	
p-Dichlorobenzene	2070	380	480	20	

Table 10.4 (*Continued*)

Compound	Basin Load	Upstream After	On-Site 3D, After	On-Site 1B, After	Downstream After
o-Dichlorobenzene	1120	270	240	120	
α-Terpineol	1270				
Decanol		40	60	60	
Naphthalene	P[a]	50	220	70	30
Ethylbenzaldehyde		80	330	140	
Triethylphosphate		40		70	
t-Butylphenol	1150	P	610		P
Ferrocene		100	200	10	
Indole					20
Di-t-butyl-p-phenol	530			730	
Diphenylmethane	(200)[c]	(90)[c]	260		40
Di-t-butyl-p-cresol			P	790	P
Methoxynaphthalene	550	11			
3,4-Dichloroanaline	770		150	220	
Dimethyl phthalate			90		
1,1,3,3-Tetramethyl-butylphenol		530	3540	110	60
Diethyl phthalate or tributylphosphate	700	260	290		10
Benzophenone		460	1600		
tris(2-Chloroethyl)phosphate	1520	350	P	560	
Butylbenzenesulfonamide	3330	240	300	190	70
Dibutyl phthalate	29,460	240	2130	520	5
Bromacil				990	
1,1-bis(Chlorophenyl)ethanol		220	160		80
Diphenamid	7370	2640	3600	(2210)[c]	390
Benzyl phthalate	70			X[d]	120
bis(2-Ethylhexyl)phthalate	950	90	20	80	60
tri(o-Cresol)phosphate	370	50	10		

[a] The identity of all compounds reported was confirmed by GC retention time matching.

[b] Indicates that a compound was present, but could not be quantitated.

[c] Concentrations in parentheses are the sum of the two compounds in the list and were not resolvable by GC.

[d] The resin column blank contained this compound at equal or greater concentration than the sample.

tration with time before, during, or after inundation ($P = 0.909$), (3) the 3D-on-site and the upstream samples were from different aquifer sources ($P > 0.999$), and (4) leakage is not occurring down the on-site well casings ($P = 0.629$). The lack of variation in concentration of ground water samples over short time periods is consistent with that found at the Phoenix, Arizona, site.

Lubbock, Texas, Gray Farm

Frank Gray's farm is located 2 miles northeast of Lubbock, Texas. About 20% of his land is used for detention and storage of waste water for up to 3 months per year (refer to Table 10.1 for site characteristics). Lubbock is near the southern edge of the Ogalala aquifer; north of town there is a producing aquifer at about 90 ft, but a few miles south of town potable water is several hundred feet deep. The background of Gray's farm and the value of water in this area is interesting, and the reader is referred to Loehr (1977) and Jeffrey (1978).

Numerous compounds have been identified in samples from ground water wells on Frank Gray's farm. A representative list of C-TLOs found in these samples is:

Chloroform	Phenanthrene
1,1,1-Trichloroethane	Naphthalene
Trichloroethylene	2,4,4-Trimethylhexane
Dibromochloromethane	Diisooctylphthalate
Tetrachloroethylene	Octane
Toluene	Trimethylbenzene
Chlorobenzene	Dichlorobenzene
3-Ethyl-3-methylhexane	Xylene
2-Methyl-2H-benzotriazole	Numerous Alkanes

It was not possible to lower a glass pump or Teflon tube into the wells, so water samples were obtained from surface spigots at each well. For this reason and other analytical difficulties, no quantitation is reported, and the compounds listed above may be artifacts of the sampling regime. Further research on C-TLOs in the ground water at this site is in progress by another research group associated with Lubbock Christian College.

Before work was stopped on the site, soil samples from one of Frank Gray's irrigation plots were collected at various depths down to the caliche layer at 5 ft. Soil samples were extracted with ether, analyzed by GC, and the chromatographic traces in Figures 10.5a and b were obtained. Note the scaling factor by which each chromatogram must be multiplied for direct comparison. Though the use of capillary GC/MS, many of these peaks were isolated and identified to be fatty acids such as octadecanoic, decanoic, and

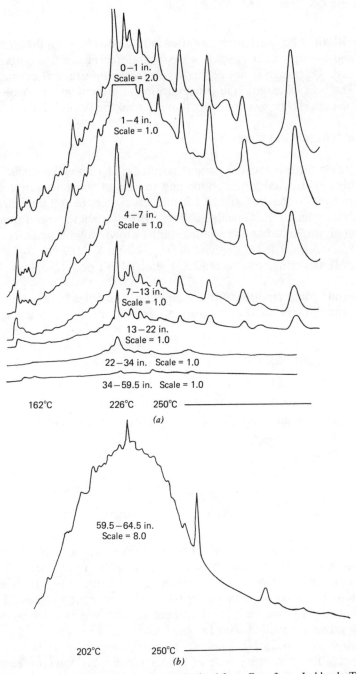

Figure 10.5. GC traces from extracts of soils obtained from Gray farm, Lubbock, Texas. (*a*) Reduced and superimposed GC traces of extracts from soil at various depths (as noted on figure). The size of the peaks in the top chromatogram should be multiplied by 2 for quantitative comparison. (*b*) Reduced chromatogram of soil extract from 59.5 to 64.5 in. This corresponds to the top 4 in of the caliche layer found below the soil in *a*. To compare peak sizes with those in *a*, multiply these peaks by 8.

tetradecanoic acids. Most of the peaks in Figure 10.5a can be fitted to a common exponential decay:

$$C = C_0 \exp(-kx) \qquad (10.1)$$

where x is depth in inches and $k = 0.15 \pm 0.01$ inches^{-1} with an average correlation coefficient of 0.96. The calculated C_0 values are reasonable estimates of C at the surface. At 5 ft a hard caliche layer was encountered, and the chromatogram in Figure 10.5b was obtained. Clay lenses are quite common in the subsurface, and the possibility that clay lenses may act to concentrate C-TLOs as ground water passes around them could be significant to predicting the fate of organics in the subsurface.

Ft. Polk, Louisiana

In order to treat further the secondary waste water from Ft. Polk, the largest rapid infiltration site in the United States was built. The ground water from the site area was sampled and analyzed for C-TLOs before operation was to have begun, and generally the area ground water was free of C-TLO contamination. Therefore, a laboratory study was undertaken to investigate the potential for ground water contamination by trace organics during operation of this system. Four columns were packed with topsoil obtained from one of the basins at the site. Unchlorinated secondary effluent was also obtained from the site and used as feed solutions for the columns. Soil columns were operated for a period of four months on a 2-day flooding/12-day drying cycle to parallel design operation of the rapid infiltration system. Feed solution and column effluents were monitored for 22 trace organics by reverse ion search using capillary GC/MS. Each of nine target trace organics found consistently in the feed solution, with the exception of p-dichlorobenzene, was detected in the column effluents during the first inundation cycle; all target trace organics were detected in the column effluents during subsequent cycles. The concentrations of tetrachloroethylene, p-dichlorobenzene, acetophenone, 2(methylthio)benzothiazole, and bis(2-ethylhexyl) phthalate in the feed solution were generally reduced by passage through the soil column, where diethyl phthalate, benzophenone, N-butylbenzenesulfonamide, and dibutyl phthalate concentrations were unaffected or enhanced. Addition of mercuric chloride as a bacteriocide during the seventh and eighth inundation cycles had varying effects for the different compounds. These preliminary results indicate that ground water contamination at the Ft. Polk rapid infiltration site may occur once the system begins operation, although concentrations of specific trace organics may be significantly reduced. Further details of these column simulation studies have been published elsewhere (Hutchins et al., 1983). Once operation of the Ft. Polk rapid infiltration site begins, the ground water will be monitored versus time for C-TLOs. This should produce valuable information on C-TLOs and soil adaption during start-up of a rapid infiltration system.

DISCUSSION

Proposed Simplified Measures of Organics in Ground Water

Extraction and GC/MS analysis of C-TLOs in ground water are demanding in terms of time and technical skill. Two alternate approaches to characterizing organics in ground water have been explored: (1) the use of indicator compounds, and (2) an operational definition of percent C-TLO removal using the relative areas under the GC traces.

1. If one or a few individual organic compounds could be identified as "indicators" of anthropogenic pollutants, then a specific analytical screening program could possibly be developed using simple and inexpensive tests. Table 10.5 contains a list of C-TLOs that have been found in the ground water at more than one site. Once an indicator compound or series of indicator compounds has been identified, it may be possible to develop an immunoassay for that compound or group (Gerba, private communication). If this can be done, the cost per analysis may be negligible.

2. It may be possible to characterize the change in the trace-organic water quality associated with a point source of contamination such as an infiltration site, a hazardous-waste site, or a septic tank by using the relative areas under chromatograms from the waste water and the ground water, when extracted in a prescribed manner. Such a parameter can be defined as follows:

$$\Delta\text{C-TLO\%} \equiv \frac{\left(\begin{array}{c}\text{Area under}\\\text{influent GC trace}\end{array}\right) - \left(\begin{array}{c}\text{Area under}\\\text{output GC trace}\end{array}\right)}{\begin{array}{c}\text{Area under}\\\text{influent GC trace}\end{array}}$$

Such a parameter could be used either to characterize the relative effectiveness of treatment regimes designed for trace-organic removal or to characterize the changing water quality of a single system. The ΔC-TLO%'s for several waste water treatment plants have been determined and vary from about 80% for classical primary treatment to 96.6% for the Ft. Devens rapid infiltration system.

It is expected that the prime use of such a water quality parameter may be in characterizing chemical treatment plants, because it may take 20–40 years for typical trace level organics to reach the sampling wells at the edge of a land application site. Therefore, the overall removal, ΔC-TLO%, may not indicate the *potential* of these sites to contaminate the ground water in the

Table 10.5   Ground Water Contaminants Found More Than Once

Tetrachloroethylene
Toluene
m-Xylene
Styrene
m-Dichlorobenzene
p-Dichlorobenzene
o-Dichlorobenzene
2,3-Dihydro-1-methyl-1H-indene
Naphthalene
p-*t*-Butylcyclohexanone
m-*t*-Butylphenol
Diphenylmethane
2,6-di-*t*-Butyl-*p*-cresol
2-Methyl-2H-benzotriazole
2,6-di-*t*-Butyl-*p*-benzoquinone
3,4-Dichloroaniline
2-(Methylthio)benzothiazole
Indole
Dimethyl phthalate
p-*t*-Pentylphenol
1,5-bis(*t*-Butyl)-3,3-dimethylbicyclo(3.1.0)hexan-2-one
o-Phenylphenol
5-Methyl-*t*-phenyl-2-hexanone
Diethyl phthalate
Benzophenone
Tributyl phosphate
p-(1,1,3,3-Tetramethylbutyl)phenol
N-Butylbenzenesulfonamide
Simazine
tris(2-Chloroethyl)phosphate
Dibutyl phthalate
bis(2-Ethylhexyl) phthalate

future. What is needed instead is a mechanistic understanding of the processes important to the transport and fate of C-TLOs in the subsurface in combination with a material balance for specific organic compounds.

Model of the Fate of C-TLOs

Degradation and adsorption are probably the major mechanisms responsible for the apparently retarded rate of movement of C-TLOs in ground water relative to a conservative tracer such as chloride. Before quantitative generalizations for biological or chemical degradation are possible, much more research is needed, but progress has been made toward predicting the physical fate of C-TLOs.

Recently, Karickhoff et al. (1979) made a significant advance toward understanding the role of adsorption in the retardation of organics of widely differing water solubilities in soils of different organic carbon (OC) content. In general, the parameter needed to predict the average rate of movement of an organic compound in ground water (V_c) relative to the average movement of the water (V_w) is the partition coefficient of the compound, Kp:

$$Kp = C_s/C_w \tag{10.2}$$

where C_s is the mass of C-TLO per gram of soil (ng/g_s) and C_w is the concentration of C-TLO in the water phase (ng/g_w). The relative velocity of contaminant and water is given by Freeze and Cherry (1979):

$$\frac{V_w}{V_c} = 1 + \frac{\rho_b}{n} Kp \tag{10.3}$$

where ρ_b is the bulk density $(\sim 2 \text{ g/cm}^3)$ and n is the porosity, generally 0.3–0.5. Karickhoff et al. (1979) found that

$$Kp = 0.63 \, K_{ow} \, (OC) \tag{10.4}$$

and

$$K_{ow} = 10^5 \, S^{-0.67} \tag{10.5}$$

where K_{ow} is the octanol-water partition coefficient for a compound $[K_{ow} = C_o/C_w$, similar to Eq. (10.2)] and S is the water solubility of the compound $(\mu mol/L)$. Substituting Eqs. (10.4) and (10.5) into Eq. (10.3), taking the logarithm of both sides and neglecting the "1" produces:

$$\log V_w/V_c = \log(0.63 \times 10^5 \, \rho_b/n) + \log(OC) - 0.67 \log S \tag{10.6}$$

For a particular soil, the first term on the right-hand side of Eq. (10.6) is constant. A series of $\log V_w/V_c$ vs. $\log S$ isopleths for various OC values is

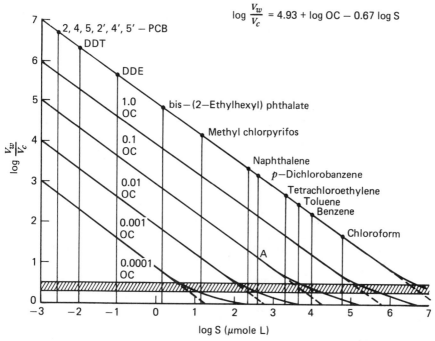

$$\log \frac{V_w}{V_c} = 4.93 + \log OC - 0.67 \log S$$

Figure 10.6. Log (V_w/V_c) vs. log S (μmol/L) constant OC isopleths for typical soils and C-TLOs—see Eq. (10.6).

plotted in Figure 10.6 along with the solubility of a few typical C-TLOs for comparison.

The isopleths in Figure 10.6 can be used to help correlate some of the concentrations of C-TLOs in ground water reported in this chapter. For discussion purposes "significant retardation" will be taken to be log $(V_w/V_c) = 0.3-0.5$ (i.e., $V_w/V_c = 2-3$) and larger, as indicated by the hashed lines in Figure 10.6. Typical fertile surface soils and rapid infiltration basin topsoils have 1–10% organic carbon—OC = 0.01–0.10 (Loehr et al., 1979 and Satterwhite et al., 1976). From Figure 10.6 it can be seen that all compounds of water solubility equal to or less than benzene's would be significantly retarded in an 0.01 OC soil. For example, for an 0.01 OC aquifer, p-dichlorobenzene would be predicted to have a relative velocity $V_w/V_c = 10^{1.15} = 14$ (Figure 10.6), that is, p-dichlorobenzene would be expected to move in such an aquifer only 1/14th as fast as the water. Many aquifers have very low OC values; OC values of 0.001 and less are common. For such an aquifer, all compounds with a water solubility equal to or greater than naphthalene would *not* be expected to be significantly retarded. Bis-(2-ethylhexyl) phthalate has the lowest water solubility (1 μmol/L) of all the compounds in ground water reported herein for rapid infiltration sites. If OC and water solubility were the only important factors, it would be expected that the order of relative breakthrough concentrations for compounds would follow their water solubilities. To further test this notion, additional sampling trips to Ft. Devens, Massachusetts, were made and concentrations of the more water-soluble compounds such as chloroform were measured. Treatment of the experimental data to predict breakthrough concentrations required the use of rather complicated numerical models with adjustable dispersion coefficients and is beyond the scope of the present chapter, but the reader is referred to Bedient et al. (1982, 1983) for further information. Essentially it was concluded that, if large dispersion coefficients are used, the above retardation formalism [Eqs. (10.2)–(10.6)] is basically correct. McCarty et al. (1979) studied the reinjection of treated waste water with C-TLOs such as chlorobenzene and obtained quantitative agreement with the retardation predictions, but for biodegradable C-TLOs such as naphthalene, greater removal was observed. These field observations suggest that for moderately soluble refractory C-TLOs, the adsorption relations in Eqs. (10.2)–(10.6) and Figure 10.6 may be usable to estimate the fate of C-TLOs once the problems of dispersion are treated explicitly as was done by Bedient et al. (1982), or implicitly by reactor theory as was done by McCarty et al. (1979).

SUMMARY

In summary, the potential for land application of municipal waste water to contaminate ground water with C-TLOs such as dichlorobenzenes or phthalate esters has been assessed. Results from three rapid infiltration sites, one irrigation site, and one laboratory column study have been reported. Two simplified measures of C-TLO removal have been proposed. Average removal efficiencies, ΔC-TLO%, for land application were generally 90% or better, with concentrations of C-TLOs in the sewage generally <1 μg/L and in the ground water generally <0.1 μg/L. A retardation model based upon the organic carbon content, OC, of the soil and the water solubility, S, of a compound has been proposed to explain these apparent removal efficiencies. Owing to the slow movement of ground water and the slower movement of C-TLOs in ground water, these ΔC-TLO% values may not express the potential of a site to contaminate ground water in future years. To date, no single C-TLO has been found in ground water that should be reason for immediate health concern to an individual drinking that water.

Unfortunately, the effectiveness of land application to remove C-TLOs has been deduced based upon our "failure" to measure various C-TLOs in ground water samples relative to sewage input samples. Several mechanisms (e.g., biodegradation, physical–chemical reaction, volatilization, adsorption, or dispersion) are consistent with such limited observations. With some of these mechanisms, a compound is destroyed, but with others it is still present as a potential ground water contaminant. For these reasons it would be useful to perform sampling and analyses sufficient in time and space to produce a mass balance for a few compounds. Such a mass balance would probably allow one to eliminate most of the above mechanisms and hence to better assess the long range potential for ground water contamination by C-TLOs. To perform such a thorough mass balance for a complete field site over several years might be prohibitively expensive using present normal field and laboratory methodologies, but several research groups are working on promising new remote sensing and rapid analytical methods to overcome these difficulties. Once developed, these procedures should enable researchers to better assess the long term potential for ground water contamination with C-TLOs by land application of waste water.

REFERENCES

Bedient, P. B., Springer, N. K., Baca, E., Bouvette, T. C., Hutchins, S. R., and Tomson, M. B. (1983). Ground water transport from wastewater infiltration. *J. Environ. Engr. ASCE* **109**(2):485–501.

Bedient, P. B., Springer, N. K., Cook, C. J., and Tomson, M. B. (1982). Modeling chemical reactions and transport in ground-water systems: A review. In: K. L. Dickson, A. W. Maki, and J. Cairns, Jr. (Eds.), *Modeling the Fate of Chemicals in the Environment.* Ann Arbor Science, Ann Arbor, MI, pp. 215–245.

Bouwer, E. J., McCarty, P. L., and Lance, J. C. (1981). Trace organic behavior in soil columns during rapid infiltration of secondary wastewater. *Water Res.* **15**(1):151–159.

Bouwer, H., Lance, J. C., and Escarcega, E. D. (1974). A high-rate land treatment system for renovating secondary sewage effluent: The Flushing Meadows Project. *J. Water Pollut. Control Fed.* **5**:835–843.

Carlson, R. R., Lindstedt, K. D., Bennett, E. R., and Hartman, R. B. (1982). Rapid infiltration treatment of primary and secondary effluents. *J. Water Pollut. Control Fed.* **54**(3):210–280.

Curran, C. M. and Tomson, M. B. (1983). Leaching of trace organics into water from five common plastics. *Ground Water Monitoring Review* **3**:68–71.

Doull, John, Klaassen, Curtis D., and Amdur, Mary O. (Eds.) (1980). *Casarett and Doull's Toxicology: The Basic Science of Poisons.* Macmillan, New York.

Dunlap, W. J., Shew, D. C., Scalf, M. R., Cosby, R. L., and Robertson, J. M. (1976). Isolation and identification of organic contaminants in groundwater. In: L. H. Keith (Eds.), *Identification and Analysis of Organic Pollutants in Water.* Ann Arbor Science, Ann Arbor, MI, pp. 453–477.

Freeze, R. A. and Cherry, J. A. (1979). *Groundwater.* Prentice-Hall, Englewood Cliffs, NJ.

Hutchins, S. R., Tomson, M. B., and Ward, C. H. (1983). Trace organic contamination of ground water from a rapid infiltration site: A laboratory field coordinated study. *Environ. Toxicol. Chem.* **2**:195–216.

Jeffrey, D. A. (1978). Hydrogeological significance of the application of sewage effluent to agricultural land on the Frank Gray farm, Lubbock County, TX. M. S. Thesis, West Texas State University, Canyon, TX.

Junk, G. A., Richard, J. J., Grieser, M. D., Witiak, J. L., Arguello, M. D., Vick, R., Svec, H. J., Fritz, J. S., and Calder, G. V. (1974a). Contamination of water by synthetic polymer tubes. *Environ. Sci. Technol.* **8**:1100–1106.

Junk, G. A., Richard, J. J., Grieser, M. D., Witiak, J. L., Arguello, M. D., Vick, R., Svec, H. J., Fritz, J. S., and Calder, G. V. (1974b). Use of macroreticular resins in the analysis of water for trace organic contaminants. *J. Chromatogr.* **99**:745–762.

Karickhoff, S. W., Brown, D. S., and Scott, T. A. (1979). Sorption of hydrophobic pollutants on natural sediments. *Water Res.* **13**:241.

Keswick, B. H. and Gerba, C. P. (1980). Viruses in ground water. *Environ. Sci. Technol.* **14**:1290–1297.

Leach, Lowell E., Enfield, Carl G., and Harlen, Curtis C., Jr. (1980). Summary of long-term infiltration system studies. U.S. Environmental Protection Agency, Ada, OK, EPA-600/2-80-165.

Loehr, R. C., Jewell, W. J., Novak, J. D., Clarkson, W. W., and Friedman, Gerald S. (1979). *Land Application of Wastes,* Vols. I and II. Van Nostrand Reinhold Company, New York.

Loehr, R. C. (Ed.) (1977). *Land as a Waste Management Alternative.* Ann Arbor Science, Ann Arbor, MI.

Majeti, Vimalla A. and Clark, E. Scott (1980). Potential health effects from persistent organics in wastewater and sludges used for land application. U.S. Environmental Protection Agency, Cincinnati, EPA-600/1-80-025.

McCarty, P. L., Rittmann, B. E., and Reinhard, M. (1979). Processes affecting the movement and fate of trace organics in the subsurface environment. In: T. Asano and P. V. Roberts (Eds.), *Proceedings of Symposium, California State Polytechnic University,* Pomona, Calif., Sept. 5–7.

Merck and Co., Inc. (1976). Martha Windholz (Ed.), *The Merck Index,* Rahway, NJ.

Miller, G. D. (1982). Uptake and release of lead, chromium, and trace level volatile organics exposed to synthetic well casings. In: *Proceedings of the Second National Symposium on Aquifer Restoration and Ground Water Monitoring,* May 26–28, Columbus, OH, pp. 236–245.

Pendleton, James and Miller, S. G. (1980). Public Utilities, Boulder, CO, personal communication.

Rice, R. C. and Gilbert, R. G. (1978). Land treatment of primary sewage effluent: Water and energy conservation. In: *Hydrology and Water Resources in Arizona and the Southwest* (Proceedings, Arizona Section of the American Water Resources Assn. and the Hydrology Section of the Arizona Academy of Science, Flagstaff, AZ, April 14–15), Vol. 8, pp. 33–36.

Satterwhite, M. B., Condime, B. J., and Stewart, G. L. (1976). Rapid infiltration of primary sewage effluent at Fort Devens, Massachusetts, Corps of Engineers, U.S. Army, Cold Regions Research and Engineering Laboratory (CRREL) Reports 76-48, 76-49.

Sax, N. Irving (1979). *Dangerous Properties of Industrial Materials.* Van Nostrand Reinhold, New York.

Tomson, M. B., Dauchy, J., Hutchins, S., Curran, C., Cook, C. J., and Ward, C. H. (1981). Ground water contamination by trace level organics from a rapid infiltration site. *Water Res.* **15:**1109–1116.

Tomson, M. B., Hutchins, S., King, J. M., and Ward, C. H. (1980). A nitrogen powered continuous delivery, all-glass-Teflon pumping system for groundwater sampling from below 10 meters. *Ground Water* **18:**444–446.

Tomson, M. B., Hutchins, S. R., King, J. M., and Ward, C. H. (1979). Trace organic contamination of groundwater: Methods for study and preliminary results. In: *Papers, Third World Congress on Water Resources,* Mexico City, Mexico, Vol. 8, pp. 3701–3709.

U.S. Dept. of Health, Education, and Welfare (HEW) (1977). Edwin Fairchild (Ed.), *Registry of Toxic Effects of Chemical Substances,* Vols. 1 and 2, Cincinnati.

U.S. Environmental Protection Agency (EPA) (1977). Process design manual for land treatment of municipal wastewater, EPA-625/1-77-008, Washington, D.C., pp. 7-44 to 7-52.

U.S. Environmental Protection Agency (EPA) (1980a). Ambient water quality criteria for dichlorobenzenes, EPA-440/5-80-039, Washington, D.C.

U.S. Environmental Protection Agency (EPA) (1980b). Ambient water quality criteria for naphthalene. EPA-440/5-80-059, Washington, D.C.

U.S. Environmental Protection Agency (EPA) (1980c). Ambient water quality criteria for phthalate esters. EPA-440/5-80-067, Washington, D.C.

U.S. Environmental Protection Agency (EPA) (1980d). Ambient water quality criteria for tetrachloroethylene. EPA-440/5-80-073, Washington, D.C.

U.S. Environmental Protection Agency (EPA) (1980e). Ambient water quality criteria for toluene. EPA-440/5-80-075, Washington, D.C.

PART TWO

METHODS FOR GROUND WATER QUALITY RESEARCH

11

OVERVIEW OF METHODS FOR GROUND WATER INVESTIGATIONS

C. P. Young
K. M. Baxter

Groundwater, Environmental Protection
Water Research Centre, Medmenham Laboratory
United Kingdom

Until relatively recently in Europe, studies of anthropogenically induced changes in ground water quality have formed adjuncts to ground water resource assessment investigations. Within the last decade or so, increasing recognition of the polluting potential of point sources such as solid waste disposal sites and effluent discharges, and of dispersed sources (chiefly agricultural), and of the possible geochemical effects of artificially recharging aquifers with waters of a different composition to the host ground water, has led to the development of techniques for obtaining a more complete characterization of the subsurface environment.

The investigatory techniques selected are determined very largely by local hydrological conditions. In this report the United Kingdom may, perhaps, be considered atypical of Europe in that its principal water supply aquifers, the Cretaceous Chalk and the Triassic Sandstone, are consolidated formations with water tables often at some tens of meters depth below ground level. Although comparable hydrogeological conditions are present in the Chalk of northern France, a high proportion of ground water abstracted for supply in Europe comes either from unconsolidated sedimentary deposits with shallow water tables or from karstified limestone, both of which are of only minor importance in Britain.

The optimization of ground water quality investigations requires forward planning. Before any decisions are made on the situating of investigatory boreholes and techniques to be employed for sampling, analysis, and monitoring, a thorough appraisal should be made of all available information, with the objective of defining as fully as possible the dynamic flow systems and geochemical environments of the target area.

The starting point for any investigation should be an examination of well, borehole, and spring records, which may include not only details of lithological successions but also the results of pumping tests, natural ground water level fluctuations, and water quality analyses. The fullest use should be made of centralized data banks, now often computerized, such as the Hydrogeology Well Records Data Bank (Horder, 1981) maintained in the United Kingdom by the Institute of Geological Sciences. However, experience has shown that such record collections are rarely complete, and valuable additional information may often be obtained from the records of drilling contractors operating within the study area.

In those cases in which ground water quality has been degraded as a result of waste disposals, information on the type of wastes tipped, and on the quantities and dates, may be obtainable from a number of sources, including the site owner/or operator, the statutory authority responsible for licensing the site (County or District Councils in the case of the United Kingdom), and the companies whose wastes are known to have been deposited. Information may also be volunteered by local sources, by hearsay, or from memory, but such information should always be accepted with reserve. If the contamination is considered to be from a dispersed source, such as agriculture, examination of local farming records and consultation with agricultural advisers familiar with the area will be necessary in order to gain a knowledge of the range of farming regimes practiced in the area, especially the rates of fertilizer applications and irrigation demands. Particular attention should be paid to past changes in farming patterns and practices; sequential land-use surveys and aerial photographic cover, if available, may be useful in this respect.

The assessment of available information constitutes only the first of several stages of a ground water quality investigation. In a detailed consideration of the problems associated with the study of contamination arising from waste disposal sites, Naylor et al. (1978) stated that:

> The object of any such investigation should be to define the extent and type of pollution associated with a landfill and to relate these to the history of tipping at the site, modes of operation, types of waste, local hydrogeological and climatic conditions, in such a way that predictions may be made of future concentrations and movement of the pollutants. The rates of production and movement of leachates under most conditions are generally slow. The three-dimensional picture of pollution levels within the landfill, in the unsaturated zone and below the water table, built up from a site study lasting typically 3–6 months, is therefore often in the nature of a "snap shot" of a dynamic situation. Con-

tinued monitoring of the pollution, both using permanent sampling boreholes and drilling additional boreholes, is therefore strongly recommended so that prediction may be checked and refined.

MATERIALS AND METHODS

Drilling Methods

The choice of drilling technique is controlled largely by the material to be drilled and the requirements of the program of investigation. The most important considerations are (1) if samples are required they should be representative of the depth of sampling, and (2) extraneous fluids should not be introduced into the samples or the formation. For this latter reason water- or mud-flush techniques are not generally acceptable.

Flight-auger drilling has been found to be satisfactory for preliminary investigation and for supplementing more detailed work. Auger drilling is rapid and inexpensive when drilling through waste material, superficial deposits, chalk, and soft sandstone. Augered samples of sandstone have often been warm on extraction due to frictional heating, which is a disadvantage since it could seriously reduce the volume of pore water available for extraction. Care must also be taken to ensure that the flights are clean in order to minimize cross-contamination.

For detailed investigations, the constraints of the sampling program restrict drilling techniques to cable-tool (percussion) and air-flush rotary methods. Cable-tool drilling is generally used in softer geological formations and for shallower (<100 m) holes. Direct air-flush rotary methods are employed in sandstone and hard rock formations, generally for deeper boreholes. Through layered strata a combination of methods may be required. During air-flush rotary drilling, any water encountered is violently expelled from the borehole. A "side-exit" flange is usually fitted at the top of the borehole casing to direct the discharged water into a settlement lagoon.

To prevent the contamination of the recovered samples, or of the aquifer via the borehole, particular attention should be paid to cleanliness. Temporary casing must be inserted into the top 5 m or so of each borehole to prevent surface water from entering the borehole. All drilling equipment must, whenever possible, be free of oil and grease.

An indication of the cost of drilling by both cable-tool and air-flush rotary methods is given in Table 11.1, assuming a 150 mm diameter borehole from which continuous undisturbed samples are required. Where only "bulk" samples or "open-hole" drilling are required, the cost is reduced to two-thirds of that given (Young and Gray, 1978; Naylor et al., 1978).

Sample Recovery

During any site investigation, solid and liquid samples should be collected systematically for both physical and chemical analysis. The first stage is to

Table 11.1   Approximate Costs for Drilling a 150 mm Diameter Borehole by Cable-Tool and Air-Flush Rotary Methods[a]

Depth (m)	Cable-Tool (£)	Air-Flush Rotary (£ 000)
10	0.7	1.2
20	1.4	1.9
40	2.3	3.7
60	4.5	5.8

[a] 1981 prices, £ sterling.

prepare a lithological log on-site as the samples are being recovered. During auger drilling recovered material is carried upwards to be sampled at the surface; the precise depth from which the material was obtained is therefore not known with any accuracy. More accurate sampling is achieved by with-drawing the augers at selected depth intervals.

Undisturbed U100 or rotary cores may be produced during percussion and air-flush rotary drilling, respectively. The U100 cores are usually ex-truded on-site from the core barrel by a hydraulic pusher into 250 gauge polyethylene "lay flat" tubing, which is heat-sealed at each end after as much air as possible has been expelled. The sample is then labeled and enclosed in a second wrapping. Much the same treatment is given to the rotary cores, except that the core is protected in the core barrel by a flexible liner, which is removed prior to wrapping or left intact if the core is simply to be stored in a core-box.

The frequency of sampling and the volume of sample required will depend on the physical and chemical results required but may be limited by the drilling method, the lithology of the site being drilled, or simply the analyti-cal load/cost that can be managed. Approximately 200 g of solid material every 2 m may be sufficient during the exploratory stage of an investigation, but 0.5 m of core, or 1–2 kg of bulk sample, every meter may be more appropriate during the detailed study.

Fluid samples can be collected from the bailer during cable-tool drilling or from the borehole head during air-flush rotary drilling. In most cases a 1-L sample every 2–3 m is sufficient for most analytical requirements.

Sample Handling, Storage, and Interstitial Water Extraction

Alteration of both solid and liquid sample composition in the time between recovery and analysis can occur by evaporative loss, chemical reaction, adsorption, or biochemical degradation. It is important, therefore, that sam-ples collected for chemical or microbiological analysis are packed and stored in such a manner that as little change as possible occurs.

All solid samples should be double sealed in polyethylene to reduce the loss of volatile compounds. Liquid samples should be filtered to remove any sediment and acidified on-site. Where appropriate, liquid samples are acidified to minimize the adsorption of heavy metals by the walls of the sample bottle. Chemical biodegradation may also be prevented by the use of biocidal preservatives. Careful note of the composition of any additives must be made so as to minimize any interference with other constituents. The choice of material for the sample bottle must also be appropriate to the type of analyses to be undertaken (Naylor et al., 1978).

Core samples can be transported and stored under different temperature conditions prior to the extraction of interstitial water. The effects of storage under freezing ($-18°C$) conditions on the chemical constituents in the porewater compared with those of storage under ambient and refrigerated conditions were investigated by Young and Gray (1978). A site in southern England, with calcareous, solifluction deposits overlying Chalk, was chosen and three 7-m deep boreholes drilled. Samples from each borehole were stored together, one in ambient conditions, one frozen in dry ice on-site and moved to a freezer each day, and the third frozen in wet ice and moved to a refrigerator each day. Five centimeter subsamples were taken from each set of cores and centrifuged after 2, 6, 14, 30, and 60 days. The extracted water was refrigerated and analyzed for nitrate within 3 days of centrifuging.

The 5-cm subsamples were taken sequentially from the core and could therefore be plotted in their correct stratigraphic sequence. The results (Figure 11.1) show that the greatest scatter of nitrate concentration is seen in the samples stored at ambient temperatures, while the least scatter was observed from the deep-frozen samples. The greatest change was noted in the upper part of the profile at about 1.2–1.5 m depth, where the nitrate concentrations of the ambient samples were almost twice those refrigerated or frozen. This difference may be attributed to the mineralization of organic residues derived from the overlying soil and incubated at laboratory temperature. An attempt to quantify the changes of nitrate concentration with time has been made by integrating the nitrate values under each of the five curves that may be drawn for each borehole, assuming a constant rock moisture content. The values are strongly influenced by individual values rather than the overall trend and hence subject to error. The results of this analysis (Table 11.2) show that the only real changes in nitrate content occur between 2 and 6 m depth.

Extraction of pore waters from consolidated materials may be made most effectively by high speed centrifugation (Edmunds and Bath, 1976; Barber et al., 1977). Experiments reported by Barber et al. (1977), on core samples from the Chalk and Triassic Sandstone aquifers, suggest that optimum recovery of pore water from both formations is achieved by centrifuging at about 6000 rpm for a minimum period of 30 min (Figure 11.2). Comparisons have been made between the compositions of interstitial pore waters and mobile ground waters obtained from the same depths. Barber et al. (1977)

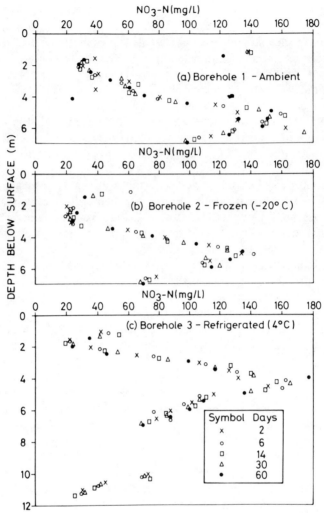

Figure 11.1. Changes in interstitial nitrate concentrations during storage—Chalk.

Table 11.2 Change in Integrated Nitrate Concentrations Under Different Storage Conditions[a]

Storage Conditions	Base (2 days)	Change			
		6 days	14 days	30 days	60 days
Ambient	100	106	108	111	110
Refrigerated	100	104	108	105	104
Frozen	100	98	96	96	111

[a] Samples from between 2- and 6-m Coombe deposits over Chalk, July 1976 (all figures in %).

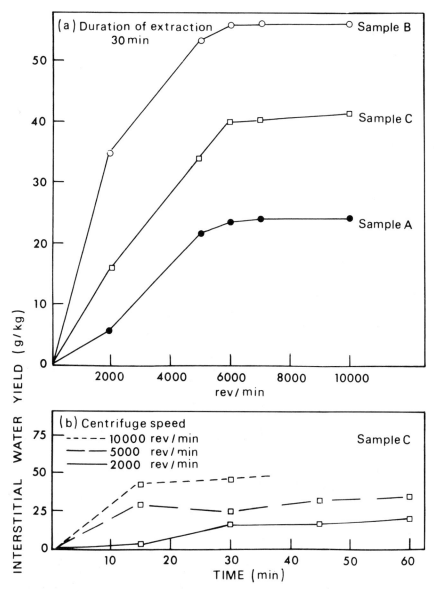

Figure 11.2. Variation in yield of interstitial water from sandstone samples with (*a*) duration of extraction at the stated speed and (*b*) centrifuge speed.

report that sodium, potassium, ammoniacal nitrogen, and total organic carbon were present at essentially similar concentrations in both sets of samples, but that zinc was present in interstitial water at concentrations about 10 times that of adjacent mobile waters, indicating partitioning. It was also found that chloride concentrations in mobile water from the Chalk aquifer were higher than those from pore water, possibly due to anionic exclusion. The possibility that concentration gradients of solutes such as nitrate are present within pores and that incomplete extraction of the water by centrifuging would provide an unrepresentative sample was investigated by Young et al. (1976). Comparisons of measurements of nitrate concentration in waters extracted by high-speed centrifuge with those made following complete disaggregation and leaching of the duplicate samples of Chalk revealed no significant differences (Figure 11.3), thereby indicating a lack of partitioning between finer and coarser pores.

Borehole Construction

Initial Investigations at Landfills

Auger holes drilled within landfills should be terminated at the base of the fill material because temporary casing is not normally used and transfer of leachate into the underlying aquifer could occur if deeper drilling were carried out. The holes may be either backfilled and sealed with a basal plug of cement grout or retained to monitor leachate quality. Construction for moni-

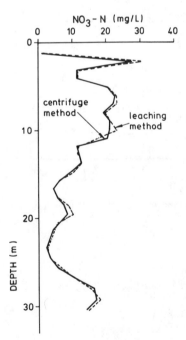

Figure 11.3. Comparison of nitrate concentrations, centrifuged and leached samples.

toring purposes may involve the installation of piezometer type tubes or *in situ* samples (see below). Holes drilled outside the landfill area to penetrate below the level of the local water table may serve as permanent ground water monitoring points. The construction, whether to accommodate piezometers or *in situ* samplers or to form open holes, will depend on the formation conditions.

Detailed Investigations

For the detailed stage of an investigation, several types of borehole construction need to be considered. The form of construction chosen for Type A boreholes will depend on the site conditions, the initial sampling requirements, and the long-term monitoring program. Examples of the different forms of Type A borehole construction are described in more detail in this section and illustrated in Figure 11.4; some have been used in the Water Research Centre's landfill investigations (Naylor et al., 1978).

In all cases the boreholes are drilled initially to the base of the landfill. Temporary steel casing is used in the construction to prevent borehole collapse and surface and subsurface leakage of contaminants into the boreholes. To reduce disturbance in the fill material and to allow representative samples to be taken, the base of the casing should be as close as possible to the base of the borehole at each stage of boring. Surging of the casing should be avoided, but this practice may be necessary if obstructions are encountered. Once the base of the landfill is reached, the casing should be driven into the underlying formation to act as a temporary seal against the movement of fluids from the base of the landfill into the borehole.

If the intention is not to drill deeper, but to use the borehole to monitor landfill fluids, construction may be of the type shown in Figure 11.4a. A cement grout plug approximately 2 m thick should be placed at the base of the borehole as a seal against movement of landfill fluids into the underlying formation, and permanent lining, slotted or perforated, should be installed. The base of the lining should be lowered into the cement grout before it has completely set; gravel or suitable material should be inserted into the annular space to prevent clogging of the slotted section of the lining, and the temporary casing then removed. A concrete platform should be constructed at ground level around the borehole to prevent the direct channeling of surface contaminants down the gravel pack.

If the intention is to drill into the underlying formation, with completion in the unsaturated zone, and to use the borehole to monitor landfill fluids, the construction should be as shown in Figure 11.4b. The landfill/formation interface should be sealed with a 2-m thick plug of either bentonite/cement mix or cement grout, depending on whether percussion or air-flush rotary methods, respectively, are to be employed. The seal and underlying formation should then be drilled at a reduced diameter to the required depth using temporary casing inserted into the upper part of the formation and also into the lower part if the formation conditions require it. The borehole should be

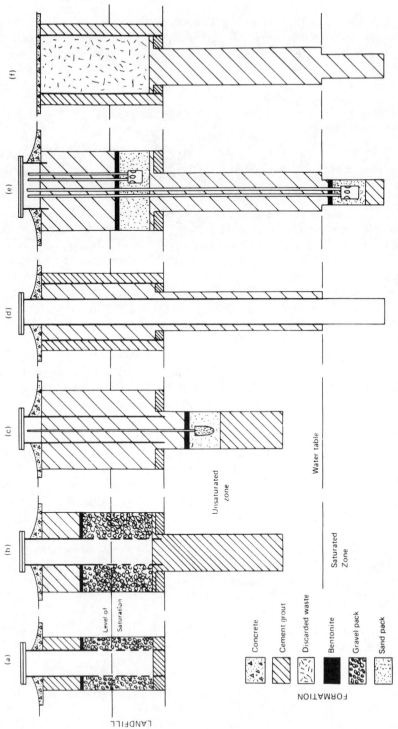

Figure 11.4. Construction of Type A boreholes.

228

backfilled with cement grout to 2 m above the landfill/formation interface and the temporary casing removed. The procedure for the construction of a landfill fluid monitoring borehole should then be followed as outlined above. Where fluids or gases from the unsaturated zone are to be monitored, suction probes may be installed (Figure 11.4c). This method of unsaturated zone sampling is suitable only in granular aquifers and is not applicable in the Chalk.

Boreholes drilled beneath the landfill into the ground water zone may be used to monitor ground water beneath the landfill or both ground water and landfill fluids. If it is intended only to monitor ground water, permanent solid lining should be installed and grouted in for the full depth of fill material before drilling deeper to the ground water zone (Figure 11.4d). The permanent lining installed into the formation may be required for the full depth, in which case the lining should be slotted in the ground water zone and surrounded with a suitable pack material.

If both ground water and landfill fluids are to be monitored, the drilling procedure should be as outlined in the second case. Small-diameter piezometer tubes or *in situ* samplers should be installed in both monitoring horizons, surrounded by sand packs and bentonite seals, and isolated by cement grout in the intervening levels (Figure 11.4e). Ground water samplers designed for this type of installation have been developed by the Water Research Centre (Joseph, 1980).

Where a Type A borehole, drilled to any depth, is not required for monitoring purposes, construction should be as shown in Figure 11.4f. Permanent lining should be installed to the base of the landfill and the base sealed before deeper drilling. Once the required depth has been reached, the borehole should be backfilled with cement grout to 2 m above the landfill/formation interface. Discarded drilling materials may be used to backfill from this level to the surface, where a concrete cover can be placed to act as a surface seal.

In the above type of construction, it is important to commence drilling at a diameter that will allow several diameter reductions with depth. Drilling through the landfill at 300 or 250 mm diameter provides sufficient flexibility for reductions in the underlying formations and has been found to be adequate for the recovery of solid and liquid samples and the installation of monitoring probes.

Type B boreholes serve as permanent observation points to monitor changes in ground water quality. The finished borehole construction (Figure 11.5) will be determined by the existing hydrogeological conditions and the form of contaminant movement. In simple, single homogeneous aquifer systems, standard open-hole construction, as shown in Figure 11.5a, will usually be suitable. Permanent, solid lining should be installed to the highest expected ground water level. In unstable formations lining may be required for the full borehole depth; it must be perforated or slotted in the ground water zone. The permanent lining may be of steel or uPVC, the latter having

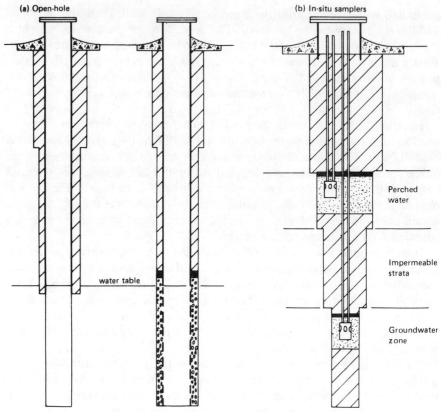

Figure 11.5.   Construction of Type B boreholes.

the advantage of being less than 50% of the cost of steel. The finished internal diameter of the borehole should be large enough to allow geophysical logging, sampling, and water level monitoring equipment to be inserted. An internal diameter of 150 mm is regarded as suitable to accommodate this equipment.

In heterogeneous aquifers, vertical variations in permeability may produce layering effects in the aquifer. In such cases, the presence of "perched" water, highly permeable zones, or multilobate contamination plumes will involve monitoring from specific levels. This may be achieved either by drilling separate boreholes, lined to each layer, or by installing a nest of sampling tubes within a single borehole. Construction to accommodate the installation of *in situ* samplers is shown in Figure 11.5*b*.

Geophysical Techniques

The use of surface geophysical techniques in ground water studies in the United Kingdom has been recently reviewed by Griffiths et al. (1981). The method most commonly used is that of surface resistivity measurements, which have been found to be capable of broad definition of zones of mineralized ground water, including chloride-rich pollution plumes originating from discharges of effluents, at sites with homogenous or uniformly layered substrates. More recently, the induced polarization method has been applied successfully to the tracing of the movement of chloride-rich leachate, derived from colliery spoil tips, through vertical fracture systems in the Triassic Sandstone of Nottinghamshire (Finch and Griffiths, 1977).

The most consistent contribution made by geophysics to ground water investigations has been to provide information on the dynamics of ground water flows by repeated logging of observation boreholes with suites of tools; loggings typically include temperature and differential temperature, fluid conductivity, flowmeter, and caliper. An idealized example of the type of information that may be gained by combination of the logs provided by the various tools is given in Figure 11.6. The velocities of ground water flows in boreholes under nonpumping conditions are often small; an important advance in measuring these velocities has been the development of a heat-pulse flowmeter by Dudgeon et al. (1975), which has a threshold velocity an order of magnitude lower than that of conventional impeller-type flowmeters (1–3 mm/sec vs. 30–50 mm/sec).

Indirect sensing of fluctuations of moisture content in the unsaturated zones of aquifers has formed an important part of ground water resource studies since development of the manually operated neutron probe (Bell, 1969). An increased interest in the mechanisms of the transfer of both water and potential pollutants through unsaturated media has resulted in increased use of pressure transducer tensiometers (Watson, 1967) to monitor soil moisture tensions and allow the calculation of moisture fluxes. Both instruments were originally designed in manually operated forms, but the increased pace of research into the physics and chemistry of water and solute transfer through partially saturated media has led to the development of automated versions of both tools (Kitching et al., 1981). The neutron probe operates automatically to depths of 2 m at present, and series of the tensiometers have been installed successfully to depths of 12.5 m in unsaturated Chalk. Both tools input to magnetic tape loggers.

Ground Water Sampling Techniques

The monitoring of ground water quality in the saturated zone at various depths below the water table from a number of different locations requires various techniques.

The base line from which monitoring and interpretation of the analytical results normally proceed is provided by the data obtained from solid, gaseous, and liquid samples during a site investigation.

Liquid and gaseous samples can normally be obtained from within the landfill and unsaturated zone by means of boreholes, small-diameter tube-

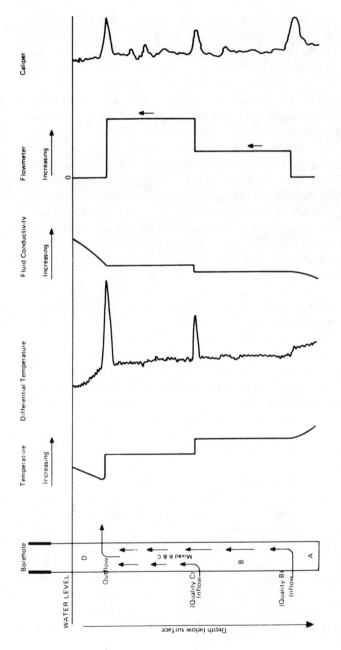

Idealised logs showing mixing within the water column

Inflows of water of differing quality entering the borehole are shown by the step form of the Temperature, Differential Temperature and Fluid Conductivity logs. The points of lateral inflow and outflow and vertical fluid movement are shown on the Flowmeter log and may be matched against fissure-like features on the Caliper log.

Confirmation that mixing is taking place is provided by the zero geothermal gradient (Temperature log) and constant conductivity between the step changes (Fluid Conductivity log).

Figure 11.6. Downhole geophysical logging.

wells, and suction probes. The routine collection of solid samples, if necessary, may be achieved only by repeated drilling.

Boreholes serving as permanent ground water monitoring points generally fulfill two functions: to provide access for the collection of ground water quality samples and to allow measurement of ground water levels.

The collection of representative ground water samples can be by open-hole or *in situ* sampler methods. In open borehole construction (Figure 11.5*a*), samples from the upper part of the water column can be taken using a simple "bucket on a string" type of sampler (Figure 11.7). "Surface" samples may be important where the contamination is less dense than the host ground water. They will, however, not usually be representative of the water column as a whole because contact with the atmosphere may increase gas exchange across the surface. A more positive approach to the problem of representative sampling is to take samples from various levels in the borehole. A depth sampler (Figure 11.7) can be lowered to the required depth, "triggered" to retain a water sample from that depth, and recovered.

The above types of sampling are appropriate at sites where the aquifer is homogeneous and isotropic and where rapid vertical changes in ground water composition are absent. Where mixing within the water column occurs, the samples obtained cannot be considered representative of local ground water composition and can therefore give misleading results. Mixing may be caused by vertical flow due to head differences within the aquifer, adverse ground water density gradients, diffusion along concentration gradients, or convection produced by the natural geothermal temperature gradient (Sammel, 1968). Identification of movement within boreholes and determination of whether mixing is taking place can be provided by downhole geophysical methods (see above).

A borehole packer method of sampling may be used to overcome the problem of mixing in open boreholes (Price et al., 1977). By this method fresh samples of formation water may be obtained by isolating specific sections of the borehole. Water is drawn from the selected length of the formation by means of a small submersible pump positioned between the inflatable packer seals (Figure 11.7). A considerable quantity of water (four or five times the volume of water contained between the packers) must be pumped to waste before the sample is collected. The packer system is probably an efficient method of sample recovery from unlined boreholes with smooth, firm walls—in consolidated aquifers with predominantly horizontal, small fissures, for example. It has been used successfully in boreholes with irregular, soft walls, but may not provide an efficient seal in hard, vertically fissured formations. Normally, boreholes containing slotted casing cannot be sampled using packers, unless the formation is of high permeability (e.g., gravels). Downhole television logging may locate suitable positions for placing the packers. A specially designed packer system has been used by the Water Research Centre landfill research programs, but in general such systems are not available commercially.

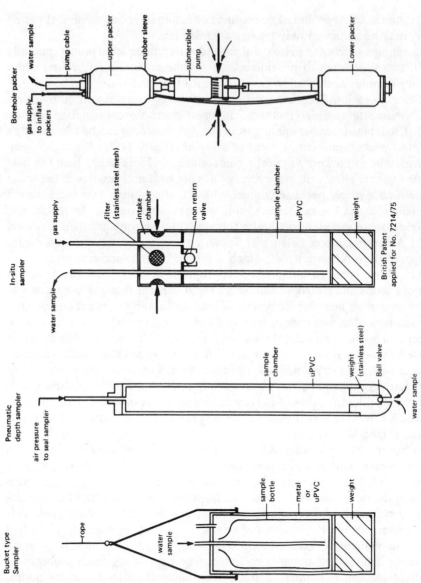

Figure 11.7. Ground water sampling techniques.

Borehole packer
- water sample
- gas supply to inflate packers
- pump cable
- upper packer
- rubber sleeve
- submersible pump
- Lower packer

In-situ sampler
- gas supply
- filter (stainless steel mesh)
- intake chamber
- non return valve
- sample chamber
- uPVC
- weight
- water sample
- British Patent applied for No. 7214/75

Pneumatic depth sampler
- air pressure to seal sampler
- sample chamber
- uPVC
- weight (stainless steel)
- Ball valve
- water sample

Bucket type Sampler
- rope
- sample bottle
- metal or uPVC
- weight
- water sample

234

An alternative method of obtaining representative samples from specific levels in the aquifer is to use *in situ* samplers (Brown, 1977). *In situ* sampling devices can be installed within a borehole and sealed in place to prevent vertical flow and to maintain the natural ground water quality stratification within the aquifer (Figure 11.5*b*). In this method the sample collector and reservoir, which are permanently sealed in the borehole, are connected to the surface by thin plastic tubes (Figure 11.7). Samples may be removed from shallow depths by suction and from deep installations by using compressed gas. The water within the sampler should be removed and discarded before obtaining a fresh sample representative of ground water composition. An automated version of the sampler has been developed (Kitching et al., 1981), which is capable of ejecting water samples at preselected time intervals. For reasons of economy and durability, early versions of the *in situ* samplers were manufactured from unplasticized polyvinyl chloride components with nylon operating tubes, and have been found to be entirely satisfactory where monitoring of gross inorganic contamination is required. However, significant concentrations (up to 50 mg/L) of organic contaminants may be leached from the apparatus, especially from the nylon tubes, and versions possessing acceptable leaching characteristics for monitoring trace organic pollution have been produced in stainless steel with teflon operating tubes. Samplers of the type described above have been installed and operated successfully to depths of up to 120 m and against operating levels of up to 75 m (Young and Gray, 1978). Care during installation and during sampling is essential, as sampler failure may require the drilling of a replacement hole, often at considerable expense. In situations where monitoring of microbially labile contaminants is to be carried out, particular care should be taken to ensure sterility of the apparatus and of surrounding gravel packs and sealing layers. The results of failure to do so have been reported by Young and Gray (1978), who described a site where samplers operated in the Chalk, with compressed nitrogen used as the ejector gas, became infected with denitrifying bacteria from a contaminated gravel pack. A rapid decrease in nitrate concentrations, which only slowly returned to background levels as food sources in the gravel were consumed, resulted (Figure 11.8). The use of compressed air as the ejector gas is now recommended for situations where the native ground water is aerobic.

Ground water level measurements in boreholes are carried out to observe fluctuations in the water table with time in the vicinity of the site. Measurement may be made manually at selected time intervals, using a simple borehole "dipper," or automatically by the use of a continuous, permanently installed water level recorder. The choice of method should be related to the type of aquifer system to be monitored. At sites with dominantly intergranular flow, the water level hydrograph, typically smooth and sinusoidal in form (Figure 11.9*a*), can be characterized by regular manual measurement. In fissure flow aquifers, however, large, irregular fluctuations in water level are likely to occur, so the form of the resultant hydrograph (Figure 11.9*b*) can be described only by continuous readings.

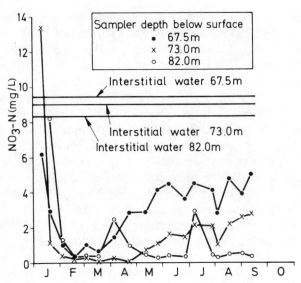

Figure 11.8. Long-term monitoring—*in situ* samples in Chalk.

At sites where surface leachates appear, monitoring of leachate composition and flow rate may be made from any drainage channels. Flow rates may be estimated from simple weir type structures, determined by regular readings from suitable staff gauge, or continuously recorded by a float-operated automatic water level indicator. Flow measurement stations often provide convenient sites for the manual collection of routine water quality samples or the installation of automatic water quality sampling equipment.

The frequency of monitoring a site will depend on the anticipated rate of movement of the leachate plume and the position of ground water and surface water sources. Where ground water development or surface sources in the area are thought to be at risk, weekly monitoring may be required, whereas 4-month or longer monitoring intervals may be suitable in "low risk" areas. Regular monitoring should include both the taking of water samples for laboratory determinations and the on-site measurement of unstable parameters such as dissolved oxygen content, pH, and temperature. As a check against the levels recorded by automatic equipment at the site, ground water levels should be measured manually on the occasions when quality samples are obtained.

Artificial Tracers

The injection of artificial tracers into ground water systems has been carried out in the United Kingdom in pursuit of three distinct objectives: to determine ground water flow paths and ground water residence time, to map and characterize karstic systems, and to provide data with which to estimate

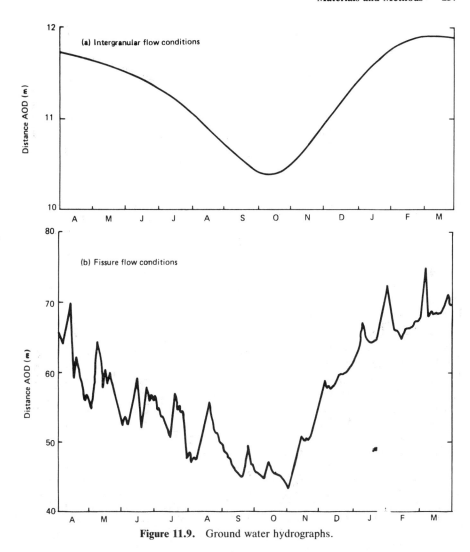

Figure 11.9. Ground water hydrographs.

aquifer parameters. A potentially wide range of radioactive, chemical, fluorescent dye, bacterial, and particulate substances are available for water-tracing studies. Those most suitable, by virtue of their mobility, low toxicity, and ease of detection, for ground water studies have been reviewed by White (1977)—who also suggested the use of short-lived isotopes created by neutron activation of elements above fluorine in the Periodic Table—and more recently by Atkinson and Smart (1981).

Mather and Smith (1973) made use of tritium to determine the contribution to spring and borehole yields of ground water present in a coal mine in South Wales. Tritium mixed with water in the mine workings was monitored at surrounding springs and borehole discharges, and it was possible to estimate the potential effect direct abstraction from the mine workings had on their yields. Proposals to discharge road drainage into a natural swallow hole in the Chalk of Hampshire were investigated by Atkinson and Smith (1974), who found evidence of a rapid, probably fissure-flow, connection with a large spring used for public supply at 5 km distance. The measured dispersion of the tracer emerging at the spring enabled an estimate to be made of the potential impact on water quality that would occur if a road tanker were to spill its load directly into the swallow hole.

The mapping and characterization of karstic factors in the Jurassic Limestone of the Bath region of England has been carried out by Smart (1976a), who was able to identify separate catchments contributing to individual springs. The principal karst development in the United Kingdom is in the Carboniferous Limestone of the Mendip Hills, Somerset, and the Malham area of Yorkshire. Extensive tracer studies of the Mendip area, especially at Atkinson (1971, 1977), have shown the presence of discrete branching networks of conduits feeding each spring. The work suggested that the conduits act as the main transmitting paths for ground water, but that there is substantial storage in a plexus of fine, intersecting fissures, which during ground water recession periods release their water in a way analogous to bank storage in rivers.

The single borehole dilution technique has been used by Smart (1976b) to measure a hydraulic conductivity of 47 m/day in fractured Jurassic Limestone in England, and more recently the same technique has been applied to a study of zones of high fissure permeability in the Chalk of the East Anglian region (Atkinson and Smart, 1981). A two-well pulsed tracer technique described by Ivanovitch and Smith (1978) was applied by them to a study of the interrelationship between ground water movement and storage either in enlarged fractures and micropores in fine fissures and micropores in the Chalk aquifer of Dorset, England. Their results indicated fracture water velocities of about 2.4 m/hr, with a storage coefficient of 0.005, compared with velocities of 0.4 m/hr and storage coefficient of 0.02 for the fine fissures. Measurements of the dispersion coefficients of Triassic Sandstone in Nottinghamshire, England, using Br-82, rhodamine WT, lithium chloride, and potassium iodide as tracers have been made by Oakes and Edworthy (1977). The studies were carried out at an experimental artificial recharge well, surrounded by observation wells; measurements were made during both recharge and abstraction. Data analyzed by comparison with a numerical model yielded values for the dispersion coefficient of 0.60 m for the full depth of flow, with individual beds giving values down to 0.16 m.

CONCLUSIONS

A wide range of investigating techniques have been developed to suit the predominant hydrogeological conditions present in the United Kingdom.

The high capital costs of borehole drilling have encouraged the development of more sophisticated sampling methods and of the increased use of repeatable, geophysical borehole logging techniques. The use of artificial tracers in ground water studies is increasing and promises to be a useful tool in elucidating the ground water flow relationships in dual porosity (matrix vs. fractures) media.

ACKNOWLEDGMENTS

This paper is published with the permission of the Director, Environmental Protection, Medmenham Laboratory, Water Research Centre, United Kingdom.

REFERENCES

Atkinson, T. C. (1977). Diffuse flow and conduit flow in limestone terrain in the Mendip Hills, Somerset (Great Britain). *J. Hydrol.* **35**:93–110.

Atkinson, T. C. (1971). The dangers of pollution of limestone aquifers. *Proceedings of University of Bristol Spelaeological Society* **12**:281–290.

Atkinson, T. C. and Smart, P. L. (1981). Artificial tracers in hydrogeology. In: *A Survey of British Hydrogeology 1980*, Royal Society of London, pp. 173–190.

Atkinson, T. C. and Smith, D. I. (1974). Rapid groundwater flow in fissures in the Chalk: an example from South Hampshire. *Q. J. Eng. Geol.* **7**:197–205.

Barber, C., Maris, P. J., and Knox, K. (1977). Groundwater sampling: The extraction of interstitial water from cores of rock and sediments by high-speed centrifuge. Water Research Centre Technical Report No. TR.54 (Sept.), The Centre, U.K.

Bell, J. P. (1969). A new design principle for neutron soil moisture gauges: The "Wallingford" Neutron Probe. *Soil Sci.* **108**:160–164.

Brown, S. L. (1977). Groundwater quality surveillance near a landfill site—comparative study of groundwater sampling techniques. In: *Proceedings of a Conference on Groundwater Quality, Measurement, Prediction and Protection*, Reading, U.K., September 1976, Water Research Centre, U.K., pp. 385–404.

Dudgeon, C. R., Green, M. J., and Smedmor, W. J. (1975). Heat-pulse flowmeter for boreholes. Water Research Centre Technical Report No. TR.4 (March), The Centre, U.K.

Edmunds, W. M. and Bath, A. H. (1976). Centrifuge extraction and chemical analysis of interstitial waters. *Environ. Sci. Technol.* **10**:467–472.

Finch, J. W. and Griffiths, D. H. (1977). The detection of contaminated groundwater in the Nottinghamshire Trias by geophysical techniques. In: *Proceedings of a Conference on Groundwater Quality, Measurement, Prediction and Protection*, Reading, U.K., September 1976, Water Research Centre U.K., pp. 364–372.

Griffiths, D. H., Barker, R. D., and Finch, J. W. (1981). Recent applications of the electrical resistivity and induced-polarization methods to hydrogeological problems. In: *A Survey of British Hydrogeology 1980,* Royal Society of London, pp. 85–96.

Horder, M. F. (1981). The use of databanks and databases within the Institute of Geological Sciences. *J. Geol. Soc. London* **138**(5):575–582.

Ivanovitch, M. and Smith, D. B. (1978). Determination of aquifer parameters by a two-well pulsed method using radioactive tracers. *J. Hydrol.* **36**:35–45.

Joseph, J. B. (1980). Installation and operating manual for the WRC "in situ" sampler. Report No. 27-M, Water Research Centre, U.K.

Kitching, R., Bell, J. P., Edworthy, K. J., and Tate, T. K. (1981). New instrumentation in hydrogeology. In: *A Survey of British Hydrogeology 1980,* Royal Society of London, pp. 113–124.

Mather, J. D. and Smith, D. B. (1973). Thermo-nuclear tritium—its use as a tracer in local hydrogeological investigations. *Proceedings of the Society of Water Treatment & Examination* **17**:187–196.

Naylor, J. A., Rowland, C. D., Young, C. P., and Barber, C. (1978). The investigation of landfill sites. Water Research Centre Technical Report No. TR.91 (Oct.), The Centre, U.K.

Oakes, D. B. and Edworthy, K. J. (1977). Field measurements of dispersion coefficients in the United Kingdom. In: *Proceedings of a Conference on Groundwater Quality, Measurement, Prediction and Protection.* Reading, U.K., Sept. 1976, Water Research Centre, U.K., pp. 274–297.

Price, M., Robertson, A. S., and Foster, S. S. D. (1977). Chalk permeability: A study of vertical variation using water injection tests and borehole logging. *Water Serv.* **81**(980):603–610.

Sammel, E. A. (1968). Convective flow and its effects on temperature logging in small diameter wells. *Geophysics* **33**(6):1004–1012.

Smart, P. L. (1976a). Catchment delimitation in karst areas by the use of quantitative tracer methods. In: *Proceedings of the Third International Symposium of Underground Water Tracing,* Bled, Yugoslavia, pp. 291–298.

Smart, P. L. (1976b). The use of optical brighteners for water tracing. *Trans. Br. Cave Res. Assoc.* **3**:62–76.

Watson, K. K. (1967). A recording field tensiometer with rapid response characteristics. *J. Hydrol.* **5**:33–39.

White, K. E. (1977). Tracer methods for the determination of groundwater residence time distributions. In: *Proceedings of a Conference on Groundwater Quality, Measurement, Prediction and Protection.* Reading, U.K., Sept. 1976, Water Research Centre, U.K., pp. 246–273.

Young, C. P. and Gray, E. M. (1978). Nitrate in groundwater—the distribution of nitrate in the Chalk and Triassic Sandstone aquifers. Water Research Centre Technical Report No. TR.69 (Jan.), The Centre, U.K.

Young, C. P., Hall, E. S., and Oakes, D. B. (1976). Nitrate in groundwater—studies on the Chalk near Winchester, Hampshire. Water Research Centre Technical Report No. TR.31 (Sept.), The Centre, U.K.

12

BIOCHEMICAL METHODS FOR DETECTION OF SUBSURFACE CONTAMINATION/BIOMASS

Franklin R. Leach
Jenq C. Chang
Jeffrey L. Howard
JoAnn J. Webster
Andrea B. Arquitt
RosaLee Merz
Elizabeth R. Doyel
Phyllis T. Norton
Ginger J. Hampton
Jeralyn Z. Jackson

Department of Biochemistry
Oklahoma State University
Stillwater, Oklahoma

Astounding progress has been made in the biochemical sciences during the last four decades. Avery et al. (1944) reported evidence that a nucleic acid of the deoxyribose type was the fundamental unit of the transforming principle of *Pneumococcus* Type III—in other words, that DNA was the repository of genetic information. Now we are able to decipher genetic messages hidden in base sequences and even to synthesize a gene chemically. Things undreamed of even by the science fiction writers of that era are now commonplace.

This biological revolution has attracted some of the best scientific minds and much of the support available for science. Many techniques have been developed for application to small-sized samples. Many of the procedures developed during these studies are recorded in methods publications such as *Methods in Enzymology* (a continuing series now containing 106 volumes), *Methods of Biochemical Analysis* (27 volumes), and the journal of *Analytical Biochemistry*.

Are there any techniques adaptable for use in detecting and quantitating sparse populations of microorganisms? Can the methods be utilized in the more applied fields to improve the quality of life? In this category we include fields with fewer immediate and obvious benefits than medical care. When a life is on the table in an emergency or hospital room, we are inclined to spare no expense in saving and/or prolonging that life. But what about the wider range of public health functions such as water quality and sewage disposal? How much are we willing to pay for the necessities of life that do not attract as much serious attention these days because lives are not seen to be in immediate jeopardy?

Both living organisms and their metabolic products are of potential contaminatory significance. Detection and quantitation of sparse populations of living organisms in environmental samples is a germane problem that can be approached using the methods of biochemical analysis. The literature relevant to such an approach has been reviewed by Dermer et al. (1980). The inherent sensitivity in various methods of biochemical analysis was discussed. Table 12.1 illustrates typical results for DNA and ATP. By measuring either the ultraviolet absorption at 259 nm or the color reaction obtained with diphenylamine, 1 μg of DNA can be detected using standard laboratory instrumentation. A biological assay for the transforming activity of *Bacillus subtilis* DNA can readily detect as little as 100 ng of DNA. Using enzymatic end labeling techniques and high-specific activity radioactive precursors, the amount of polynucleotide detectable is 3 ng. This is a procedure that is used in the determination of base sequences, but in general there must be an enrichment of that sequence of genetic material by cloning or growing large amounts of the organism. A fluorometric assay using DAPI gives the greatest current sensitivity, 0.4 ng.

Adenosine triphosphate (ATP) performs a central role in cellular energy exchange first recognized by Lipmann (1941) and by Kalckar (1941). ATP occurs ubiquitously in living organisms and functions as an allosteric effector, as a group-carrying coenzyme, and as a substrate and product. Atkinson (1977) found that the ratio of the adenylates was a controlling factor in many metabolic sequences and proposed the concept of energy charge to account for these observations. Thus the concentration of ATP is fairly constant within an organism, and the amount of ATP would reflect the biomass (Holm-Hansen and Booth, 1966). ATP absorbs light at 259 nm with an extinction coefficient of 15.4×10^8 and can therefore be determined spectrophotometrically. A coupled enzymatic assay with a fluorometric determina-

Table 12.1   Inherent Sensitivity of DNA and ATP Measurements

Assay	Method	Minimum Detectable Amount in Ordinary Instrumentation (ng)	(pg)
DNA			
Spectrophotometric	Direct	1000	
Spectrophotometric	Deoxyribose via diphenylamine	1000	
Fluorometric	DAPI	0.4	
Transformation	*B. subtilis*	100	
End labeling for sequencing	[^{32}P]	3	
ATP			
Spectrophotometric	Direct	100	
Fluorometric	Coupled enzyme		50
Bioluminescent			0.05
Double enzymatic cycling			0.005

tion of the pyridine nucleotide produced increases the sensitivity of quantitation (Table 12.1). The direct enzymatic assay of ATP uses firefly luciferase which consumes ATP as a substrate and produces an easily measured photon of light (quantum yield 0.9). Advantages of this method (to be discussed in detail later) include simplicity, specificity if the enzyme is pure enough, sensitivity, and the commercial availability of both instrumentation and reagents. Enzymatic cycling reactions can be used to detect even smaller amounts of ATP, but these require sophisticated microtechniques and are in the realm of only an accomplished enzymologist.

A substance may be determined stoichiometrically where the quantity being measured is either equal to or some fraction (depending upon the extent of reaction, such as derivatization) of the amount of the original substance. It is possible to increase greatly the sensitivity when the substance or something derived from it is used catalytically. For example, the fluorometric limit of detection of NADH is 60 ng, and that of the protein alcohol dehydrogenase is also 60 ng. If the alcohol dehydrogenase is used to reduce NAD$^+$ to NADH at a rate of 1000 moles of NADH produced per minute of incubation by each molecule of alcohol dehydrogenase, a 10-min incubation would yield a 10,000-fold increase in the sensitivity of measuring alcohol dehydrogenase over just a protein determination.

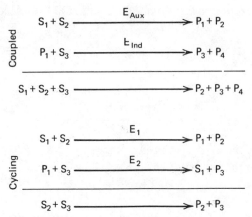

Figure 12.1. Coupled and cycling assays: S, substrate; P, product; E, enzyme; Aux, auxiliary; Ind, indicator.

Sometimes it is advantageous to convert the substance being measured to a substance that can be measured more easily or with a greater sensitivity. Figure 12.1 shows how coupled enzymatic assays can be used to accomplish this. A simple chemical example of this is the conversion by treatment with 10 N NaOH of NAD^+ or $NADP^+$ to a reaction product with a 10-fold increase in fluorescence.

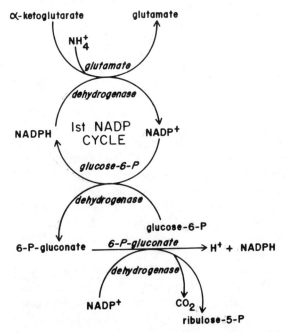

Figure 12.2. A $NADP^+$ cycle. NADPH and $NADP^+$ are the reduced and oxidized form of nicotinomide adenine dinucleotide phosphate, respectively. P, phosphate.

Firefly

$$LH_2 + Lu + MgATP \rightleftharpoons Lu - LH_2 - AMP + MgPP$$

$$Lu - LH_2 - AMP + O_2 \longrightarrow Lu + AMP + CO_2 + OL + Light$$

Bacterial

$$BLU + FMNH_2 + O_2 \longrightarrow BLU - FMNHOOH$$

$$BLU - FMNHOOH + RCHO \longrightarrow BLU + RCOOH + FMN + H_2O + Light$$

Figure 12.3. Bioluminescent assay system: LH_2, firefly luciferin; LU, firefly luciferase; OL, oxidized luciferin; BLU, bacterial luciferase; $FMNH_2$, reduced flavin adenine dinucleotide; RCHO, a long-chain aldehyde.

Lowry et al. (1961) introduced an amplification technique called enzymatic cycling which has been further developed into a system for flexible enzymatic analysis (Lowry and Passonneau, 1972). In this technique (also illustrated in Figure 12.1) two enzymatic reactions having a common substrate—such as $NADP^+(S_1)$—generate as much product P_3 as would a 20,000-fold amplification of the $NADP^+$. If two enzymatic cycles are coupled, the 20,000-fold amplification of one cycle is multiplied by the 20,000-fold amplification of the second cycle to produce a final amplification of 400 million-fold over the amount originally present. Figure 12.2 shows the detailed mechanism involved in a $NADP^+$ cycle. There is the possibility of stacking of catalytic functions so that the signal is amplified through a cascade of reactions, as occurs in blood clotting and in the metabolism of glycogen.

We applied standard biochemical determinations of DNA, RNA, protein, and organic phosphate and determined the sensitivity achievable with typical samples. Catalytic assays using enzymatic and/or cycling procedures were used to determine enzymes, cofactors, or lipopolysaccharides. The conditions we found that optimize assay sensitivity will be reported. Particular emphasis was placed on bioluminescent (see Figure 12.3) and chemiluminescent methods because of their inherent sensitivity.

MATERIALS AND METHODS*

Materials

Chemicals

EDTA, Folin reagent, hydrogen peroxide, INT, and lactic acid were from Fisher. Armour was the source of the bovine serum albumin, and Aldrich

---

* Abbreviations: EDTA—ethylenediaminetetraacetic acid; DAPI—4′,6-diamidino-1-phenylindole 2 HCl; DABA—3,5-diaminobenzoic acid; TCA—trichloroacetic acid; DMSO—dimethyl sulfoxide; INT—2-(p-iodophenyl-3-p-nitrophenyl-5-phenyltetrazolium chloride; INTF—the formazan of ?-p-iodophenyl-3-p-nitrophenyl-5-phenyltetrazolium chloride; DTT—dithiothreitol.

Chemical supplied the bromosulfalein, DABA, diphenylamine, ethidium bromide, and orcinol. ADP, ATP, Coomassie Blue, DNA, FMN, glucose, hemoglobin, 2-mercaptoethanol, NAD$^+$, NADP$^+$, p-nitrophenyl disodium phosphate, 6-phosphogluconate, o-phthalaldehyde, pyruvic acid, and RNA were all from Sigma. DAPI was from Serva. Luciferin was synthesized in Dr. K. D. Berlin's laboratory or obtained commercially from Calbiochem or Boehringer-Mannheim. Other acids, bases, buffers, salts, and solvents were reagent grade.

Enzymes

Adenylate kinase, alkaline phosphatase, catalase, diaphorase, lactate dehydrogenase, glucose-6-phosphate dehydrogenase, glutamate dehydrogenase, hexokinase, 6-phosphogluconate dehydrogenase, and ribonuclease were from Sigma. Firefly luciferase was obtained from Analytical Luminescence Laboratory, Boehringer-Mannheim, Calbiochem, DuPont, LKB, Lumac, SAI Technology, and Sigma.

Limulus amebocyte lysate and *E. coli* lipopolysaccharide were obtained from Associates of Cape Cod.

Organisms

E. coli Crookes ATCC No. 8739 was obtained from the American Type Culture Collection.

Methods

Protein was determined by the following procedures: Lowry et al. (1951); McKnight (1977); McGuire et al. (1977); Kutchai and Geddis (1977); Butcher and Lowry (1976); and Schultz and Wassarman (1977). DNA was determined by the following methods: Abraham et al. (1972); Setaro and Morley (1977); Cattolico and Gibbs (1975); Kapuscinski and Skoczylas (1977); and both DNA and RNA by a modification of LePecq and Paoletti (1966); Boer (1975); Karsten and Wollenberger (1977); and El-Hamalawi et al. (1975).

RNA was also determined by the procedure of Ceriotti (1955). Organic phosphate was determined according to Going et al. (1975) and EPA (1974).

Adenylate kinase was determined as described by Sottocasa et al. (1967); Rasmussen (1978); and Lowry and Passonneau (1972). Alkaline phosphatase was measured according to Malamy and Horecker (1964); and Torriani (1968). The procedures in the *Worthington Enzyme Manual* (1972) were used to determine catalase and diaphorase. Lactic acid dehydrogenase was measured by procedures described by the following: Worthington (1972); Curl and Sandberg (1961); Wieser and Zech (1976); Lowry et al. (1961); Chi et al. (1978); and Lowry and Passonneau (1972).

The pyridine nucleotides were measured according to the following: Lowry and Passonneau (1972); Lowry et al. (1961); and Chi et al. (1978). The flavin procedure was according to Okrend et al. (1977). The iron porphyrin procedure was modified from Thomas et al. (1977); Neufeld et al. (1965); Oleniacz et al. (1968); Searle (1975); and Ewetz and Thore (1976).

The firefly luciferase procedures for ATP were as specified by Webster et al. (1979, 1980a, 1980b, 1981a, 1981b), and Webster and Leach (1980). The enzymatic cycling procedure for ATP was adapted from Lowry et al. (1961); and Lowry and Passonneau (1972). The *Limulus* amebocyte lysate assay for lipopolysaccharide was based on descriptions by Cohen (1979) and Associates of Cape Cod (1981).

RESULTS

Standard Biochemical Tests

The standard biochemical assays currently in use in most biochemistry laboratories for protein, DNA, RNA, and organic phosphate were set up and tested in this laboratory. Linear standard curves were obtained and the standards run enough times to establish both statistical and operator confidence. The assays were then performed to measure the particular substance in a culture of *E. coli* in order to allow expression of the sensitivity of the assay in terms of the number of *E. coli* cells. Table 12.2 summarizes these experiments. Under the "Complications/Remarks" column we note particular limiting factors or difficulties that we encountered. While all of these assays were satisfactory within the inherent and biochemically understandable limitations, none were of sufficient sensitivity to be applicable to the sparse populations of organisms expected in subsurface environmental samples.

Catalytic Tests

Enzyme Assays

One of the goals of this research was to increase or optimize the sensitivity of assay using normal laboratory equipment. Table 12.3 shows the sensitivity achieved in this laboratory (Howard, 1980) and the sensitivities reported by other workers with several enzymes.

Normal enzymatic assays involved the "continuous assay" of the reaction over a short time period, which takes advantage of several primary characteristics of enzymes and eliminates many potential problems. The "incubation assay," in which enzymes act over a longer time and a single evaluation of the amount of product is used, yields greater sensitivity. Difficulties such as inhibitors, instability, and changes in other parameters are, however, introduced. Figure 12.4 shows the results of determining lactate dehydrogenase by a kinetic assay (*A*) and by an incubation assay (*B*). Through the use of enzymatic cycling procedures, an even greater sensitivity of determining lactic acid dehydrogenase can be obtained (Figure 12.4*C*).

Table 12.2 Standard Biochemical Determination Methods and Their Sensitivities

Substance	Method or Principle	Reference	Specific Method	Range[a]	Number of E. coli Cells[b]	Complications/Remarks
Protein	Lowry	Lowry et al., 1951	Folin phenol, biuret	10–80 μg	10^8	Phenol, tyrosine interference
	Dye binding	McKnight, 1977	Coomassie Blue	0.5–10 μg	10^7	Not all proteins bind dyes to the same extent
	—	McGuire et al., 1977	Bromosulfalein	0.5–10 μg	10^7	
	Fluorescent	Kutchai and Geddis, 1977	Without hydrolysis	1–25 μg	2×10^5	Only amino groups
	—	Butcher and Lowry, 1976	With hydrolysis	1.5–12 ng(Ala)	—	6 ng of protein, background is the problem
	Radioisotopic	Schultz and Wassarman, 1977	Labeling amino terminal	0.08–2.5 μg	10^7	Reproducibility poor
DNA	Burton	Abraham et al., 1972	Diphenylamine	5–50 μg	6×10^7	Adeoxyribose determination

248

Substance	Reagent	Reference	Assay	Range[a]	Number[b]	Comments
	DABA	Setaro and Morley, 1977	Spectrophotometric	25–750 µg	10^8	Not as sensitive, but one can run spectrophotometric or fluorometric assay depending on the DNA content of the unknown
	—	Cattolico and Gibbs, 1975	Fluorometric	0.1–1.6 µg	10^6	
	Fluorescent	Various	Ethidium bromide	0.02–0.1 pg	6×10^5	Also react with RNA
	—	Kapuscinski and Skoczylas, 1977	DAPI	1–16 ng	3×10^4	
RNA	Orcinol	Ceriotti, 1955		1–50 µg	10^8	For pentose
	Ethidium bromide	Various	Fluorometric	5–20 ng	10^6	Also reacts with DNA
Organic Phosphate	Molybdoantimonylphosphoric acid	Going et al., 1975	Direct	40–640 ng	10^6	
	—	EPA, 1974	Extraction	0.2–6.4 ng	10^5	6 fg ATP/E. coli cell

[a] Range found in this laboratory to yield linear and accurate results.

[b] The number of E. coli cells required to yield the minimum amount of substance that is determinable by the particular methods.

Table 12.3   Comparison of Laboratory-Measured and Literature Sensitivities for Enzymatic Bioindicator Determinations

| | Limit of Detection | | |
Bioindicator	This Study	Literature	Reference
Lactic acid dehydrogenase	1.0 pg	2.0 ng	Howell et al., 1977
Alkaline phosphatase	0.1 ng	5.0 ng	Doellgast, 1977
Catalase	1.0 ng	20.0 pg	Weetall et al., 1965
Adenylate kinase	0.1 ng	0.1 ng	Brolin et al., 1979

Cofactor Assays

Similar experiments were done using assays for cofactors. Table 12.4 shows results of these studies; the improvement of sensitivity achieved was not so good as that achieved with the enzyme assay. Work done on most of these assays was not sufficient to optimize our determinations but we are confident that improvements are possible.

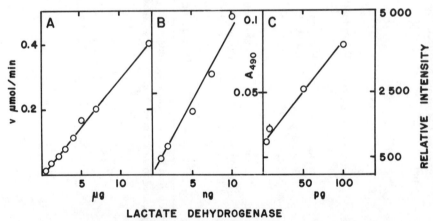

LACTATE DEHYDROGENASE

Figure 12.4.   Assay of lactate dehydrogenase. (*A*) Kinetic assay. The utilization of NADH was measured at 340 nm. The minimum absorbance change measurable was 0.005. The reagent consisted of 4 mM NADH and 10 mM pyruvate in 30 mM phosphate buffer, pH 7.4. The reaction volume was 1 mL and the sample holder was maintained at 30°C. (*B*) Incubation assay. The reagent contained 0.14 mM NAD$^+$, 50 mM lactate, 40 μg/mL INT, and 1 U/mL diaphorase in 0.2 M tris-acetate, pH 8.0. A 0.1-mL sample was incubated with 0.9 mL reagent at 37°C for 1 hr. The INTF formation was determined at 490 nm after extraction into tetrachloroethylene : acetate (1.5 : 1). (*C*) Enzymatic cycling assay. Lactate dehydrogenase was measured by enzymatic cycling of NAD$^+$ produced by a 1-hr incubation of a 20-μL sample with 10 μL containing 10 nmol NADH and 30 nmol pyruvate in 0.09 M phosphate buffer, pH at 37°C. Unreacted NADH was destroyed by treatment with 10 μL 1 N HCl, and the NAD$^+$ was measured by enzymatic cycling.

Table 12.4 Comparison of Laboratory-Measured and Literature Sensitivities for Respiratory Cofactor Bioindicator Determinations

	Limit of Detection		
Bioindicator	This Study	Literature	Reference
Pyridine nucleotides	37.0 pg	15.0 pg	Chi et al., 1978
ATP	100.0 pg	0.2 pg	Howard et al., 1979
FMN	1.0 ng	0.4 ng	Stanley, 1971
Iron Porphyrins			
Hemoglobin	2.0 pg	6.0 pg	Neufeld et al., 1965
E. coli	1000 cells	3400 cells	Oleniacz et al., 1968

Sensitivities of Catalytic Assays in Terms of Number of E. coli Cells

Enzymes and cofactors were extracted from *E. coli* cells (Howard, 1980) and the amounts of selected ones determined. Table 12.5 summarizes the limit of detection found, which is related in turn to the number of *E. coli* cells required to yield the minimum detectable amount of that specific component.

Table 12.5 Sensitivities of Catalytic Assays for Bioindicators in Terms of *E. coli* Cell Number

Bioindicator and *E. coli* Number	Assay Method	Limit of Detection	Minimum Cell Concentration (m/L)
Lactate dehydrogenase	Continuous	0.1 μg	
1.1.1.27	Incubation	1.0 ng	1×10^6
	Cycling	1.0 pg	5×10^5
Alkaline phosphatase	Continuous	20.0 ng	
3.1.3.1	Incubation (1 hr)	1.0 ng	8.5×10^6
	Incubation (24 hr)	0.09 ng	
Catalase	Continuous	0.2 μg	
1.11.1.6	Incubation	1.0 ng	1×10^7
Adenylate kinase	Continuous	3.0 ng	
2.7.4.3	Incubation	0.1 ng	4×10^5
Pyridine nucleotides	Cycling		5×10^6
	NADP$^+$	37.0 pg	—
	NAD$^+$	3.0 ng	—
Adenosine triphosphate	Coupled enzyme	100.0 ng	—
	Cycling	100.0 pg	5×10^5
Flavin mononucleotide	Bacterial luciferase	0.25 ng	—
Iron porphyrin	Luminol	2.0 pg	1×10^3

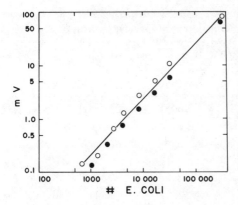

 # E. COLI

Figure 12.5. *E. coli* determination using luminol. Results from two separate experiments are shown. The logarithm of the peak height voltage read from a recorder trace is plotted against the logarithm of the number of *E. coli* cells. The reagent contained 0.25 m*M* luminol, 6.3 m*M* EDTA, and 29 m*M* hydrogen peroxide in 0.75 *N* potassium hydroxide. The reagents were added in a specific order with a 50-μL sample and the peak height of light production determined on a Picolite photometer.

A typical result for the luminol determination with two different *E. coli* samples is shown in Figure 12.5.

ATP Determinations

Firefly Luciferase

The reactions catalyzed by firefly luciferase have already been shown in Figure 12.3. Table 12.6 summarizes the properties of the firefly enzyme. It is

Table 12.6   Properties of Firefly Luciferase *E. coli* No. 1.13.12.7[a]

	Binding Sites	
Site Definition	Other Compounds Bound	Number/1 $\times$ 10^5
Substrate		
MgATP	AMP, ATP	1
Luciferin	Dehydroluciferin, luciferyl adenylate, 2-cyano-6-chlorobenzothiazole, naphthalene dyes.	2
	Dehydroluciferyl adenylate, oxyluciferin	1
Product		
Mg pyrophosphate		1
Oxyluciferin		1
Others		
ATP	AMP	1
Anion		1
Essential groups		
$-$SH	*p*-Chloromercuribenzoate	2

[a] Photinus luciferin: oxygen 4-oxidoreductase (decarboxylating, ATP-hydrolyzing) Euglobulin, crystalline and homogeneous; M_r 100,000; two subunits 50,000 pI 6.2–6.3; optimum temperature 23–25°; optimum pH 7.8; quantum yield 0.9; large conformational change; 25 msec lag, maximum emission 290 msec, light of 590 nm; K_m in Tricine ATP, 125 μM; Mg^{2+}, 315 μM; Luciferin, 125 μM.

a very complex one, with several substrates, several binding sites, large conformation changes, fast and slow steps, and complicated kinetics. Besides, the measurements being made yield a rate directly and are thus unlike those of most enzyme assays. In spite of being so complicated, the enzyme is exquisitely useful for sensitive and specific determinations of ATP.

The commercial availability of various firefly luciferase preparations has enhanced the facility of using this ATP determination method. We have evaluated the commercial luciferase preparations (Webster et al., 1979; Webster et al., 1981b). Table 12.7 summarizes some of the most important findings related to using the reagents. The inherent light production, that is, the background light emission without added ATP, is a significant limiting factor, which can sometimes be partially reduced by incubation of the preparation prior to use. As shown in Table 12.8, the response of most of the commercial preparations is stimulated by the addition of extra luciferin; this factor may be significant for obtaining the maximum sensitivity. Contaminating enzymes such as adenylate kinase and/or nucleoside diphosphate kinase give light production through formation of ATP from nucleotides other than ATP. A measure of purity of the various preparations is given by the light output obtained from 225 μg of ADP (that is a measure of the adenylate kinase).

When we reviewed the literature on firefly luciferase, we found that many

Table 12.7 Firefly Luciferase Preparations

		Light Units				
		Inherent Light Emission		1 ng ATP		225 μg ADP
Type	Manufacturer	(None)	(Luciferin)	(None)	(Luciferin)	(Luciferin)
Crude	Calbiochem	0.1	33	1.8	292	8.9
	Sigma FLE-50	0	44.6	0.8	254	151
Partially purified	Analytical Luminescence	0	0.2	20.6	86.8	0.4
	DuPont	0	2.0	5.0	64	0.1
	LKB	0.6	0.8	41.9	49.4	0
	Lumac HS	0	0.3	0.4	8.8	0
	Lumac PM	0	0.1	0.2	3.4	0
	Packard Picozyme	0	1.8	2.3	23.8	14
	SAI	0	0.2	14.9	77.9	1.9
Crystalline	Boehringer Mannheim	0	0	0	21.8	
	Sigma Type IV	0	2.1	0	43.1	
	This lab	0	1.9	0	166.8	(5.9)[a]

[a] 1 ng ADP.

Table 12.8 Stimulation of Commercial Firefly Luciferase Preparations by Additional Luciferin[a]

Preparation	Fold Stimulation
Crude Calbiochem, Sigma	30–165
Group I Lumit HS, DuPont, Picozyme	10–25

Group II
  Lumit PM, SAI, Firelight, LKB,
  Sigma Type IV, Boehringer
	1–8

[a] The apparent requirement for additional luciferin is also related to the volume of the luciferin–luciferase preparation used (see Webster and Leach, 1980).

Table 12.9 Comparison of the Catalytic Activity of Firefly Luciferase in Various Buffers[a]

Buffer	pK_a	Luciferase Activity Relative to Tricine
Mopso	6.95	1.00
Mops	7.20	0.52
Phosphate	7.21	0.07
Tes	7.50	0.43
Hepes	7.55	0.75
Dipso	7.6	1.06
Tapso	7.7	0.84
Popso	7.85	0.47
Heppso	7.9	1.05
Hepps	8.00	0.54
Tricine	8.15	1.00
Glycinamide	8.20	0.64
Tris	8.30	0.80
Tris-tricine	8.30–8.15	0.90
Bicine	8.35	0.56
Taps	8.40	1.06
Glycylglycine	8.40	0.57

[a] Buffers obtained from Research Organic or Sigma.

different conditions were specified for the assay and that 64% of the more than 100 sampled used arsenate, which inhibits firefly luciferase. We decided that it was necessary to optimize the reaction conditions. Table 12.9 shows the influence of various buffers and buffer combinations on firefly luciferase activity. Additional studies on buffer combinations did not yield any enhancement of luciferase activity. We selected Tricine as the buffer for our assay system. It was shown (Webster et al., 1980a) that the buffer influences the conformation of luciferase, and we concluded that Tricine yields a more active conformation.

Table 12.10 shows the requirements for the luciferase reaction, that is, the effect of addition or of omission of various components on luciferase activity. Through a number of experiments the conditions were optimized (Webster and Leach, 1980). A summary of major factors involved in optimization is shown in Table 12.11.

We have also investigated the stability of the reagents because such factors could be important for the practical utility of the assay. Table 12.12 shows that we defined conditions under which the reagents were sufficiently stable to allow utilization of the technique.

The standard curve for a sensitive determination of ATP is shown in Figure 12.6. We can detect as little as 50 fg of ATP (Webster et al., 1980b).

Table 12.10   Requirements for the Firefly Luciferase Assay[a]

	Light Units (1 ng ATP)
Omissions	
None	20.6
Less 5 mM MgSO$_4$	0
Less 0.5 mM EDTA	12.6
Less 50 μg Luciferin (LH$_2$)	5.2
Less 0.5 mM DTT	13.6
Less 50 μg Bovine serum albumin (BSA)	20.0
Additions	
None	0.1
MgSO$_4$	3.7
LH$_2$	0.5
DTT	1.1
BSA	0.3
MgSO$_4$ + LH$_2$	13.8
MgSO$_4$ + LH$_2$ + DTT	13.6

[a] This experiment was performed with Firelight enzyme.

Table 12.11 Improvements in the Firefly Luciferase Assay

Factor	Change	Fold Increase
Reaction vessel	Glass → Plastic	2
Reaction volume	0.5 mL → 0.2 mL	5
Luciferin	Commercial → Synthetic	2
Buffer	Phosphate → Tricine	3
Additives	None → Complete	10
Enzyme	Crude → Partially purified	4
Light effect on glass reaction vessel (Picolite)	Fluoro light → Dark maintained	$(-9)^a$
Instrument	Initial → Microprocessor	8

[a] This is a decrease in the inherent light emission from glass tubes (the only kind available for the Picolite).

Table 12.12 Stability of Firefly Luciferase and Related Reagents

Substance	Temperature (°C)	Time (days)	Condition	Stability (% activity remaining)
Luciferase	4	10	Tricine	96
		12	Phosphate	86
	−15	4	Repeated freeze-thaw	92
	−15	56	Single freeze-thaw	91
Luciferin	−15	200	N_2 (in dark)	100
ATP	−15	730	<10 μg/mL concentrations	100

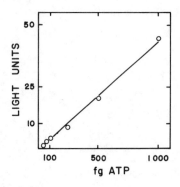

Figure 12.6. Standard curve for ATP determination by firefly luciferase. Light production was determined using a Picolite photometer maintained at 25 ± 0.1°C. The light production without added ATP was subtracted. The reaction system (0.2 mL) contained 0.025 M Tricine buffer, pH 7.8, 5 mM MgSO$_4$, 0.5 mM EDTA, 0.5 mM DTT, 100 μg bovine serum albumin, and 50 μg luciferin. Various amounts of ATP contained in 20-μL samples were injected into the cuvette, and the light emission was counted for 30 sec after a 1-sec delay.

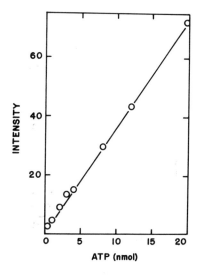

Figure 12.7. Coupled enzyme assay of ATP. ATP was measured by reaction with hexokinase and glucose-6-phosphate to produce NADPH, which was measured fluorometrically (excitation 340 nm, 1-mm slit; emission 455 nm, 2-mm slit). ATP samples (0.1 mL) were mixed with the coupled enzymes reagent (0.9 ml) and incubated 15 min at 37°C.

Other Enzymatic Measures of ATP

Coupled Enzyme Assay

A coupled enzyme reaction of hexokinase and glucose-6-phosphate dehydrogenase was used to measure ATP. The product actually determined fluorometrically was NADPH. Tests of the assay using either ATP or glucose-6-phosphate standards showed the reaction was completed in less than 5 min at 37°C. A typical standard curve is shown in Figure 12.7. The limit of detection was about 0.1 μg of ATP and the range of the assay was up to 10 μg of ATP.

Enzymatic Cycling

The sensitivity of detection of ATP was improved by enzymatic cycling of the NADPH produced by the coupled assay described above. Excess NADP$^+$ in the coupled enzyme assay reagent was destroyed by alkali treatment. A 1-hr incubation of samples at pH 12.5 at 65°C starting with 1×10^{-9} mol reduced the NADP$^+$ content to 5×10^{-13} mol (a 99.99% destruction). The undestroyed NADP$^+$ was the major contributor to the blank fluorescence and largely determined the sensitivity of ATP detection (Howard, 1980).

ATP was measured by the above-cited enzymatic cycling procedures. The limit of detection was 0.1 pg and the range was 0.1–10 pg. Figure 12.8 shows this standard curve. NADP$^+$ standards produced a parallel curve but with a lower blank and thus a greater sensitivity. The limit of detection of ATP by enzymatic cycling was improved 1000-fold over the coupled enzyme assay.

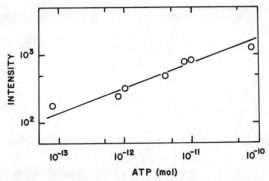

Figure 12.8.   Enzymatic cycling assay of ATP. ATP was measured by enzymatic cycling of the NADPH produced by the coupled enzyme reactions described in Figure 12.7. Samples (20 µL) were incubated with 10 µL of coupled enzymes reagent for 15 min, treated with NaOH, incubated for 1 hr at 65°C, neutralized, and then treated by the standard NADP+ cycling procedure.

Comparison of Enzymatic Cycling and Firefly Luciferase

Table 12.13 compares the results we obtained using the enzymatic cycling with those from firefly luciferase procedures for the determination of ATP. The parameters compared were range, sensitivity, cost, productivity (numbers of samples measured, not counting standards), inhibitors, enzyme, equipment, turn-around time (time required to get a result and repeat an assay), and specificity. The firefly luciferase is clearly as good or better in every category except susceptibility to inhibitors. That fact might be significant with environmental samples. However, internal standards could probably correct for the inhibitions that might be found in either system.

Environmental Samples

Measurement of the ATP present in solutions or in simple experimental material such as bacteria, plant, or animal cells subjected to various treatments is straightforward. Difficulties arise when a marked concentration of the sample is required or when environmental samples containing sparse microbial populations are the subjects of investigation. ATP, being so important metabolically, is a substrate for many enzymes that can destroy it if those enzymes are not inactivated. Since the phosphate groups on ATP are highly charged, ionic interactions with material from the environment could occur. In the structure of DNA the basestacking (interaction of the planar aromatic rings) gives as much stabilization as the hydrogen bonds. The base could stack on a similar aromatic surface in the environmental sample and/ or the ATP could be held by hydrogen bonding. All of these factors should be taken into account in devising extraction procedures.

We tested many of the published extraction procedures to measure recov-

Table 12.13   Comparison of Enzymatic Cycling and Firefly Luciferase Determinations of ATP

	Assay	
Parameter	Firefly Luciferase	Enzymatic Cycling
Range[a]	0.2–100 pmol	0.3–10 pmol
Sensitivity[b]	0.1 fmol (50 fg)	0.1 pmol (50 pg)
Cost[c]	6¢/assay	9.5¢/assay
Productivity	25/hr or 200/8-hr day	96/5 hr or 192/8-hr day
Inhibitors	Metal ions, PO_4^{3-}	None encountered to date[d]
Equipment	Photometer	Fluorometer
Technical competence required	Technicians	Enzymologist
Turnaround time	<30 min	5 hr
Specificity	ATP only	ATP, NADH, NADPH

[a] Useful range of ATP which can be routinely measured

[b] Smallest amount of ATP detected by the assay.

[c] Based on 1979 prices when the experiment was done. Current cost for doing with Firelight luciferase is 32¢/assay.

[d] An optimistic statement.

eries under our laboratory conditions. These results are shown in Table 12.14. An extensive review of extraction methods has been made by Karl (1980).

Experiments on optimization of the extraction are in progress. An indication of the direction of these studies is shown in Table 12.15. Experiment A shows that the presence of sulfuric acid—in addition to phosphoric acid, EDTA, and adenosine—does not increase the yield of ATP extracted from either cells or soil. Because we knew of the inhibitory effect of phosphate on the activity of firefly luciferase in the enzyme assay (see Table 12.9), we had avoided using phosphoric acid or phosphate in extraction. Then we decided that a little of these substances might aid in soil extraction, and upon testing we found that phosphoric acid was a good extracting agent. Experiment B of Table 12.15 shows that both EDTA and adenosine aid in extracting ATP from soil but not from cells. There is an interesting differential effect of various phosphoric acid concentrations on the extraction of ATP from cells and from soil. Decreasing phosphoric acid concentrations yield increased ATP from cells (Experiment C). In contrast, increasing phosphoric acid concentrations increased ATP extracted from soil.

Table 12.14 Comparison of Extraction Procedures

Agent	Conditions	Standard Sample	Reference	Recovery (%)
Buffers (mM)				
Tris (50)	100°C, 2 min	Solution	Holm-Hansen and Booth, 1966	65
Tricine (50)	100°C, 90 sec	E. coli	This lab	93
Tricine (50) + EDTA (2)	100°C, 2 min	E. coli	This lab	75
Tricine (50 + (10)Mg + EDTA (2)	100°C, 2 min	E. coli	Lundin and Thore, 1975	84
Acids (N if not specified)				
HNO$_3$ (0.1)	4°C, 10 min	E. coli	Chappelle et al., 1975	25
H$_2$SO$_4$ (0.2)	4°C, 10 min	E. coli	Lee et al., 1971	50
H$_2$SO$_4$ (0.6)	4°C	Solution	Karl and LaRock, 1975	21
H$_2$SO$_4$ (1.5)	4°C, caution exchange	Soil	Eiland, 1979	12
H$_2$SO$_4$ (1.5)	4°C, Polytron, shaking	Soil	This lab	19
H$_2$SO$_4$ (1.5) + EDTA (20 mM)	4°C, Polytron, shaking	Soil	This lab	55
TCA (510 mM) + EDTA (17 mM)	4°C	E. coli	Lundin and Thore, 1975	48
Organics				
Butanol–octanol 1 : 8	25°C	E. coli	Sharpe et al., 1970	98
DMSO 80%	25°C	E. coli	Mathis and Brown, 1976	31
		Solution	Gabridge and Polisky, 1977	59

Table 12.15 Toward Optimization of Extraction

	Relative ATP Yield	
	Cells	Soil
Experiment A		
1.5N H₂SO₄ + 0.5N H₃PO₄ + EDTA[a] + Adenosine[a]	1	1
1N H₂SO₄ + 0.5N H₃PO₄ + EDTA + Adenosine	0.96	0.86
1N H₂SO₄ + 0.5N H₃PO₄ + EDTA + Adenosine	0.87	0.98
0.5N H₃PO₄ + EDTA + Adenosine	1	1
Experiment B		
0.5N H₃PO₄ + EDTA + Adenosine	1.0	1.0
0.5N H₃PO₄ + EDTA	0.97	0.54
0.5N H₃PO₄ + Adenosine	0.98	0.76
Experiment C		
2N H₃PO₄ + EDTA + Adenosine	0.65	1.0
1N H₃PO₄ + EDTA + Adenosine	0.63	0.90
0.5N H₃PO₄ + EDTA + Adenosine	0.93	0.72
0.1N H₃PO₄ + EDTA + Adenosine	1.0	0.26

[a] 0.02 M EDTA, 2 mg adenosine.

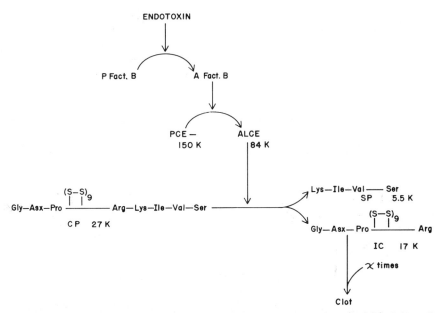

Figure 12.9. The schematic basis of the LPS assay using LAL: Endotoxin, LPS; P Fact. B, pro factor B; A Fact. B, active factor B; PCE, proclotting enzyme; ALCO, activated *Limulus* clotting enzyme; CP, clotting protein; SP, soluble peptide; IC, insoluble coagulogen. The numbers give molecular weight in thousands.

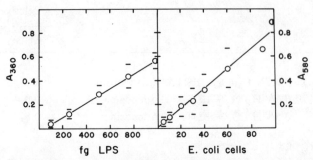

Figure 12.10. The LPS and *E. coli* standard curves using LAL. LAL was reconstituted in sterile water and centrifuged to remove any insoluble material. An aliquot of this LAL (0.2 mL) was added to a 1-mL sample that had been diluted in artificial sea water in 10 m*M* imidazole, pH 7.6. Left graph: After 1 hr incubation at 37°C the absorbance was determined at 360 nm. Right graph: For a more sensitive determination the reaction was carried on as described above and the gelled protein sedimented by centrifugation. The protein was determined by the bromosulfalein procedure.

Limulus Amebocyte Lysate Assay for Lipopolysaccharide—a Cascaded Assay

Some details of the cascaded reaction sequence involved in the activation of the preclotting enzyme by lipopolysaccharides and the ultimate formation of a clot are shown in Figure 12.9. This scheme is based on the work of Tai and Liu (1977); Tai et al. (1977); Liang et al. (1980); and Ohki et al. (1980).

The clotting reaction can be followed by measuring the increase in turbidity at 360 nm as the clot forms or at the end of a defined incubation. This is shown in Figure 12.10. A more sensitive assay is achieved by sedimenting the clot by centrifugation and by determining the protein using the bromosulfalein dye-binding procedure described earlier. Application of this procedure to determine the number of *E. coli* cells present in a sample is also shown in Figure 2.10. As few as five *E. coli* cells can be determined.

DISCUSSION

The ordinary analytical procedures that have been developed for biochemical use are sufficient for most purposes. There have been some improvements in the details of the methods, but the standard methods have been modified little since the middle 1950s. The Lowry protein procedure has been little modified since 1951, the diphenylamine DNA determination has been only slightly improved since 1956, and the orcinol is of 1955 vintage. This stability of basic methods is not due to a lack of ingenuity on the part of biochemists; there have been some improvements in dye-binding protein assays and in fluorometric determinations. This lack of change means that the methods are good and meet the everyday needs of biochemists. These methods also approach the inherent sensitivity of the physical measurements made in the currently available instruments. If marked improvements are made, they will have to be innovative departures from the current art.

In most cases the amount of sample available is not limiting, and the standard methods are therefore sufficient. For example, in most microbial experiments the amplification of the genetic information and its products brought about a growth sufficient in a relatively short time to yield enough material for determination. Besides, less sensitive methods are not influenced so much by extraneous materials as are extremely sensitive methods. The increased population size required for measurement by the standard methods, in fact, adds a degree of reliability to the measurements.

The analysis of materials available only in small amounts requires approaches different from the ordinary analytical procedures described above. With microorganisms the usual procedures involve cultivation in the proper medium, and when appropriate conditions are known, it is easy to obtain a sufficient quantity of cells for analysis. The difficulty arises from environmental samples derived from sources where proper cultural conditions are not known and where the proper medium has not been established; this is the situation with subsurface environmental samples (McNabb and Dunlap, 1975).

There are certain aspects of analytical biochemistry that apply directly. For example, Rotman (1961) developed a microscopic, fluorometric technique that allows the measurement of the activity of a single enzyme molecule. The enzyme and substrate were sprayed in droplets 0.1–40 μm in diameter onto a microscope slide coated with a silicone. The slide was covered with a cover glass and incubated 15 hr at 36°C; then the fluorescence was determined with a Zeiss microscope and an Aminco photomultiplier microphotometer over an area 20 μm in diameter. Similar techniques were applied by Collins et al. (1964) to the estimation of penicillinase in single bacterial cells. More recent optical developments such as the use of fiber optics (Vurek and Bowman, 1969) enable measurements in submicroliter samples. While these techniques are important for investigating specific problems in the research laboratory, they are not widely applicable to routine laboratory investigations.

We have already discussed and reviewed at some length the literature related to biochemical indicators of subsurface pollution (Dermer et al., 1980) and will not repeat those discussions and arguments here.

The developments that have occurred recently in analytical technique are based on (1) fluorometric techniques, (2) uses of modern physical developments such as lasers, and (3) cascade, cyclic, or catalytic procedures. While the application of cascade and cyclic assay procedures is of fairly recent vintage, the concepts have been around as long as life itself since these techniques are those normally functional in the biochemistry practiced by living systems.

Male and female fireflies have been using bioluminescence to find each other for eons, while it has been used by analysts only during the last quarter of a century. Commercial developments of reagents and instrumentation has made ATP analysis using firefly luciferase a broadly applicable technique throughout many disciplines. Our studies on the optimization of the assay should aid in increasing the applicability and dependability of the firefly luciferase ATP assay.

Enzyme analysis and enzymatic cycling procedures are useful in the research laboratory setting. Most require sophisticated equipment and technical procedures found only in laboratories capable of biochemical research. These procedures have been discussed in detail by Lowry and Passonneau (1972).

Lust et al. (1981) recently described enzymatic methods for the determination of ATP and other adenine nucleotides and p-creatine in biological samples. Their methods are similar to those that we used. Theirs included (1) a direct fluorometric procedure for the measurement of 0.1–10 nmol using hexokinase and glucose-6-P-dehydrogenase, (2) an enzymatic cycling procedure with a sensitivity of 1–50 pmol, and (3) the measurement of light emission in the luciferin–luciferase system with a sensitivity of 0.1–80 pmol.

Figure 12.7 shows that we can measure approximately the same range in the coupled assay, 0.2–20 nmol. The enzymatic cycling procedure yielded an effective range of 0.1 pmol to 10 nmol (Figure 12.8). Therefore, our procedures yield results equivalent to those obtained in the laboratory of a recognized leader in the application of enzymatic cycling procedures. Our ATP analysis using firefly luciferase has been optimized to yield an effective range of 0.1–2 fmol.

The utility of using light-emitting reactions for analysis has been stressed recently (DeLuca, 1978; Schram and Stanley, 1979; DeLuca and McElroy, 1981). Since many examples are presented in those references, we will not discuss the general applications further.

The sensitivity of the *Limulus* amebocyte lysate assay for lipopolysaccharides is great because of the cascaded nature of the reactions involved. The application of this assay to subsurface material samples will depend upon the development of extraction procedures.

CONCLUSIONS

The standard assays used in biochemical laboratories for DNA, RNA, protein, and organic phosphate, while excellent for most determinations, do not possess the sensitivity required with sparsely populated environmental samples. Assays with sufficient sensitivity for enzymes and cofactors could be developed using enzymatic cycling procedures. However, the sophisticated equipment and biochemical expertise required for these procedures limit their applicability, in general, to research laboratories. Assays based on bioluminescence, chemiluminescence, and cascaded enzymatic reactions are presently applicable to many environmental samples. By further optimization of conditions and additional study, the range of application of these techniques could be increased. Some simplification and better definition of parameters is also required.

RECOMMENDATIONS

This research has demonstrated that the inherent sensitivity of bioluminescent, chemiluminescent, and cascaded (or cycled) reactions is great enough for application to the sparse populations of organisms expected in subsurface environmental samples. Further work is required on optimization of assays, perfection of extraction procedures, and elimination of or correction for inhibitors found in environmental samples. Several different environmental samples must be evaluated to allow selection of conditions that are widely applicable.

ACKNOWLEDGMENTS

This research was supported in part by EPA Grant R 804613, the National Center of Ground Water Research, and Oklahoma Agricultural Experiment Station Projects No. 1640 and No. 1806. This is manuscript J-4042 of the Oklahoma Agricultural Experiment Station and was critically read by Drs. O. C. Dermer and E. C. Nelson. Taken in part from the thesis of J. L. Howard, submitted to the Oklahoma State University Graduate College (1980) in partial fulfillment of the Ph.D. requirements.

REFERENCES

Abraham, G. N., Scaletta, C., and Vaughan, J. H. (1972). Modified diphenylamine reaction for increased sensitivity. *Anal. Biochem.* **49**:547–549.

Associates of Cape Cod, Inc. (1981). *Limulus* Amebocyte Lysate (Pyrotell) for the Detection and Quantitation of Endotoxins. Associates of Cape Cod, Inc., Woods Hole, MA.

Atkinson, D. E. (1977). *Cellular Energy Metabolism and Its Regulation.* Academic Press, New York.

Avery, O. T., MacLeod, C. M., and McCarty, M. (1944). Studies on the chemical nature of the substance inducing transformation of Pneumococcal Types: Induction of transformation by a desoxyribonucleic acid fraction isolated from Pneumococcus Type III. *J. Exp. Med.* **79**:137–158.

Boer, G. J. (1975). A simplified microassay of DNA and RNA using ethidium bromide. *Anal. Biochem.* **65**:225–231.

Brolin, S. E., Borglund, E., and Agren, A. (1979). Photokinetic microassay of adenylate kinase (EC 2.7.4.3) using the firefly luciferase reaction. *J. Biochem. Biophys. Methods* **1**:163–170.

Butcher, E. C. and Lowry, O. H. (1976). Measurement of nanogram quantities of protein by hydrolysis followed by reaction with ortho phthalaldehyde or determination of glutamate. *Anal. Biochem.* **76**:502–523.

Cattolico, R. A. and Gibbs, S. P. (1975). Rapid filter method for the microfluorometric analysis of DNA. *Anal. Biochem.* **69**:572–582.

Ceriotti, G. (1955). Determination of nucleic acids in animal tissues. *J. Biol. Chem.* **214**:59–70.

Chappelle, E. W., Picciolo, G. L., Curtis, C. A., Knust, E. A., Nibley, D. A., and Vance, R. B. (1975). Laboratory Procedures Manual for the Firefly Luciferase Assay for Adenosine Triphosphate (ATP). (NASA-TM-X-70926) X-726-75-1 X75-28159. Goddard Space Flight Center, Greenbelt, MD.

Chi, M. M.-Y., Lowry, C. V., and Lowry, O. H. (1978). An improved enzymatic cycle for nicotinamide-adenine dinucleotide phosphate. *Anal. Biochem.* **89**:119–129.

Cohen, E. (Ed.) (1979). Biomedical Applications of the Horseshoe Crab (*Limulidae*). In: *Progress in Clinical and Biological Research,* Vol. 29. Alan R. Liss, New York.

Collins, J. F., Mason, D. B., and Perkins, W. J. (1964). A microphotometric method for the estimation of penicillinase in single bacteria. *J. Gen. Microbiol.* **34**:353–362.

Curl, H., Jr. and Sandberg, J. (1961). The measurement of dehydrogenase activity in marine organisms. *J. Mar. Res.* **19**:123–138.

DeLuca, M. A. (Ed.) (1978). Bioluminescence and Chemiluminescence. In: *Methods in Enzymology,* Vol. 57. Academic Press, New York.

DeLuca, M. A. and McElroy, W. D. (Eds.) (1981). *Bioluminescence and Chemiluminescence.* Academic Press, New York.

Dermer, O. C., Curtis, V. S., and Leach, F. R. (1980). *Biochemical Indicators of Subsurface Pollution.* Ann Arbor Science, Ann Arbor, MI.

Doellgast, G. J. (1977). Immunoquantitation of human placental alkaline phosphatase using radial immunodiffusion. *Anal. Biochem.* **82**:278–288.

Eiland, F. (1979). An improved method for determination of adenosine triphosphate (ATP) in soil. *Soil Biol. Biochem.* **11**:31–35.

El-Hamalawi, A.-R., Thompson, J. S., and Barker, G. R. (1975). The fluorometric determination of nucleic acids in pea seeds by use of ethidium bromide complexes. *Anal. Biochem.* **67**:384–391.

Environmental Protection Agency (1974). *Methods for Chemical Analysis of Water and Waste.* Washington, D.C.

Ewetz, L. and Thore, A. (1976). Factors affecting the specificity of the luminol reaction with hematin compounds. *Anal. Biochem.* **71**:564–570.

Gabridge, M. D. and Polisky, R. B. (1977). Intracellular levels of adenosine triphosphate in hamster trachea organ cultures exposed to *Mycoplasma pneumoniae* cells or membranes. *In Vitro* **13**:510–515.

Going, J., Wenzel, S., and Thompson, J. (1975). Spectrophotometric determination of phosphate by extraction of reduced molybdoantimonylphosphoric acid with acetophenone-chloroform. *Microchem. J.* **20:**126–131.

Holm-Hansen, O. and Booth, C. R. (1966). The measurement of adenosine triphosphate in the ocean and its ecological significance. *Limnol. Oceanogr.* **11:**510–519.

Howard, J. L. (1980). Development of Sensitive Assays for Bioindicators in Ground Water. Ph.D. Thesis, Oklahoma State University, Stillwater, OK.

Howard, J. L., Webster, J. J., and Leach, F. R. (1979). Comparison of the luciferin-luciferase and enzymatic cycling procedures for the measurement of picogram amounts of ATP. *Fed. Proc.* **38:**670.

Howell, B. F., McCune, S., and Schaffer, R. (1977). Influence of salts on Michaelis-constant values for NADH. *Clin. Chem.* **23:**2231–2237.

Kalckar, H. M. (1941). The nature of energetic coupling in biological synthesis. *Chem. Rev.* **28:**71–178.

Kapuscinski, J. and Skoczylas, B. (1977). Simple and rapid fluorimetric method for DNA microassay. *Anal. Biochem.* **83:**252–257.

Karl, D. M. (1980). Cellular nucleotide measurements and applications in microbial ecology *Microbiol. Rev.* **44:**739–796.

Karl, D. M. and LaRock, P. A. (1975). Adenosine triphosphate measurement in soil and marine sediments. *J. Fish. Res. Board Can.* **32:**599–607.

Karsten, U. and Wollenberger, A. (1977). Improvements in the ethidium bromide method for direct fluorometric estimation of DNA and RNA in cell and tissue homogenates. *Anal. Biochem.* **77:**464–470.

Kutchai, H. and Geddis, L. M. (1977). Determination of protein in red cell membrane preparations by *o*-phthalaldehyde fluorescence. *Anal. Biochem.* **77:**315–319.

Lee, C. C., Harris, R. F., Williams, J. D., Armstrong, D. E., and Syers, J. K. (1971). Adenosine triphosphate in lake sediments. I. Determination. *Soil Sci. Soc. Am. Proc.* **35:**82–86.

LePecq, J.-B. and Paoletti, C. (1966). A new fluorometric method for RNA and DNA determination. *Anal. Biochem.* **17:**100–107.

Liang, S.-M., Sakmar, T. P., and Liu, T.-Y. (1980). Studies on Limulus amoebocyte lysate III: Purification of an endotoxin-binding protein from *Limulus* amoebocyte membranes. *J. Biol. Chem.* **255:**5586–5590.

Lipmann, F. (1941). Metabolic generation and utilization of phosphate bond energy. In: F. F. Nord and C. H. Werkman (Eds.), *Advances in Enzymology,* Vol. 1. Interscience, New York, pp. 99–162.

Lowry, O. H. and Passonneau, J. V. (1972). *A Flexible System of Enzymatic Analysis.* Academic Press, New York.

Lowry, O. H., Passonneau, J. V., Schultz, D. W., and Rock, M. K. (1961). The measurement of pyridine nucleotides by enzymatic cycling. *J. Biol. Chem.* **236:**2746–2755.

Lowry, O. H., Rosebrough, N. J., Farr, A. L., and Randall, R. J. (1951). Protein measurement with the folin phenol reagent. *J. Biol. Chem.* **193:**265–275.

Lundin, A. and Thore, A. (1975). Comparison of methods for extraction of bacterial adenine nucleotides determined by firefly assay. *Appl. Microbiol.* **30**:713–721.

Lust, W. D., Feussner, G. K., Barbehenn, E. K., and Passonneau, J. V. (1981). The enzymatic measurement of adenine nucleotides and *p*-creatine in picomole amounts. *Anal. Biochem.* **110**:258–266.

Malamy, M. H. and Horecker, B. L. (1964). Purification and crystallization of the alkaline phosphatase of *Escherichia coli. Biochemistry* **3**:1893–1897.

Mathis, R. R. and Brown, O. R. (1976). ATP concentration in *Escherichia coli* during oxygen toxicity. *Biochim. Biophys. Acta* **440**:723–732.

McGuire, J., Taylor, P., and Greene, L. A. (1977). A modified bromosulfalein assay for the quantitative estimation of protein. *Anal. Biochem.* **83**:75–81.

McKnight, G. S. (1977). A colorimetric method for the determination of submicrogram quantities of protein. *Anal. Biochem.* **78**:86–92.

McNabb, J. F. and Dunlap, W. J. (1975). Subsurface biological activity in relation to ground-water pollution. *Ground Water* **13**:33–44.

Neufeld, H. A., Conklin, C. J., and Towner, R. D. (1965). Chemiluminescence of luminol in the presence of hematin compounds. *Anal. Biochem.* **12**:303–309.

Ohki, M., Nakamura, T., Morita, T., and Iwanaga, S. (1980). A new endotoxin sensitive factor associated with hemolymph coagulation system of horseshoe crab (*Limulidae*). *FEBS Lett.* **120**:217–220.

Okrend, H., Thomas, R. R., Deming, J. W., Chappelle, E. W., and Picciolo, G. L. (1977). Methodology for photobacteria luciferase FMN assay of bacterial levels. In: G. A. Borun (Ed.), *Second Bi-Annual ATP Methodology Symposium.* SAI Technology Co., San Diego, pp. 525–546.

Oleniacz, W. S., Pisano, M. A., Rosenfeld, M. H., and Elgart, R. L. (1968). Chemiluminescent method for detecting microorganisms in water. *Environ. Sci. Technol.* **2**:1030–1033.

Rasmussen, H. N. (1978). Preparation of partially purified firefly luciferase suitable for coupled assays. In: M. A. DeLuca (Editor), *Methods in Enzymology*, Vol. 57. Academic Press, New York, pp. 28–36.

Rotman, B. (1961). Measurement of activity of single molecules of β-D-Galactosidase. *Proc. Natl. Acad. Sci. USA* **47**:1981–1991.

Schram, E. and Stanley, P. (Eds.) (1978). *International Symposium on Analytical Applications of Bioluminescence and Chemiluminescence.* State Publishing and Printing, Westlake Village, CA.

Schultz, R. M. and Wassarman, P. M. (1977). [³H]Dansyl chloride: A useful reagent for the quantitation and molecular weight determination of nanogram amounts of protein. *Anal. Biochem.* **77**:25–32.

Searle, N. D. (1975). Applications of chemiluminescence to bacterial analysis. In: E. W. Chappelle and G. L. Picciolo (Eds.), *Analytical Applications of Bioluminescence and Chemiluminescence.* NASA SP-388, National Aeronautics and Space Administration, Washington, D.C., pp. 95–103.

Setaro, F. and Morley, C. D. (1977). A rapid colorimetric assay for DNA. *Anal. Biochem.* **81**:467–471.

Sharpe, A. N., Woodrow, M. N., and Jackson, A. K. (1970). Adenosinetriphosphate (ATP) levels in foods contaminated by bacteria. *J. Appl. Bacteriol.* **33**:758–767.

Sottocasa, G. L., Kuylenstierna, B., Ernster, L., and Bergstrand, A. (1967). Separation and some enzymatic properties of the inner and outer membranes of rat liver mitochondria. In: R. W. Estabrooks and M. E. Pullman (Eds.), *Methods in Enzymology*, Vol 10. Academic Press, New York, pp. 448–463.

Stanley, P. E. (1971). Determination of subpicomole levels of NADH and FMN using bacterial luciferase and the liquid scintillation spectrometer. *Anal. Biochem.* **39:**441–453.

Tai, J. Y. and Liu, T.-Y. (1977). Studies on *Limulus* amoebocyte lysate, isolation of pro-clotting enzyme. *J. Biol. Chem.* **252:**2178–2181.

Tai, J. Y., Seid, R. C., Jr., Huhn, R. D., and Liu, T.-Y. (1977). Studies on *Limulus* amoebocyte lysate II. Purification of the coagulogen and the mechanism of clotting. *J. Biol. Chem.* **252:**4773–4776.

Thomas, R. R., Picciolo, G. L., Chappelle, E. W., Jeffers, E. L., and Taylor, R. E. (1977). Use of the luminol assay for the determination of bacterial iron porphyrins: Flow techniques for wastewater effluent. In: G. A. Borun (Ed.), *Second Bi-Annual ATP Methodology Symposium*. SAI Technology Co., San Diego, pp. 569–599.

Torriani, A. (1968). Alkaline phosphatase of *Escherichia coli*. In: L. Grossman and K. Moldave (Eds.), *Methods in Enzymology*, Vol. 12B. Academic Press, New York, pp. 212–218.

Vurek, G. G. and Bowman, R. L. (1969). Fiber-optic colorimeter for submicroliter samples. *Anal. Biochem.* **29:**238–247.

Webster, J. J., Chang, J. C., Howard, J. L., and Leach, F. R. (1979). Some characteristics of commercially available firefly luciferase preparations. *J. Appl. Biochem.* **1:**471–478.

Webster, J. J., Chang, J. C., Howard, J. L., and Leach, F. R. (1981b). A comparison of commercial firefly luciferases. In: M. A. DeLuca and W. D. McElroy (Eds.), *Bioluminescence and Chemiluminescence*. Academic Press, New York, pp. 755–762.

Webster, J. J., Chang, J. C., and Leach, F. R. (1980b). Sensitivity of ATP determination. *J. Appl. Biochem.* **2:**516–517.

Webster, J. J., Chang, J. C., Manley, E. R., Spivey, H. O., and Leach, F. R. (1980a). Buffer effects on ATP analysis by firefly luciferase. *Anal. Biochem.* **106:**7–11.

Webster, J. J. and Leach, F. R. (1980). Optimization of the firefly luciferase assay for ATP. *J. Appl. Biochem.* **2:**469–479.

Webster, J. J., Taylor, P. B., and Leach, F. R. (1981a). Studies on the sensitivity of ATP determination using commercial firefly luciferase. In: M. A. DeLuca and W. D. McElroy (Eds.), *Bioluminescence and Chemiluminescence*. Academic Press, New York, pp. 497–504.

Weetall, H. H., Weliky, N., and Vango, S. P. (1965). Detection of microorganisms in soil by their catalytic activity. *Nature* **206:**1019–1027.

Wieser, W. and Zech, M. (1976). Dehydrogenases as tools in the study of marine sediments. *Mar. Biol.* **36:**113–122.

Worthington Biochemical Corp. (1972). Worthington Enzyme Manual. Freehold, NJ pp. 41–42.

13

METHODS FOR THE ASSESSMENT OF GROUND WATER POLLUTION POTENTIAL

L. W. Canter

National Center for Ground Water Research
University of Oklahoma
Norman, Oklahoma

There are multiple sources of ground water pollution within the United States, including illegal dumps, sanitary and chemical landfills, septic tank systems, municipal and industrial waste water ponds, and oil and gas field activities. With increasing emphasis on ground water quality protection, there is a growing need to evaluate man-made pollution sources systematically. Site selection is continuing for activities that could create future pollution sources if appropriate control measures are not implemented. Ground water pollution potential should be an important consideration in that selection process. Additionally, there is a critical need for the planned monitoring of existing and potential sources. Since the costs for extensive monitoring can be prohibitive, focus must be given to those areas, or "hot spots," with greater potential for ground water pollution.

The objectives of this paper are threefold: (1) to present a brief summary of three methodologies for empirically assessing ground water pollution potential; (2) to describe the application of one existing methodology to waste water ponds, septic tank systems, and landfills; and (3) to describe the development and application of two methodologies for oil and gas field activities. Empirical assessment methodologies are intended to be simple approaches for development of numerical indices of the ground water pollution potential of man's activities. The existing methodology and the two new methodologies have been applied to pollutant sources in the Garber-Wellington aquifer area in central Oklahoma. The two new methodologies are related to evaluation of oil and gas wells and dry holes, and evaluation of salt water disposal wells and waterflooding operations.

EMPIRICAL ASSESSMENT METHODOLOGIES

Table 13.1 summarizes the features of empirical assessment methodologies for evaluating the ground water pollution potential of waste water ponds and sanitary and chemical landfills. Methodologies typically focus on a numerical index, with larger numbers used to denote greater ground water pollution potential; however, some methodologies encourage the grouping or ranking of pollution potential without extensive use of numerical indicators. Methodologies typically consider several factors for evaluation, with the number, type, and importance weighting of those factors varying from methodology to methodology. Methodologies also include descriptions of measurement techniques for the factors and provide information on the scaling of importance weights (points). Final integration of information may involve summation and/or multiplication of factor scores.

Empirical assessment methodologies should be utilized for relative evaluations and not as absolute considerations of ground water pollution. Considerable professional judgment is needed in the interpretation of results. However, they do represent approaches that can be used, based on minimal data input, to provide a structured procedure for source evaluation, site selection, and monitoring planning. Three empirical assessment methods will be discussed—surface impoundment assessment (U.S. Environmental Protection Agency, 1978), waste–soil–site interaction matrix (Phillips et al., 1977), and site rating system (Hagerty et al., 1973). The surface impoundment assessment method was applied to waste water ponds, septic tank systems, and landfills in the Garber-Wellington study area; accordingly, more information will be presented on this method.

Table 13.1   Summary Features of Empirical Assessment Methodologies

Numerical indices of ground water pollution potential

Multiple factors and relative importance weighting

Measurement techniques for factors and scaling (scoring) of importance weights

Indices based on summation of factor scores or products of scores

Need for careful interpretation with professional judgment

Table 13.2 Rating of the Unsaturated Zone in the SIA Method[a]

Earth Material Category	I	II	III	IV	V	VI
Unconsolidated rock	Gravel, medium to coarse sand	Fine to very fine sand	Sand with <15% clay, silt	Sand with >15% but ≤50% clay	Clay with <50% sand	Clay
Consolidated rock	Cavernous or fractured limestone, evaporites, basalt lava fault zones	Fractured igneous and metamorphic (except lava) sandstone (poorly cemented)	Sandstone (moderately cemented) fractured shale	Sandstone (well cemented)	Siltstone	Unfractured shale, igneous and metamorphic rocks
Representative permeability						
gpd/ft²—	>200	2–200	0.2–2	<0.2	<0.02	<0.002
cm/sec—	$>10^2$	10^4–10^2	10^5–10^4	$<10^5$	$<10^6$	$<10^7$
Rating Matrix						
Thickness of the unsaturated zone (m)						
>30	9A	6B	4C	2D	0E	0F
>10 ≤ 30	9B	7B	5C	3D	1E	0G
>3 ≤ 10	9C	8B	6C	4D	2E	0H
>1 ≤ 3	9D	9F	7C	5D	3E	1F
>0 ≤ 1	9E	9G	9H	9I	9J	9K

[a] U.S. Environmental Protection Agency, 1978.

272

Surface Impoundment Assessment

The surface impoundment assessment (SIA) method is based on work by LeGrand (1964). The method was developed for evaluating waste water ponds (U.S. Environmental Protection Agency, 1978), and it yields a sum index with numerical values ranging from 1 to 29. The index is based on four factors—the unsaturated zone, the availability of ground water (saturated zone), ground water quality, and the hazard potential of the waste material. Numerical values for the unsaturated zone range from 0 to 9, for the availability of ground water from 0 to 6, for ground water quality from 0 to 5, and for hazard potential of waste from 1 to 9.

The unsaturated zone rating is based on earth material characteristics as well as zone thickness. Table 13.2 provides the basis for the evaluation, with the categories of earth materials based primarily upon permeability and secondarily upon sorption character. In a particular locality where hydrologically dissimilar layers exist, the waste is more likely to move through the more permeable zones and avoid the impermeable zones; in such cases the earth material should be rated as the more permeable of the two or more existing layers.

Table 13.3   Rating Ground Water Availability in the SIA Method[a]

Earth Material Category	I	II	III
Unconsolidated rock	Gravel or sand	Sand with $\leq 50\%$ clay	Clay with $<50\%$ sand
Consolidated rock	Cavernous or fractured rock, poorly cemented sandstone, fault zones	Moderately to well cemented sandstone, fractured shale	Siltstone, unfractured shale and other impervious rock
Representative permeability			
gpd/ft^2	>2	0.02–2	<0.02
cm/sec	$>10^{-4}$	10^{-6}–10^{-4}	$<10^{-6}$
Rating Matrix			
Thickness of saturated zone (m)			
≥ 30	6A	4C	2E
3–30	5A	3C	1E
≤ 3	3A	1C	0E

[a] U.S. Environmental Protection Agency, 1978.

The availability of ground water factor considers the ability of the aquifer to transmit ground water and is thus dependent upon aquifer permeability and saturated thickness. Table 13.3 provides information on the types of earth material and thicknesses for various ratings. The letters accompanying the rating matrices in Tables 13.2 and 13.3 are for the purpose of identifying the origin of the rating and documenting the process.

The ground water quality factor is based upon criteria associated with the Underground Injection Control program of the U.S. Environmental Protection Agency. Table 13.4 contains information on the rating (U.S. Environmental Protection Agency, 1978). If ground water has high total dissolved solids (TDS), the rating is lower since potential ground water uses would be limited. If the ground water is serving as a drinking water supply, the rating is 5 regardless of the TDS concentration.

The waste hazard potential factor is associated with the potential for causing harm to human health. Examples of hazard potential ratings of waste materials classified by source are in Table 13.5. The ratings consider toxicity, mobility, persistence, volume, and concentration. Table 13.5 includes a range of ratings for several sources, the rating at the bottom of the range being reserved for cases in which there is considerable pretreatment. The waste hazard potential rating based on wastes classified by type can also be used (U.S. Environmental Protection Agency, 1978).

Summation of the ratings for each of the four factors in the SIA method yields an overall evaluation for the source. An additional consideration is the degree of confidence of the investigator as well as data availability for the specific site. An overall evaluation of the final rating is suggested, with the ratings being either A, B, or C. A rating of A denotes high confidence and is given when the data used has been site-specific. Ratings of B and C denote moderate and low confidence, respectively, and are given when data have been obtained from a generalized source or extrapolated from adjacent sites.

Table 13.4   Rating Ground Water Quality in the SIA Method[a]

Rating	Quality
5	≤500 mg/L TDS or a current drinking water source
4	>500–≤1000 mg/L TDS
3	>1000–≤3000 mg/L TDS
2	>3000–≤10,000 mg/L TDS
1	>10,000 mg/L TDS
0	No ground water present

[a] U.S. Environmental Protection Agency, 1978.

Table 13.5   Examples of Contaminant Hazard Potential Ratings of Waste Classified by Source in the SIA Method[a]

SIC Number	Description of Waste Source	Hazard Potential Initial Rating
02	Agricultural Production—Livestock	
021	Livestock, except dairy, poultry and animal specialties	3 (5 for feedlots)
024	Dairy farms	4
025	Poultry and eggs	4
13	Oil and gas extraction	
131	Crude petroleum and natural gas	7
132	Natural gas liquids	7
1381	Drilling oil and gas wells	6
20	Food and kindred products	
201	Meat products	3
202	Dairy products	2
203	Canned and preserved fruits and vegetables	4
204	Grain mill products	2
28	Chemicals and allied products	
2812	Alkalies and chlorine	7–9
2813	Industrial gases	
2816	Inorganic pigments	3–8
2819	Industrial inorganic chemicals, not elsewhere classified	3–9
29	Petroleum refining and related industries	
291	Petroleum refining	8
295	Paving and roofing materials	7
299	Miscellaneous products of petroleum and coal	7

[a] U.S. Environmental Protection Agency, 1978.

Waste–Soil–Site Interaction Matrix

The waste–soil–site interaction matrix was developed for assessing industrial solid or liquid waste disposal on land (Phillips et al., 1977). The method involves summation of the products of various waste–soil–site considerations, with the resultant numerical values ranging from 45 to 4830. The methodology includes 10 factors related to the waste and seven factors associated with the site of potential waste application. Table 13.6 contains a

Table 13.6 Waste Factors in Waste–Soil–Site Interaction Matrix[a]

Group	Factor
Effects	Human toxicity (Ht)—ability of a substance to produce injury once it reaches a susceptible site in or on the body. Based on severity of effect, all substances are grouped into those with no toxicity, slight toxicity, moderate toxicity, and severe toxicity. The Ht values range from 0 (no toxicity) to 10 (maximum toxicity).
	Ground water toxicity (Gt)—related to minimum concentration of waste substance in ground water that would cause damage or injury to humans, animals, or plants. The Gt value is a function of the lowest concentration that would cause damage or injury to any portion of the ecosystem; the Gt values range from 0 (nontoxic) to 10 (very toxic).
	Disease transmission potential (Dp)—considers mode of disease contraction, pathogen life state, and ability of the pathogen to survive. Disease contraction includes direct contact, infection through open wounds, and infection by vectors (usually insects). Pathogen life state includes pathogenic microorganisms with more than one life state (virus and fungi), one life state (vegetative pathogens), and those which cannot survive outside their host. The ability of the pathogen to survive includes survival in air, water, and soil environments. The Dp values range from 0 (no effect) to 10 (maximum effect).
Behavioral performance	Chemical persistence (Cp)—related to the chemical stability of toxic components in the waste. Consideration is given to the concentration of toxic components after 1-day and 6-day contact with soil from potential disposal site. The Cp values range from 1 (very unstable toxic component) to 5 (very stable toxic component).
	Biological persistence (Bp)—related to the biodegradability of the waste as determined by biochemical oxygen demand (BOD) and theoretical oxygen demand (TOD). The Bp values range from 1 (very biodegradable) to 4 (nonbiodegradable).
	Sorption (So)—related to the mobility of the waste in the soil environment. Consideration is given to initial concentration of toxic component(s) in waste as well as 1-day concentration following mixing with soil from potential disposal site. The So values range from 1 (very strong sorption) to 10 (no sorption).
Behavioral properties	Viscosity (Vi)—related to the flow of the waste toward the water table. Consideration is given to the waste viscosity measured at the average maximum temperature of the site during its proposed months of use. The Vi values range from 1 (very viscous) to 5 (viscosity of water).

Table 13.6   (*Continued*)

Group	Factor
	Solubility (Sy)—along with sorption, solubility relates to the mobility of the waste in the soil environment. Waste solubility is measured in pure water at 25°C and pH of 7. The Sy values range from 1 (low solubility) to 5 (very soluble). In case the waste is miscible with water, Sy is equal to 5.
	Acidity/basicity (Ab)—considers the influence of acidic or basic wastes on the solubility of various metals. Acidic wastes tend to solubilize metals, whereas basic wastes tend to immobilize metals through precipitation. The Ab values range from 0 (no effect) to 5 (maximum effect).
Capacity rate	Waste application rate (Ar)—related to the volumetric application rate of the waste at the site, the sorption characteristics of the site (NS discussed in Table 13.7), and the concentration of toxic component(s) in the waste. The Ar values range from 1 (low volumetric application rate of a low concentration waste to a site having high sorptive properties) to 10 (high volumetric application rate of a high concentration waste to a site having low sorptive properties).

[a] Phillips et al. (1977).

description of the waste factors and their numerical scores, and Table 13.7 lists the soil–site factors with their associated weights. Table 13.8 represents a sample interaction matrix resulting from this methodology, with the total summation of the products being 990. Ten classes used for interpretation are as follows—Class 1 (45–100 points), Class 2 (101–200), Class 3 (201–300), Class 4 (301–400), Class 5 (401–500), Class 6 (501–750), Class 7 (751–1000), Class 8 (1001–1500), Class 9 (1501–2500), and Class 10 (greater than 2500). Classes 1–5 are considered acceptable, and classes 6–10 unacceptable.

Site Rating System

A methodology for chemical landfill site selection/evaluation was developed by Hagerty et al. (1973). The methodology also includes the rating of waste materials, with the factors being human toxicity, ground water toxicity, disease transmission potential, biological persistence, and waste mobility. The site rating system considers the influence of 10 factors described in Table 13.9. Infiltration potential, bottom leakage potential, and ground water velocity are factors affecting waste transmission and range in point scores from 0 to 20. Filtering capacity and adsorptive capacity affect waste trans-

Table 13.7 Soil-Site Factors in Waste–Soil–Site Interaction Matrix[a]

Group	Factor
Soil	Permeability (NP)—relates to permeability of site materials. Clay is considered to have poor permeability, fine sand moderate permeability, and coarse sand and gravel good permeability. The NP values range from 2.5 (low permeability) to 10 (maximum permeability).
	Sorption (NS)—relates to sorption characteristics of site materials. The NS values range from 1 (high sorption) to 10 (low sorption).
Hydrology	Water table (NWT)—considers the fluctuating boundary free water level and its depth. The zone of aeration occurs above the water table and is important to oxidative degradation and sorption. The NWT values range from 1 (deep water table) to 10 (water table near surface).
	Gradient (NG)—relates to the effect of the hydraulic gradient on both the direction and rate of flow of ground water. The NG values range from 1 (gradient away from the disposal site in a desirable direction) to 10 (gradient toward point of water use).
	Infiltration (NI)—relates to the tendency of water to enter the surface of a waste-disposal site. Involves consideration of the maximum rate at which a soil can absorb precipitation or water additions. A site with a large amount of infiltration will have greater ground water pollution potential. The NI values range from 1 (minimum infiltration) to 10 (maximum infiltration).
Site	Distance (ND)—relates to the distance from the disposal site to the nearest point of water use. The greater the distance, the less chance of contamination because waste dilution, sorption, and degradation increase with distance. The ND values range from 1 (long distance from disposal site to use site) to 10 (disposal site close to use site).
	Thickness of porous layer (NT)—refers to porous layer at the disposal site. The NT values range from 1 (about 100 ft or more of depth) to 10 (about 10 ft of depth).

[a] Phillips et al. (1977).

mission after contact with water, and they are assigned points ranging up to 16. The organic content and buffering capacity of ground water are associated with present conditions, and they are assigned points ranging from 0 to 10. Potential travel distance, prevailing wind direction, and population are factors outside the immediate disposal site and related to potential human exposure. These three factors are assigned numerical scores from 0 to 7. Composite scores for a given site can range up to 129.

Table 13.8. Example of Waste–Soil–Site Interaction Matrix[a]

		Soil	Soil Group		Hydrology Group			Site Group		
Waste		P[b] / P[b]	Perme-ability ($2\frac{1}{2}$–10)	Sorption (1–10)	Water Table (1–10)	Gradient (1–10)	Infil-tration (1–10)	Distance (1–10)	Thickness of Porous (1–10)	Total
		5	4	5	2	6	7	1	30
Effects Group	Human toxicity (0–10)	8	40	32	40	16	48	56	8	240
	Groundwater toxicity (0–10)	5	25	20	25	10	30	35	5	150
	Disease transmission potential (0–10)	0								
Behavioral Group — Behavioral Performance Subgroup	Chemical persistence (1–5)	3	15	12	15	6	18	21	3	90
	Biological persistence (1–4)	4	20	16	20	8	24	28	4	120
	Sorption (1–10)	5	25	20	25	10	30	35	5	150
Behavioral Properties Subgroup	Viscosity (1–5)	2	10	8	10	4	12	14	2	60
	Solubility (1–5)	1	5	4	5	2	6	7	1	30
	Acidity/basicity (0–5)	1	5	4	5	2	6	7	1	30
Capacity-Rate Group	Waste application rate (1–10)	4	20	16	20	8	24	28	4	125
	Total	33	165	132	165	66	198	231	33	990

[a] Phillips et al., 1977.

[b] P = point score.

Table 13.9 Factors in Site Rating System[a]

Group	Priority[b]	Factor
Soil	1	Infiltration potential (Ip)—relates to the potential for water to enter a waste deposit. The parameter is a function of the infiltration rate of the cover soil and the field capacity and depth of the cover soil. The Ip values range from 0.02 (most desirable) to 20 (least desirable) points.
	1	Bottom leakage potential (Lp)—relates to the water passing through the soil layer beneath the waste site and into the ground water system. The parameter is a function of the bottom soil permeability and thickness of the bottom soil layer. The Lp values range from 0.02 (most desirable) to 20 (least desirable) points.
	2	Filtering capacity (Fc)—relates to the ability of bottom soils to remove solid particles traveling downward in a fluid suspension. The parameter is dependent upon the pore spaces between individual soil grains, and is inversely proportional to the average grain size (particle diameter) in the soil stratum. The Fc values range from about 0 (good filtering capacity) to 16 (poor filtering capacity).
	2	Adsorptive capacity (Ac)—relates to the ability of bottom soils to remove organic and inorganic minerals from suspension and solution in a migrating fluid by adsorption. The parameter is dependent on the organic content and cation exchange capacity of the bottom soil. The Ac values range from 0 (good adsorptive capacity) to 16 (poor adsorptive capacity).
Ground water	3	Organic content (Oc)—relates to the role of the organic content of ground water as a substrate for reproduction of pathogenic organisms. The parameter is dependent on the biochemical oxygen demand of the ground water. The Oc values range from 0 (minimal potential for pathogenic microorganism reproduction) to 10 (maximal potential for pathogenic microorganism reproduction).
	3	Buffering capacity (Bc)—relates to the ability of the ground water system to neutralize any acidic or basic characteristics of entering waste materials. The parameter is related to the ground water pH, acidity, and alkalinity. The Bc values range from 0 (strong buffer) to 10 (weak buffer).

Table 13.9 (*Continued*)

Group	Priority	Factor
	4	Potential travel distance (Td)—relates to the potential for dispersal of pollutants through a ground water system. The parameter is related to the distance of travel from a point directly beneath the waste site through the ground water and surface water systems to the sea. The greater the distance, the greater the potential number of water uses. The Td values range from 0 (less than 500 ft) to 5 (more than 50 mi).
	1	Ground water velocity (Gv)—relates to the time of waste transmission through the ground water system. This parameter is a function of the coefficient of permeability and the hydraulic gradient. The Gv values range from 9 (minimal velocity and waste transmission) to 20 (maximal velocity and waste transmission).
Air	4	Prevailing wind direction (Wp)—relates to the transport of any airborne toxics or pathogens from the waste site in relation to the distribution of the human population around the site. The parameter considers the prevailing wind direction in four quadrants along with the population and population nodes in the quadrants. The area of influence encompasses a 42 km radius around the site. The Wp values vary from 0 (wind direction primarily away from population centers) to 5 (wind direction primarily toward population centers).
	4	Population factor (Pf)—relates to the human population around the site that could be adversely affected by escaping hazardous materials. The parameter is related to the total population within a 42 km radius of the site. The Pf values range from 0 (no population exposed) to 7 (10,000,000 people exposed).

[a] Hagerty et al. (1973).

[b] Priority 1 denotes factors that would immediately affect waste transmission. Priority 2 denotes factors that would affect waste transmission after contact with water. Priority 3 denotes factors representing the present conditions of receiving ground water. Priority 4 denotes factors outside the immediate disposal site.

STUDY AREA IN CENTRAL OKLAHOMA

The main aquifer in this study of empirical assessment methodologies was the Garber-Wellington aquifer in central Oklahoma. The surface area bounding the outcrop and underlying portions of the aquifer includes Oklahoma and Cleveland Counties as well as portions of Logan, Lincoln, Pottawatomie, McClain, Canadian, and Kingfisher Counties. This area is shown in Figure 13.1, and it includes 80 townships and 2880 sections. The Garber-

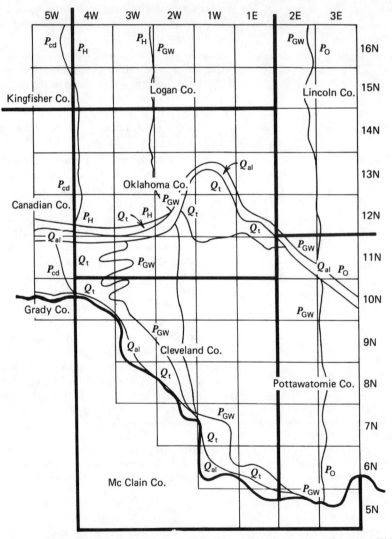

Figure 13.1. Surface geology of study area. Wavy lines denote outcrop interface. P_{cd}, El Reno group; P_H, Hennessey group; P_{GW}, Garber-Wellington formation; P_O, Oscar group; Q_{al}, alluvial deposit; Q_t, terrace deposit.

Wellington aquifer contains over 50 million acre/ft of fresh water, with approximately two-thirds potentially available for development. The thickness of the fresh water zone ranges from about 50 to 275 m. Water well depths range from about 75 to 325 m, with the deeper wells located in the western half of the study area.

There are additional aquifers potentially influenced by waste sources in the study area. For example, in Canadian County the Garber-Wellington aquifer is overlain by Permian-age rock formations such as the Hennessey shale and the El Reno group (Mogg et al., 1969). Saturated zone thicknesses are generally greater than 35 m. In Logan, Oklahoma, and Cleveland Counties, many of the potential waste sources are surrounded and underlain by alluvial or terrace deposits. Where the alluvial or terrace deposits are underlain by the Hennessey shale, as in Norman, Moore, and the western parts of Oklahoma City, the Garber-Wellington aquifer is confined (Bingham and Moore, 1975). The alluvial and terrace deposits in these areas are between 3 and 35 m thick. The eastern half of the area shown in Figure 13.1 represents the outcrop area for the Garber-Wellington aquifer (Bingham and Moore, 1975; Burton and Jacobsen, 1967).

EVALUATION OF PONDS, SEPTIC TANK SYSTEMS, AND LANDFILLS

The SIA method was applied to 77 waste water ponds, 13 septic tank system areas, and 10 landfills in the 80-township study area. The waste water ponds included 38 municipal ponds, 12 agricultural ponds, and 27 industrial liquid ponds. The waste hazard potential was based on the standard industrial classification (SIC) code for the source, with all municipal ponds assigned a rating of 5. For those five municipal waste water ponds receiving industrial discharges, the ratings were assigned based on the industrial effluents received. The waste hazard potential for the septic tank systems was the same as that for municipal ponds. Assignment of waste hazard potential ratings to landfills was based on a rating of 7, with the two landfills receiving hazardous wastes being rated 9. The rating of 7 for municipal landfills was based on the fact that approximately 10% of municipal solid waste consists of metals.

Table 13.10 displays the ground water contamination potential ratings for the 38 municipal ponds (Sammy and Canter, 1980). The highest overall score was associated with a facility in Oklahoma City which receives an industrial waste discharge. Out of the total of 38 municipal waste water ponds, 24 had ratings between 20 and 26, and 14 had ratings between 11 and 14. Those situated in the Garber outcrop area, or on alluvial or terrace deposits, all had scores in the range of 20 to 26. Lagoons west of the Hennessey-Garber outcrop interface, and primarily situated on Permian rocks of the Hennessey or El Reno groups, had overall scores in the range of 11 to 14. The greatest contributors to the differences in overall scores between the two groups were

Table 13.10 Assessment of Municipal Ponds[a]

Facility and Location (Underlying Aquifer)[b]	Unsaturated Zone Rating	Ground Water Availability Rating	Ground Water Quality Rating	Waste Hazard Rating	Overall Ground Water Contamination Potential	Average Flow (mil gal/day)
Maximum value	9	6	5	9	29	
Minimum value	0	0	0	1	1	
Confidence level	B	C	B	B	—	
Northside, OKC (A, G-W)[c]	7B	6A	5	8	26	16.8
Deer Creek, OKC (A, G-W)[c]	8B	6A	5	5	24	2.5
Coffee Creek, EDMOND (A, G-W)[c]	8B	6A	5	5	24	1.3
Deep Fork, OKC (A, G-W)[c]	8B	6A	5	5	24	0.48
LUTHER (A, G-W)[c],	8B	6A	5	5	24	0.16
Deer Creek 2, OKC (A, G-W)[c]	8B	6A	5	5	24	0.15
LEXINGTON (A, G-W)[c]	8B	6A	5	5	24	0.11
23rd Street, CHOCTAW (A, G-W)[c]	8B	5A	5	5	24	0.08
Southside, OKC (A)	6B	5A	4	8	23	26.1
Northwest, EDMOND (G-W)	7B	6A	5	5	23	0.9
HARRAH (A, G-W)	7B	6A	5	5	23	0.20
JONES (A, G-W)	7B	6A	5	5	23	0.11
SPENCER (A, G-W)	7B	6A	5	5	23	0.10
COYLE (A, G-W)	7B	6A	5	5	23	0.04
Siekel Add., CHOCTAW (A, G-W)	7B	6A	5	5	23	0.02
Ridgedale Add., CHOCTAW (G-W)	7B	6A	5	5	23	0.01
Eastwood, MIDWEST CITY (G-W)	7B	6A	5	5	23	0.01
Greenbriar, OKC (A)	8B	5A	4	5	22	0.48
NOBLE (T)	8B	5A	4	5	22	0.4
Will Rogers, OKC (A)	8B	5A	4	5	22	0.15
West Elm Creek, OKC (G-W)	6B	6A	5	5	22	0.03
Cullen's Valley Hi Add., OKC (A)	7B	5A	4	5	21	0.1
MOORE (A)	6B	5A	4	5	20	1.8
HALL PARK (G-W)	6B	5A	4	5	20	0.03

Table 13.10   (*Continued*)

Facility and Location (Underlying Aquifer)[b]	Unsaturated Zone Rating	Ground Water Availability Rating	Ground Water Quality Rating	Waste Hazard Rating	Overall Ground Water Contamination Potential	Average Flow (mil gal/day)
Maximum value	9	6	5	9	29	
Minimum value	0	0	0	1	1	
Confidence level	B	C	B	B	—	
YUKON (ER)	4D	2E	3	5	14	1.35
UNION CITY (ER)	4D	2E	3	5	14	0.05
CASHION (H)	2E	2E	4	5	13	0.05
OKARCHE (ER)	2E	2E	3	5	12	0.16
Fed. Reform, EL RENO (ER)	2E	2E	3	5	12	0.12
Spring Creek, OKC (H)	2E	2E	3	5	12	0.09
Harvest Halls Add., OKC (H)	2E	2E	3	5	12	0.09
Wilshire Hills Add., OKC (H)	2E	2E	3	5	12	0.06
Concho School, EL RENO (H)	2E	2E	3	5	12	0.04
Idelwyld Add., OKC (H)	2E	2E	3	5	12	0.04
Rambling Acres Add., OKC (H)	2E	2E	3	5	12	0.03
Willow Creek Est., OKC (H)	1E	2E	3	5	11	0.03
Embassy W. Mob. Homes, OKC (H)	1E	2E	3	5	11	0.02
Embassy W. Add., OKC (H)	1E	2E	3	5	11	0.01

[a] Sammy and Canter, 1980.

[b] A, alluvium; G-W, Garber-Wellington; T, terrace deposits; ER, El Reno group; H, Hennessey group.

[c] Sampling conducted near facility.

associated with the differences in unsaturated zone ratings and ground water quality ratings. An additional factor that could be considered when several waste water ponds have the same overall potential rating is the average inflow to the pond system. Ponds with larger inflow rates would represent larger absolute sources of potential ground water contamination. Table 13.10 also has information on the average inflows.

Ratings for the 12 agricultural ponds are shown in Table 13.11 (Sammy and Canter, 1980). As with the municipal pond systems, predominant factors

Table 13.11   Assessment of Agricultural Ponds[a]

Facility and Location (Underlying Aquifer)[b]	Unsaturated Zone Rating	Ground Water Availability Rating	Ground Water Quality Rating	Waste Hazard Rating	Overall Ground Water Contamination Potential
Maximum value	9	6	5	9	29
Minimum value	0	0	0	1	1
Confidence level	B	C	B	B	—
Barnes Ranch, GUTHRIE (G-W)	7B	6A	5	3	21
Cornett Feed Mill, OKC (A)	6B	5A	4	5	20
Tru-Fresh Egg, OKC (A)	6B	5A	4	4	19
Savon-Rich Dairy, MOORE (A)	6B	5A	4	4	19
Manning Feedlot, BANNER (ER)	4D	2E	3	5	14
Natural Feedlots, YUKON (ER)	4D	2E	3	5	14
Wilson Hog Farm, CALUMET (ER)	4D	2E	3	3	12
Merveldt Feed Yard, OKARCHE (H)	2E	2E	3	5	12
J & J Feedlot, EL RENO (ER)	2E	2E	3	5	12
Royse Bros. Feedlot, EL RENO (ER)	2E	2E	3	5	12
Experimental Farm, EL RENO (ER)	2E	2E	3	3	10
Alfadale Farm, EL RENO (ER)	2E	2E	3	3	10

[a] Sammy and Canter, 1980.

[b] G-W, Garber-Wellington; A, alluvium; ER, El Reno group; H, Hennessey group.

are the ratings of the unsaturated zone and ground water availability. The four ponds situated on alluvial deposits, or in the Garber outcrop area, have scores between 19 and 21. The eight other ponds have scores in the range of 10 to 14.

Table 13.12 displays the ratings for the 27 industrial waste water ponds (Sammy and Canter, 1980). Variations in waste hazard potential ratings were

Table 13.12 Assessment of Industrial Ponds[a]

Facility and Location (Underlying Aquifer)[b]	Unsaturated Zone Rating	Ground Water Availability Rating	Ground Water Quality Rating	Waste Hazard Rating	Overall Ground Water Pollution Potential
Maximum value	9	6	5	9	29
Minimum value	0	0	0	1	1
Confidence level	B	C	B	B	—
Nat. Guard Heliport, LEXINGTON (A, G-W)[c]	8B	6A	5	5	24
DEL CITY Paint (G-W)[c]	7B	6A	5	6	24
F & K Plating, OKC (A)	6B	5A	4	8	23
Sun Oil, MOORE (A)	6B	5A	4	8	23
Permian Corp, GUTHRIE (G-W)	7B	6A	5	4	22
Ashland Chemical, OKC (T)	7B	5A	4	6	22
Eaton Wholesale, CRESCENT (T, G-W)	8B	6A	5	2	21
Madewell and Madewell, JONES (A, G-W)	7B	6A	5	3	21
MacKlanburg-Duncan, OKC (A)	6B	5A	4	6	21
Millcon Corp., OKC (G-W)	7B	6A	5	3	21
Y Carwash, HARRAH (A, G-W)	7B	6A	5	2	20
Red Rock Petroleum, OKC (T)	7B	5A	4	4	20
Markins Concrete, NORMAN (T)	6B	5A	5	3	19
Dolese North, NORMAN (T)	6B	5A	5	3	19
Dolese South, NORMAN (T)	6B	5A	5	3	19
W & W Steel, NORMAN (T)	6B	5A	5	3	19
Thunderbird Grocery, NORMAN (T)	6B	5A	5	2	18
Cimarron Aircraft, EL RENO (ER)	2E	2E	3	8	15
Gaido-Lingo, EL RENO (ER)	2E	2E	3	8	15
MUSTANG Airport (ER)	3E	2E	4	4	13
Gemini Lacquers, EL RENO (ER)	2E	2E	3	6	13

Table 13.12   *(Continued)*

Facility and Location (Underlying Aquifer)[b]	Unsaturated Zone Rating	Ground Water Availability Rating	Ground Water Quality Rating	Waste Hazard Rating	Overall Ground Water Pollution Potential
Maximum value	9	6	5	9	29
Minimum value	0	0	0	1	1
Confidence level	B	C	B	B	—
Schwartz Refinery, OKARCHE (H)	2E	2E	3	5	12
Oklahoma Brick, UNION CITY (ER)	4D	2E	3	2	11
C RI & P Railroad, EL RENO (ER)	2E	2E	3	4	11
Highway 81 Dairy, EL RENO (ER)	2E	2E	3	4	11
Union Carbide, OKC (H)	2E	2E	3	2	9
EL RENO Biproducts (ER)	2E	2E	3	2	9

[a] Sammy and Canter, 1980.

[b] A, alluvium; G-W, Garber-Wellington; T, terrace deposits; ER, El Reno group; H, Hennessey group.

[c] Sampling conducted near facility.

greater for these ponds than for either municipal or agricultural ponds. Of 27 industrial waste water ponds, 17 were rated from 18 to 24, and 10 were in the category from 9 to 15. Ponds situated on alluvium, terrace deposits, or the Garber outcrop area had overall ratings in the range from 18 to 24.

Table 13.13 summarizes the ratings associated with septic tank systems (Carriere, 1980). As expected, the overall scores of septic tank system areas closely paralleled those of municipal ponds. Variations in the scores are primarily reflective of the geological features of the areas. Ten septic tank system areas are located on terrace deposits or in the Garber-Wellington outcrop area, and they received scores ranging from 22 to 24. The remaining three systems are located on outcrops of the Hennessey or El Reno groups, and they had scores between 12 and 15. An additional factor which could be utilized for evaluation of the pollution potential is the service area or estimated total flows into septic tank systems. Table 13.13 also contains estimates of the total waste water flows for the respective service areas.

Table 13.13   Assessment of Septic Tank Systems[a]

Facility and Location (Underlying Aquifer)[b]	Unsaturated Zone Rating	Ground Water Availability Rating	Ground Water Quality Rating	Waste Hazard Rating	Overall Ground Water Contamination Potential	System Flow (mil gal/yr)
Maximum value	9	6	5	9	29	
Minimum value	0	0	0	1	1	
Confidence level	B	C	B	B	—	
Arcadia (G-W)[c]	8B	6A	5	5	24	8
Seward (G-W)[c]	8B	6A	5	5	24	175
Arrowhead Hills (G-W)	7B	6A	5	5	23	9
Crutcho (T, G-W)	7B	6A	5	5	23	11
Forest Park (G-W)	7B	6A	5	5	23	27
Green Pastures (T, G-W)	7B	6A	5	5	23	44
Midwest City (G-W)	7B	6A	5	5	23	228
Nicoma Park (T, G-W)	7B	6A	5	5	23	57
East Norman (T, G-W)	6B	6A	5	5	22	152
Del City (G-W)	6B	6A	5	5	22	5
Sunvalley Acres (ER)[c]	5D	2E	3	5	15	3
Mustang (ER)	4D	2E	4	5	15	67
Silver Lake Estates (H)	2E	2E	3	5	12	6

[a] Carriere, 1980.

[b] G-W, Garber-Wellington; T, terrace deposits; ER, El Reno group; H, Hennessey group.

[c] Sampling conducted in area.

Table 13.14 summarizes the pollution ratings for the 10 landfills (Hajali, 1980). The ratings were generally higher than for other source types owing to the nature of the material being disposed. Hydrogeological features of individual sites were also important. Nine of the sites are located on more permeable strata and thus received overall scores between 24 and 27. One site, situated on the Hennessey outcrop, received a score of 14.

A modest field sampling program was conducted to evaluate the pollution potential predictions for 15 of the 100 waste sources in the study area. Ten waste water pond systems (eight municipal ponds and two industrial ponds), four septic tank system areas, and one landfill were included. The program consisted of locating existing wells near the 15 sources, pumping the wells for several minutes, and then collecting 1-L samples. Field measurements included pH, salinity, and conductivity. Subsequent laboratory analyses were performed for orthophosphates, total phosphorus, Kjeldahl (organic) nitrogen, nitrate-nitrogen, alkalinity, hardness, and TDS. Specific sources

Table 13.14   Assessment of Landfills[a]

Facility (Underlying Aquifer)[b]	Unsaturated Zone Rating	Ground Water Availability Rating	Ground Water Quality Rating	Waste Hazard Rating	Overall Ground Water Contamination Potential
Maximum value	9	6	5	9	29
Minimum value	0	0	0	1	1
Confidence level	B	C	B	B	—
Industrial Disposal Service Company (A, G-W)	9F	6A	5	7	27
Midwest City (A, G-W)	9F	6A	5	7	27
Norman Asphalt Disposal Co. (A)[c]	9F	5A	5	7	26
OKC Disposal Co.— Moseley Rd. (A)	7B	5A	5	9	26
Yukon (A)	9F	5A	5	7	26
John Pritner (G-W)	6B	6A	5	9	26
Lexington Treatment Center (G-W)	7B	6A	5	7	25
Del City (A, G-W)	7B	6A	5	7	25
OKC Disposal Co.— Air Depot, Depot Rd. (A)	7B	5A	5	7	24
Dick Richardson (H)	2E	2E	3	7	14

[a] Hajali, 1980.

[b] A, alluvium; G-W, Garber-Wellington; H, Hennessey group.

[c] Sampling conducted in the area.

monitored during the program are identified in Tables 13.10, and 13.12–13.14. The following criteria were used in interpreting the key analytical results (U.S. Environmental Protection Agency, 1976):

pH	value should be 6.5–8.5
Orthophosphate	6 mg/L represents weak domestic sewage
Total phosphorus	6 mg/L represents weak domestic sewage
Nitrate-nitrogen	10 mg/L is drinking water standard
TDS	500 mg/L is drinking water standard

Twenty wells were sampled near the 10 selected waste water ponds, and the results are in Table 13.15. Wells indicative of background conditions include No. 12 at Coffee Creek, No. 14 at Deer Creek-1, No. 22 at Deep Fork, and No. 31 at Lexington. Evidence of ground water pollution was primarily associated with the nitrate-nitrogen, TDS, and total phosphorus. The nitrate-nitrogen concentrations often exceeded the recommended Oklahoma standard of 10 mg/L. The TDS concentrations in several wells exceeded the U.S. Public Health Service (USPHS) drinking water standard of 500 mg/L. Total phosphorus in well No. 30 at Lexington was 59 mg/L, and this is higher than typical concentrations in untreated domestic sewage. Well No. 30 was formerly used for drinking water, but was abandoned several years ago because of taste and odor problems. In summary, all 10 pond systems examined exhibited evidence of ground water pollution.

Eleven wells were sampled in the four septic tank system areas, and the results are in Table 13.16. All wells were within the septic tank system areas, and none can be considered as background wells. Seven of the 11 wells exceeded the Oklahoma nitrate-nitrogen standard; 4 wells exceeded the USPHS TDS standard, and 9 wells had organic phosphorus concentrations of greater than 1 mg/L (weak domestic sewage has 2 mg/L). Therefore, ground water contamination is occurring in each of the four septic tank system areas.

Two wells were sampled near the landfill serving the city of Norman, with the results shown in Table 13.17. Well No. 48 is a background well, with the high nitrate-nitrogen concentration probably resulting from a nearby cattle farming operation. Measurements at Well No. 47 are indicative of the influence of leachates from the landfill, as concentrations of phosphorus, TDS, hardness, and alkalinity are elevated. Of the 15 sources sampled, all yielded indications of ground water pollution.

EVALUATION OF OIL AND GAS FIELD ACTIVITIES

Oil and gas field activities within the central Oklahoma study area constitute a major potential source of ground water pollution. Since no empirical assessment methodologies existed for these sources, prioritization methodologies were developed for oil and gas wells and dry holes, and for salt water disposal and waterflooding operations. The methodologies were developed by an interdisciplinary team consisting of four persons—an environmental engineer, a geologist, a soil microbiologist, and a natural scientist. Development involved three basic steps:

1. Identification of factors related to the drilling of oil and gas wells, their plugging or abandonment, and their production; and identification of factors related to salt water disposal and waterflooding operations.

2. Assignment of relative importance weights, or points, to each of the identified factors in both methodologies.
3. Delineation of a scaling approach that could be used to determine the fractional evaluation of each factor in each methodology.

Based on the availability of data from the Oklahoma Corporation Commission (OCC) and other state agencies, 7 factors were selected for the prioritization methodology for oil and gas wells and dry holes; an additional 7 factors, for a total of 14 factors, were selected for the prioritization methodology for salt water disposal wells and waterflooding operations. The

Table 13.15 Well Samples and Analysis for Waste Water Holding Pond Areas

Parameter	Luther		Choctaw			Coffee Creek		Deer Creek-1	
	4[a]	5	6	7	8	12	13	14	15
pH	7.65	7.45	7.85	7.50	7.85	7.75	8.05	7.95	7.65
Salinity (ppt)	0.5	0.5	0	0	0	0	0	0	0
Conductivity (μmhos/cm)	1170	1150	400	650	475	500	800	500	400
Orthophosphate (mg/L)	2.5	2.5	1.0	2.2	3.0	4.1	4.1	0.2	0
Total phosphorus (mg/L)	4.0	4.0	1.5	2.3	3.5	4.6	6.6	2.5	2.5
Total Kjeldahl nitrogen (mg/L)	<1	0.28	1.1	<1	<1	0	0.56	0	0.15
Nitrate-N (Mg/L)	16.4	15.7	3.3	17.7	4.4	5.6	3.4	5.2	3.5
Alkalinity (mg/L as $CaCO_3$)	265	265	195	248	230	285	355	250	250
Hardness (mg/L as $CaCO_3$)	240	222	166	246	188	258	150	130	118
Total dissolved solids (mg/L)	735	771	269	462	329	335	529	270	383
Depth (m)	50	65	40	27	33	140	25	30	30
Distance (m) and direction from source	170 E	370 E	120 NW	150 NW	180 NW	33 SW	17 SW	1700 W	170 S

[a] Denotes well number.

selected factors are generally applicable and not limited to central Oklahoma.

Assignment of importance weights was made by using the nominal group process technique (Voelker, 1977). To facilitate importance weight assignments, a pairwise comparison technique was utilized for the initial allocation (Dean and Nishry, 1965). Consensus point assignments for the oil and gas well and dry hole methodology, along with factor descriptions, are presented in Table 13.18. The relative importance weight total for an oil or gas well or dry hole was 100 points, and a given well or dry hole that has maximum pollution potential would receive 100 points. The minimum would be 3

Deer Creek-2		North-side OKC	Deepfork			Del City Paint		Lexington		Lexington Heliport
16	17	20	22	23	24	28	29	30	31	32
7.50	7.50	7.60	7.80	7.20	7.45	7.30	7.45	7.30	7.35	7.60
0.5	0	0	0	0	0	0	0	0	0	0.5
1050	450	370	300	550	750	850	750	1150	450	850
0.6	0	0.60	0.60	1.55	0.30	1.90	4.80	10.70	7.0	1.30
2.5	1.0	3.60	1.00	2.00	3.60	5.50	5.20	59.00	10.5	4.50
0.98	0	1.26	0.56	0.84	0.84	0.28	0.56	1.40	0.98	0.84
18.34	12.5	13.3	2.1	3.2	1.3	18.8	16.5	2.0	5.3	2.4
280	220	190	173	288	392	362	360	434	262	278
452	172	88	62	252	186	316	314	414	228	52
754	727	217	232	221	453	407	678	472	357	714
30	30	260	65	30	30	17	30	18	18	65
900	900	1700	430	330	500	100	650	100	300	900
E	E	S	W	E	SE	E	SE	NW	N	SE

Table 13.16 Well Samples and Analysis for Septic Tank System Areas

Parameter	Arcadia			Seward			Midwest City			Sunvalley Acres	
	1[a]	2	3	9	10	11	25	26	27	18	19
pH	7.45	7.75	8.00	7.60	7.70	7.55	7.30	7.55	7.35	7.60	7.65
Salinity (ppt)	<1	<1	<1	0	<1	0	0	0	0	<1	<1
Conductivity (μmhos/cm)	720	705	750	500	800	550	375	500	450	975	925
Orthophosphate (mg/L)	5.2	1.0	1.0	2.1	0.2	2.1	2.55	1.60	2.55	1.2	0.2
Total phosphorus (mg/L)	5.4	1.8	3.0	5.0	2.0	4.0	8.00	6.00	6.50	3.4	3.4
Total Kjeldahl nitrogen (mg/L)	1.4	<1	<1	1.12	0	2.8	1.12	0.56	0	0.28	0.15
Nitrate-N (mg/L)	30.1	3.3	5.2	17.6	83.7	16.8	18.0	37.6	20.5	9.7	7.9
Alkalinity (mg/L as $CaCO_3$)	349	343	376	240	285	200	149	201	217	350	345
Hardness (mg/L as $CaCO_3$)	316	322	362	256	330	240	150	212	220	322	374
Total dissolved solids (mg/L)	565	478	556	382	565	273	306	376	187	422	529
Depth (m)	<30	<30	65	25	35	20	27	45	45	31	30
Distance (m) and direction from source	8	12	25	25	20	27	25	25	25	30	25
	—	—	—	W	W	W	—	—	—	N	S

[a] Denotes well number.

Table 13.17   Well Samples and Analysis for Norman Landfill Area

	Well Numbers	
Parameter	47	48
pH	7.00	7.55
Salinity (ppt)	1.5	0
Conductivity (μmhos/cm)	2950	450
Orthophosphate (mg/L)	31.0	5.5
Total phosphorus (mg/L)	43.5	5.5
Total Kjeldahl nitrogen (mg/L)	3.85	0
Nitrate-N (mg/L)	1.64	15.6
Alkalinity (mg/L as $CaCO_3$)	817	234
Hardness (mg/L as $CaCO_3$)	1086	240
Total dissolved solids (mg/L)	2228	391
Depth (m)	10	30
Distance (m) and direction from source	100, SW	100, N

points for a well drilled recently and meeting more than the minimum requirements for several factors. The 14 factors in the methodology for salt water disposal wells and waterflooding operations are described in Table 13.19. A basic premise is that there is greater potential for ground water pollution from a salt water disposal well or waterflooding operation than there is from an oil or gas well or dry hole. This greater potential is reflected by a total of 200 points to characterize the worst case for a salt water disposal well or waterflooding operation. In cases where information was lacking for any factor in either methodology, a worst case approach was utilized; that is, maximum points were assigned for the factor. Detailed descriptions of the fractional scaling approaches for both methodologies are found elsewhere (Fairchild et al., 1981).

The 80-township study area has had a total of 14,127 oil and gas wells and dry holes drilled since record-keeping was initiated by the OCC in 1917. Of

Table 13.18   Factors in Prioritization Methodology for Oil and Gas Wells and Dry Holes[a]

Factor	Description
Years unplugged (20 points)	This factor represents the total years that the oil or gas well, or dry hole, has been unplugged from the year of initial drilling through 1980. The concept is that the greater the number of years unplugged, the greater the possibility for production, and thus the greater the potential for contaminant materials to move to water-bearing aquifers. Actual production information from each of the wells was unavailable from the OCC. This factor was calculated by considering whether or not the well had been unplugged since its initial date of drilling, or whether it had been plugged on one or more occasions since the initial drilling date.
Age category (30 points)	Oil and gas wells have been drilled in central Oklahoma for over 50 yr. As time has progressed from the initial years of drilling, the rules and regulations of the OCC have become more complete and better oriented to the production of ground water resources. Wells drilled more than 50 yr ago were not subject to the more stringent regulations of today. This factor is based on the concept that an older well, drilled under conditions of less stringent requirements, has more likelihood for causing ground water pollution than a more recently drilled well. An evaluation of the rules and regulations of the OCC since 1917 revealed that four major time periods could be identified based on increasing stringency of the regulations.
Surface casing (10 points)	This factor accounts for the casing material used near the ground surface for protection of underlying ground water aquifers. Surface casing requirements have become more stringent in recent years. The concept is that if the surface casing was inadequate in terms of the regulations in force during a given time period, the potential for ground water pollution is increased. Evaluation of surface casing compliance with regulations in existence at the time the well was drilled was based on either compliance or noncompliance.
Drilling compliance (10 points)	This factor accounts for whether the drilling activities associated with the oil or gas well or dry hole were conducted in a manner as stipulated by the rules and regulations of the OCC. Failure to comply with the requirements of the OCC is indicative of greater potential for ground water pollution. Drilling compliance evaluation was also based on regulations that existed in the time period in which the well was drilled.

Table 13.18 (*Continued*)

Factor	Description
Plugging compliance (10 points)	Dry holes or abandoned wells must meet certain OCC regulations related to plugging. Lack of compliance with regulations for plugging are indicative of greater potential for ground water pollution since pathways for pollutant movement to water-bearing aquifers can exist. The plugging compliance factor was also based on the regulations that existed when the well was plugged.
Area (10 points)	This factor is indicative of the uncemented production area of the casing exposed to the subsurface environment. The concept is that the greater the surface area of the pipe in the subsurface environment, the greater the possibility for some type of disruption of the physical integrity of the pipe, and thus the greater the potential for ground water pollution. Disruption of physical integrity can occur from corrosion, improper joints or seals, and improper seating of the production pipe.
Chemical substances (10 points)	In the drilling activities associated with an oil or gas well, drilling muds and associated chemicals are often utilized. Additional chemicals may be required for acidizing or fracturing to increase production. This factor accounts for the number of chemicals used in conjunction with drilling or production activities.

[a] Fairchild et al., 1981.

Table 13.19   Factors in Prioritization Methodology for Salt Water Disposal Wells and Waterflood Operations[a]

Factor	Description
Years unplugged (20 points)	The concept of this factor is simply that the greater the number of years that a salt water disposal well or waterflooding operation has existed in an unplugged category, the greater the potential for pollution through movement of undesirable materials into ground water aquifers. This concept is similar to that described for oil and gas wells and dry holes.
Age category (30 points)	This factor reflects the general time period that the salt water disposal well or waterflooding operation was drilled, because the rules and regulations have become more stringent over time. The general concept is that an operation initiated some years ago probably has greater potential for ground water pollution because the regulations existing at that time were not as strin-

Table 13.19 (*Continued*)

Factor	Description
	gent as those at the current time. This concept was also used for the oil and gas wells and dry holes methodology.
Surface casing (10 points)	The concept of this factor is based on the fact that there are regulations for surface casing, which have become more stringent over time. Failure to comply with the regulations for surface casing at the time of well construction is indicative of a situation which can create undesirable ground water pollution owing to the inadvertent transport of pollutant materials into fresh water aquifers. This concept is the same as that used for oil and gas wells and dry holes.
Drilling compliance (10 points)	The concept of this factor is similar to that for surface casing in that there are regulations that have existed in conjunction with drilling activities. Failure of the given salt water disposal well or waterflooding operation to comply with the drilling requirements at the time it was drilled is considered to increase the potential for ground water pollution. This concept is also the same as that used in the methodology for oil and gas wells and dry holes.
Area (10 points)	The concept of this factor is that the greater the exposed surface area of the production casing in the subsurface environment, the greater the potential for ground water pollution because of the greater area available for possible physical disruption of the casing integrity. This concept was also used in the prioritization methodology for oil and gas wells and dry holes.
Chemical substances (10 points)	This factor is indicative of the number of chemicals that may have been used in drilling or usage activities associated with the salt water disposal well or waterflooding operation. It is similar to the factor used in the oil and gas well and dry hole methodology, with the concept being that the greater the number of chemicals utilized, the greater the potential for ground water pollution.
Fresh water well location (5 points)	The presence of a fresh water well within one-half mile of the salt water disposal well or waterflooding operation represents usage of ground water from the area, and the potential for the well's pollution is greater than for fresh water wells located further distances away. The concept of the point assignment is simply whether

Table 13.19 *(Continued)*

Factor	Description
	or not a fresh water well is located within one-half mile.
Quality of ground water (5 points)	This factor is associated with the existing quality of the ground water in the vicinity of the salt water disposal well or waterflooding operation. For any fresh water wells within one mile of the operation, consideration was given to the total dissolved solids as an indicator of general ground water quality. The concept used in this factor is that the better the existing ground water quality, the greater the potential for pollution and thus reduction in the uses of the ground water.
Injection quantity (20 points)	This factor reflects the amount of brine or fresh water injected into either the salt water disposal well or waterflooding operation, respectively. The concept is that the greater the amount of fluid injected, the greater the potential for causing ground water pollution.
Injection pressure (30 points)	This factor is related to the injection pressure utilized for the salt water disposal well or waterflooding operation. The concept is that the greater the injection pressure, the greater the potential for causing ground water pollution in the area of the operation.
Distance (15 points)	This factor is indicative of the injection interval as well as the location of that interval relative to the overlying fresh water aquifer. The general concept is that the closer the injection interval to the fresh water aquifer, the greater the potential for ground water pollution to occur. Additionally, the greater the injection interval itself, the greater the potential for ground water pollution.
Quality of injected water (5 points)	This factor accounts for the quality of the injected fluid and is indicated by the total dissolved solids in the injected fluid. The concept is that the greater the total dissolved solids, the greater the potential for ground water pollution.
Cumulative injection (20 points)	This factor is reflective of the total quantity of injected fluids over the period of operation of the salt water disposal well or waterflooding operation. The concept is that the greater the quantity of injected material, the greater the potential for ground water pollution.

[a] Fairchild et al., 1981.

the total, 8460 (60%) are oil wells, 1707 (12%) are gas wells, and 3960 (28%) are dry holes. There is an average of 106 oil wells and 21 gas wells per township. The term "dry holes" refers to drilled wells found insufficient for oil or gas production. Dry holes are either plugged or abandoned, and there is an average of 50 dry holes per township. The township with the largest number of oil wells is 11N 3W, with a total of 1104 wells. This same township has 324 gas wells, and this represents the largest number of gas wells in any township in the study area.

One of the issues related to ground water pollution is the number of wells which have been plugged improperly according to the extant regulations at the time of plugging. A total of 8524 wells have been plugged or abandoned in the 80 township study area, with 6165 (72%) considered to have been improperly plugged based on the regulations that existed or the lack of plugging information.

There are 1051 salt water disposal wells and waterflooding operations within the study area. This includes 706 salt water disposal wells, 196 waterflood operations, and 149 injection wells representing either salt water disposal wells or waterfloods, the specific delineation of which is impossible due to lack of data. There are averages of nine salt water disposal wells, two waterflood operations, and two injection wells per township. Of the 706 salt water disposal wells, 17% had been plugged in 1979–1980, with 4% in use and 79% idle. Only 6% of the waterflood wells had been plugged in 1979–1980, with 12% in use. Only one injection well was in use in 1979–1980, with 64% having been plugged. Township 10N 3W has the largest number of waterflood operations with a total of 12, while Township 11N 3W has the largest number of salt water disposal wells with 74. Townships 13N 3E and 11N 3W include the largest number of injection wells, each having 20. Again, it should be noted that Township 11N 3W has the highest number of oil and gas wells.

Each of the 14,127 oil and gas wells and dry holes were subjected to the prioritization methodology described in Table 13.18. Point assignments for the wells were organized according to sections and townships to enable priority ranking based on ground water pollution potential. Each of 1051 salt water disposal wells and waterflood operations were evaluated based on the prioritization methodology described in Table 13.19. The resultant data were also organized by section and township, with priority rankings assigned based on the ground water pollution potential due to these activities.

The points for each section and township were aggregated to allow for a combined consideration of oil and gas wells and dry holes, and salt water disposal wells and waterflood operations. Table 13.20 contains the resultant aggregate points for the top 20 townships (25% of the study area townships). The top three townships have sums considerably higher than the rest of the townships. Table 13.20 also shows the percentage contribution from oil and gas wells and dry holes as well as salt water disposal and waterflood operations. The primary contributors to the point totals for the seven highest

Table 13.20   Priority Ranking of Townships Based on Composite Evaluation[a]

		Summation			Percentage Contribution	
Rank	Township	O/G/DH[b]	SWD/WF/IW[c]	Sum	O/G/DH[b]	SWD/WF/IW[c]
1	11N 3W	61,052	15,560	76,612	80	20
2	12N 3W	34,896	3,260	38,156	91	9
3	14N 4W	23,913	4,838	28,751	83	17
4	15N 3E	12,838	7,567	20,405	63	37
5	10N 3W	10,614	8,445	19,059	56	44
6	14N 3E	11,975	5,720	17,695	68	32
7	16N 3E	14,081	2,629	16,710	84	16
8	13N 3E	8,069	8,103	16,172	50	50
9	15N 4W	11,675	4,474	16,149	72	28
10	14N 2W	11,386	4,735	16,121	71	29
11	13N 4W	12,073	3,491	15,564	78	22
12	10N 2W	9,104	4,160	13,264	69	31
13	11N 2W	10,040	2,928	12,968	77	23
14	14N 2E	8,107	4,209	12,316	66	34
15	13N 1E	8,775	3,369	12,144	72	28
16	14N 3W	7,979	4,144	12,123	66	34
17	12N 2W	7,379	4,540	11,919	62	38
18	6N 3E	6,490	5,357	11,847	55	45
19	7N 3E	7,089	4,078	11,167	63	37
20	16N 5W	4,869	5,903	10,772	45	55

[a] Fairchild et al., 1981.
[b] Oil and gas wells and dry holes.
[c] Salt water disposal wells, waterflooding operations, and injection wells.

ranked townships are oil and gas wells and dry holes. Table 13.21 contains the priority ranking for the top 57 sections within the study area, with these representing 2% of the 2880 sections within the study area. The top 13 sections are all in either Township 12N 3W or 11N 3W.

A cursory field sampling program was also conducted for oil and gas field activities. A total of 14 wells in 11 sections with high ground water pollution potential (all 11 within the top 57 sections) were sampled. The specific sections are listed in Table 13.21. Existing wells were utilized, and the analytical determinations included pH, salinity, conductivity, and TDS. Two additional sections were selected for sampling (11N 3W Sections 2 and 10); however, owing to their water quality, all wells in these two sections had been previously condemned by Oklahoma City. Most of the sections selected for sampling were over the confined parts of the Garber-Wellington

Table 13.21 Priority Ranking of Sections Based on Composite Evaluation[a]

Rank	Township	Section	Summation O/G/DH[b]	Summation SWD/WF/IW[c]	Sum	Percentage Contribution O/G/DH[b]	Percentage Contribution SWD/WF/IW[c]
1[d]	12N 3W	34	9419	531	9950	95	05
2[d]	11N 3W	2	7448	888	8336	89	11
3[d]	11N 3W	10	7417	194	7611	97	03
4	11N 3W	3	7350	111	7461	99	01
5	11N 3W	36	2828	4115	6943	41	59
6	11N 3W	26	2413	3634	6047	40	60
7	11N 3W	25	4927	1113	6040	82	18
8	11N 3W	22	5321	609	5930	90	10
9[d]	12N 3W	22	5662	153	5815	97	03
10	12N 3W	15	3972	1195	5167	77	23
11[d]	11N 2W	31	3413	1639	5052	68	32
12[d]	12N 3W	27	4803	194	4997	96	04
13	11N 3W	15	3995	276	4271	94	06
14[d]	14N 3W	31	2772	1471	4243	65	35
15	11N 3W	1	2912	1025	3937	74	26
16	6N 3E	23	2568	1144	3712	69	31
17	12N 3W	35	3533	118	3651	97	03
18[d]	11N 2W	30	2445	982	3427	71	29
19	12N 3W	23	2804	411	3215	87	13
20	11N 3W	14	2993	163	3156	95	05
21	11N 3W	24	2509	618	3127	80	20
22	11N 3W	13	2638	307	2945	90	10
23[d]	14N 3E	1	1670	1107	2777	60	40
24	11N 3W	11	2329	404	2733	85	15
25[d]	14N 3E	2	1563	1158	2721	57	43
26	13N 2W	30	890	1573	2463	36	64
27	13N 3E	28	910	1479	2389	38	62
28	11N 3W	12	2239	131	2370	95	05
29	12N 3W	26	2251	0	2251	100	00
30	8N 3E	29	537	1602	2139	25	75
31	11N 2W	19	1829	307	2136	86	14
32	14N 3E	3	1439	560	1996	72	28
33	13N 1E	28	1140	809	1949	59	41
34	6N 3E	22	1475	464	1939	76	24
35	10N 2W	29	854	1041	1895	45	55
36	11N 3W	23	1375	518	1893	73	27
37	14N 1W	3	1321	541	1862	71	29

Table 13.21   (*Continued*)

Rank	Township	Section	Summation O/G/DH[b]	Summation SWD/WF/IW[c]	Sum	Percentage Contribution O/G/DH[b]	Percentage Contribution SWD/WFIW[c]
38	13N 3E	21	476	1334	1810	26	74
39	11N 3W	27	1021	731	1752	58	42
40	14N 3W	1	675	1033	1708	40	60
41	15N 3E	21	1078	596	1674	64	36
42	15N 3E	1	907	746	1653	55	45
43	10N 3W	26	794	855	1649	48	52
45	14N 2E	2	747	901	1648	45	55
45	8N 1W	15	527	1093	1620	33	67
46	10N 3W	30	1037	556	1593	65	35
47[d]	10N 2W	21	1302	266	1568	83	17
48	13N 2W	31	671	852	1523	44	56
49	13N 3W	20	659	844	1503	44	56
50	14N 3E	4	1054	425	1479	71	29
51[d]	10N 3W	29	1268	201	1469	86	14
52	13N 3E	2	637	830	1467	43	57
53	13N 1E	26	1158	292	1450	80	20
54	10N 2W	6	756	689	1445	52	48
55	15N 2E	35	1443	0	1443	100	00
56	15N 3E	12	838	604	1442	58	42
57[d]	13N 3W	6	1256	177	1433	88	12

[a] Fairchild et al., 1981.
[b] Oil and gas wells and dry holes.
[c] Salt water disposal wells, waterflooding operations, and injection wells.
[d] Denotes sampling point.

aquifer; thus the aquifers being sampled were the Permian formations, including the Hennessey shale and El Reno group. The three wells in Township 14N 3E were in the outcrop area of the Garber-Wellington aquifer.

The results of the field sampling program are in Table 13.22. Indications of possible oil and gas field-related pollution result from comparisons between wells within an area, from comparison with standards, and from verbal reports obtained during the field investigation. Perhaps the most indicative parameter is TDS; only three wells (Nos. 38, 41, and 46) had TDS concentrations below the USPHS drinking water standard of 500 mg/L. One well (No. 35) had a TDS concentration of nearly 3000 mg/L; residents living near Well No. 35 reported that most domestic wells had been abandoned because of pollution related to oil and gas field activities. In all, wells in 10 of the 11

Table 13.22   Results of Field Sampling Program for Oil and Gas Field Activities[a]

Location	Well Number	Cumulative Priority Points	Rank	Depth of Well (ft)	pH	Salinity (ppt)	Conductivity ($\mu\Omega$/cm)	Total Dissolved Solids (mg/L)
12N 3W 34	44	9950	1	NA[b]	6.95	0.5	1150	962
34	45	9950	1	54	6.80	NM[c]	920	579
11N 3W 2	—	8336	2	Verbal indication of pollution				
11N 3W 10	—	7611	3	Verbal indication of pollution				
12N 3W 22	42	5815	9	46	6.85	0.5	1600	925
31	34	5052	11	110	7.05	NM	1250	765
27	43	4997	12	40	6.90	NM	700	546
14N 3W 31	40	4243	14	NA	7.35	NM	925	688
31	41	4243	14	200	7.40	NM	550	362
11N 2W 30	35	3427	18	144	6.80	2.0	4100	2994
14N 3E 1	37	2777	23	NA	7.20	NM	800	531
2	36	2721	25	NA	6.95	NM	850	652
2	38	2721	25	140	7.10	NM	500	311
10N 2W 21	33	1568	47	NA	7.35	NM	800	554
10N 3W 29	46	1469	51	NA	7.30	NM	650	474
13N 3W 6	39	1433	57	NA	7.05	NM	1050	656

[a] Fairchild et al., 1981.
[b] NA, not available.
[c] NM, not measurable.

sections sampled exceeded 500 mg/L TDS. When considering verbal indications along with the sampling results, evidence of ground water pollution was found in 12 of 13 sections included in the survey.

SUMMARY

Empirical assessment methodologies can be useful tools for comparative evaluations of ground water pollution sources, site selection, and the planning and conduction of ground water monitoring programs. This discussion has reviewed three empirical assessment methodologies and presented information on the application of the Surface Impoundment Assessment methodology to 77 waste water ponds, 13 septic tank system areas, and 10 landfills in the central Oklahoma area overlying the Garber-Wellington aquifer. The study area included 80 townships and 2880 sections. A cursory field sampling program related to the top 10 waste water ponds, 4 septic tank system areas, and 1 landfill indicated potential evidence of ground water pollution at all 15 sources.

New prioritization methodologies were developed for oil and gas wells and dry holes as well as for salt water disposal wells and waterflood operations. These methodologies were applied to 14,127 oil and gas wells and dry holes, and 1051 salt water disposal wells and waterflood operations in the study area. A field sampling program in 13 of the top 57 sections (2% of the total number of sections) revealed that 12 exhibited evidence of increased TDS in aquifers used as drinking water sources.

Several research needs can be identified in conjunction with empirical assessment methodologies. Included are needs for field verification studies, professional input into methods development, and more structured value judgment approaches. Although several methodologies have been developed, and additional methodologies are planned, only cursory field verification of the results of methodologies has been conducted. The empirical assessment methodologies generally tend to be developed by single persons or small groups of professionals, and there is need for greater input from a wider range of professionals associated with ground water quality management. Finally, since empirical assessment methodologies require the considerable exercise of value judgment, more structured approaches for soliciting value judgments are needed.

ACKNOWLEDGMENT

This research has been sponsored by the National Center for Ground Water Research (NCGWR) through a Cooperative Agreement with the U.S. Environmental Protection Agency. The NCGWR is a consortium consisting of the University of Oklahoma, Oklahoma State University, and Rice University.

REFERENCES

Bingham, R. H. and Moore, R. L. (1975). Reconnaissance of the water resources of the Oklahoma City quadrangle, central Oklahoma. *Hydrologic Atlas 4,* Oklahoma Geological Survey, Norman, OK.

Burton, L. C. and Jacobsen, C. L. (1967). Geologic map of Cleveland and Oklahoma counties, Oklahoma. Oklahoma Geological Survey, Norman, OK.

Carriere, G. D. (1980). Priority ranking of septic tank systems in the Garber-Wellington area. NCGWR 80-32 (Nov.), National Center for Ground Water Research, Norman, OK.

Dean, B. V. and Nishry, J. J. (1965). Scoring and profitability models for evaluating and selecting engineering products. *J. Oper. Res. Soc. Amer.* **13**:550–569.

Fairchild, D. M., Hall, B. J., and Canter, L. W. (1981). Prioritization of the ground water pollution potential of oil and gas field activities in the Garber-Wellington area. NCGWR 81-4 (Sept.), National Center for Ground Water Research, Norman, OK.

Hagerty, D. J., Pavoni, and Heer, J. E., Jr. (1973). *Solid Waste Management*. Van Nostrand Reinhold, New York.

Hajali, P. A. (1980). Priority ranking of sanitary landfills in the Garber-Wellington area. NCGWR 80-35 (Nov.), National Center for Ground Water Research Norman, OK.

LeGrand, H. E. (1964). System of reevaluation of contamination potential of some waste disposal sites. *J. Am. Water Works Assoc.* **56**:959–974.

McKee, J. E. and Wolf, H. W. (1963). *Water Quality Criteria*, 2nd ed. California State Water Quality Control Board, Sacramento.

Mogg, J. L., Schoff, S. L., and Reed, E. W. (1969). Ground water of Canadian county. Bulletin 87, Oklahoma Geological Survey, Norman, OK.

Phillips, C. R., Nathwani, J. D., and Mooij, H. (1977). Development of a soil–waste interaction matrix for assessing land disposal of industrial wastes. *Wat. Res.* **11**:859–868.

Sammy, G. K. and Canter, L. W. (1980). Priority ranking of waste water pond systems in the Garber-Wellington area. NCGWR 80-31 (Nov.), National Center for Ground Water Research, Norman, OK.

U.S. Environmental Protection Agency (1976). *Quality Criteria for Water*. U.S. Government Printing Office, Washington, D.C.

U.S. Environmental Protection Agency (1978). *A Manual for Evaluating Contamination Potential of Surface Impoundments*. EPA 570/9-78-003, Office of Drinking Water, Washington, D.C.

Voelker, A. H. (1977). Power plant siting—an application of the nominal group process technique. ORNL/NUREG/TM-81, Oak Ridge National Laboratory, Oak Ridge, TN.

14

BIOCHEMICAL MEASURES OF THE BIOMASS, COMMUNITY STRUCTURE, AND METABOLIC ACTIVITY OF THE GROUND WATER MICROBIOTA

David C. White
Janet S. Nickels
Jeffrey H. Parker
Robert H. Findlay
Michael J. Gehron
Glen A. Smith
Robert F. Martz

Department of Biological Science
Florida State University
Tallahassee, Florida

Ground water is an indispensable resource providing the major supply of fresh water to communities and industry in many areas of the world. The land disposal of domestic and much industrial waste has in some celebrated cases seriously contaminated this vital resource. Consequently the understanding of how ground water can be protected, or can be purified once contaminated, necessitates the study of the unique microbial community that recent studies have begun to define in samples of the ground water substrate that have been carefully protected from contact with surface contaminants.

The application of biochemical methods has provided a means by which the microbiota can be examined without the selection and consequent bias of plating methods that require microbial growth. Lipid analysis has recently been utilized in studies of marine and estuarine microbial ecology. Lipid extraction is a relatively rapid and simple means for recovering components from the membranes of marine microbes and gaining an initial purification with a relatively easily achieved hundredfold concentration of those components. Lipids in the estuarine detritus and marine sediments have a rapid turnover (King et al., 1977; White et al., 1979c), and consequently can be utilized to measure the "viable" biomass. Since both the prokaryotes and microeukaryotes contain membrane lipids, the total membrane content of a sediment can be used as a measure of the "viable" biomass. Because of differences in lipid composition between groups of microbes, it has been possible to gain insight into the changes in community structure of these microbial assemblies. For example, the fatty acid components of the lipids of marine sedimentary microbes are readily recoverable and can be separated by glass capillary gas-liquid chromatography into at least 260 components (Bobbie and White, 1980). Many of these fatty acids can be assigned to specific groups of microorganisms. For example, the various component portions of the bacteria are enriched in short chain iso- and anteiso-branched fatty acids (Schultz and Quinn, 1973), while others form the 3-carbon cyclopropane ring from monounsaturated fatty acids esterified to phospholipids incorporated in the microbial lipids (Law, 1971). These cyclopropane fatty acids accumulate if the culture is stressed by nutrient exhaustion, decrease in terminal electron acceptors, or the presence of inhibitors (Knivett and Cullen, 1965). Certain other bacteria form the monounsaturated fatty acids during anaerobic chain elongation at the level of the decyl derivative, a formation that results in cis-vaccenic rather than oleic acid (Bloch, 1969). The microeukaryotes do not form significant amounts of these fatty acids but do form long chain (longer than 20 carbon atoms) polyunsaturated fatty acids. These long chain polyunsaturated fatty acids are not found in bacteria (Kates, 1964), with the exception of some of the morphologically complex cyanophytes, which form 18-carbon polyenoic fatty acids (Kenyon et al., 1972), and some *Flexibacteria* (Johns and Perry, 1977).

Using these simple criteria together with other lipid markers and physiologic activities of components of the marine microbial community, it has been possible to begin to define the biological structure of the attractant of grazing amphipods to the estuarine detrital microbiota (White et al., 1979d), the shift in microbial communities in the microfouling film formed on different metals (Berk et al., 1981), and to measure the efficiency of removing attached bacteria on metal surfaces to facilitate heat transfer (Nickels et al., 1981a, 1981c; White and Benson, 1984). These same criteria can be used to define the effects of light on the formation of the detrital microbiota (Bobbie et al., 1981), the effects of sand grain microtopology on the colonization of marine sands (Nickels et al., 1981b), the effects of oil and gas well drilling fluids on the colonization of marine sands (Smith et al., 1982a), the response of the sedimentary microbiota to grazing by sand dollars (White et al., 1980b), the response of the detrital microbiota to grazing by amphipods (Morrison and White, 1980), the partitioning of the detrital microbiota between sympatric amphipods (Smith et al., 1982b), and the recolonization of fecal mounds of the enteropneusts *Ptychodera bahamensis* (Fazio et al., 1982).

Several methods have been utilized to provide validation of the changes in community structure of the microbiota induced by the various treatments listed above. The first method is more a check on the analytical reproducibility than a validation of community structure. This method consists of the analysis of known mixtures of organisms. For example, as the proportion of *E. coli* was increased in constant weight mixtures containing *Neurospora crassa,* the concentrations of muramic acid, extractable lipid phosphate, and cis-vaccenic acid increased while the proportions of ergosterol, polyenoic fatty acids longer than 20 carbon atoms, the lipid glucose, and wall inositol decreased (White et al., 1980a).

A second method of validation involves the isolation of organisms from the environmental assembly and analysis of the lipids, showing the presence of particular lipids in both the mixture and the isolated monocultures. This technique has been exploited in the study of marine sediments by R. B. Johns (Perry et al., 1979) and in the isolation of estuarine fungi (White et al., 1980a). This method assumes that no major changes in lipid composition occur during the isolation procedure.

A third method of validation involves the manipulation of the microbiota by changing nutrient feed-stocks or growth conditions, or introducing antibiotics, so that the biochemistry can be compared to the predicted changes in morphology by scanning electron microscopy. In this way the estuarine detrital microbiota was shifted from a fungus "heaven" (acid pH, nutrient broth rich in sucrose with penicillin and streptomycin) to a fungus "hell" (alkaline pH, glutamine and phosphate with cyclohexylimide) for 7 days. On analysis the obvious strands of fungal mycelia were found in the "heaven" along with marked increases in the ergosterol and other steroids, the sulfolipid synthesis, the lipid glycerol, the polyenoic fatty acids, the long chain saturated fatty acids, the lipid inositol, and lysine. In the fungus "hell" there were no mycelia strands, only bacteria with increases in muramic acid, phospholipid synthesis, DNA formation from thymidine, short chain saturated and branched fatty acids, cis-vaccenic acid, cyclopropane fatty acids, lipid ethanolamine and xylose (White et al., 1980a).

The fourth method of validation involves use of specific inhibitors such as molybdate for the sulfate reducing bacteria and chloroform for the methane reducing bacteria. These inhibitors affect the rates of sulfate reduction or methane generation in expected ways in marine sediments (Oremland and Taylor, 1978). The sulfate reducing bacteria have been shown to contain unique ester-linked extractable hydroxy fatty acids, and H. L. Fredricksor from this laboratory has shown a correlation between sulfate reduction and the presence of the extractable ester-linked hydroxy fatty acids, which are stimulated by the addition of sulfate and inhibited by molybdate (Fredrickson, 1981).

The fifth method for validating the assignment of "signature" lipids is to exploit the known specificity in the feeding behavior of detrital grazers or to deposit feeding animals in the sediments. The microbiota is exposed and the residue examined for the changes expected. Gammaridean amphipods graze the detrital microbiota, and their removal of the slower growing algae and fungi leaves a residue enriched in rapidly growing bacteria and diatoms (Morrison and White, 1980; Smith et al., 1982b). The sand dollar *Mellita quinquiesperforata* specifically harvests the non-photosynthetic microeukaryotes from the sedimentary microbiota (White et al., 1980b).

METHODS

Samples

Aquifer sediments were recovered with the autoclavable drill bits described by Wilson et al. in this symposium volume (Chapter 26); the sediments were frozen in dry ice and shipped to our laboratory for analysis.

Extraction

The sediments were lyophilized, the dry weight determined and then extracted by the modified Bligh and Dyer single phase chloroform-methanol method (White et al., 1979c). The lipid fraction, the aqueous fraction, and the lipid-extracted residue were separated and subjected to the analytical sequence illustrated in Figure 14.1.

Analysis of the Lipids

The lipid fraction was dehydrated by passage through Whatman 2v filter paper and then fractionated into neutral and phospholipids by chromatography on silicic acid columns (King et al., 1977). The neutral lipids are particularly valuable in the determination of the community structure of the microeukaryotes by analysis of the steroids (White et al., 1980a) and also of the nutritional status by examination of the triglyceride glycerol (Gehron and White, 1982). The neutral lipids, when fractionated in this way, also contain the prokaryotic endogenous storage lipid poly-β-hydroxy butyrate, which has been shown to form during conditions of unbalanced growth in environmental samples (Nickels et al., 1979; Findlay and White, 1983). The phospholipid fraction was recovered and subjected to mild alkaline hydrolysis, which quantitatively deacylates the lipids (King et al., 1977). The fatty acids

Analysis

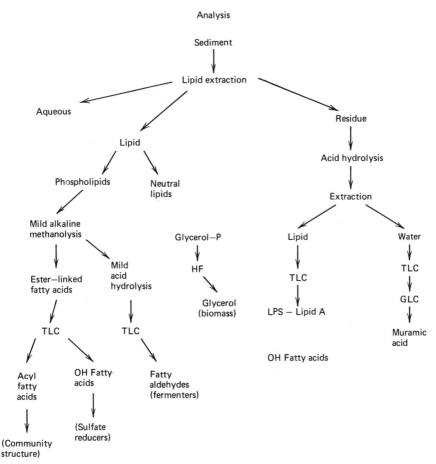

Figure 14.1.   Diagram of the analytical procedure for the microbiota of the ground water strata.

and undeacylated phospholipids were recovered in the organic phase, and the glycerol phosphoryl esters that had ester-linked fatty acids were recovered in the aqueous phase. The ground water microbiota is so relatively sparse that the colorimetric analysis following perchloric acid digestion, which is sensitive to 10 μM phosphate and has proved such a valuable measure of the "membrane" biomass (White et al., 1979a, 1979b, 1979c), could not be utilized. Consequently a new methodology was developed in which the glycerol phosphoryl esters are acid hydrolyzed to release the glycerol, ethanolamine, choline, serine, and so on, as well as cyclic glycerol phosphate. The glycerol from phosphatidyl glycerol or cardiolipin is triacetylated and measured by gas-liquid chromatography (GLC) (Gehron and White, 1982), as is the cyclic glycerol phosphate after hydrolysis by concentrated hydrofluoric acid at 0°C (Gehron and White, 1983).

The lipid fraction was taken to dryness in a stream of nitrogen and redissolved in cyclohexane; the fatty acids were extracted quantitatively into aqueous bicarbonate and then fractionated by thin layer chromatography (TLC) into acyl and hydroxy fatty acid bands. The bands were recovered, and the acyl fatty acid band was methylated and analyzed by glass capillary GLC (Bobbie and White, 1980). The hydroxy fatty acids were methylated and then acylated with heptafluorobutyrate prior to analysis by GLC.

The remaining lipids were redissolved in chloroform and subjected to mild acid hydrolysis in the presence of mercury, which quantitatively liberates the fatty aldehydes from the vinyl ether containing plasmalogens (White et al., 1979b, 1979c). The aldonitriles of the fatty aldehydes were formed from pentafluorophenyl hydroxylamine and analyzed by capillary GLC.

Water-Soluble Portion of the Lipid Extraction

The water-soluble portion of the lipid extract contains the free amino acids and nucleotides of the cell sap of the organisms in the sediments. This fraction has been utilized to measure the adenylate energy charge and the more sensitive adenosine/adenosine triphosphate ratio, which measures one of the homeostatic mechanisms by which the energy charge is maintained (Davis and White, 1980), or to measure the free amino acid pools as a response to oil and gas well drilling fluid exposure to reef-building corals (White et al., 1983a). This technique has not yet been applied to the ground water samples.

Lipid-Extracted Residue

The residue after lipid extraction was subjected to mild acid hydrolysis ($1M$ HCl, 100°C, 5 hr) and extracted. The organic phase was recovered, redissolved in cyclohexane, and the fatty acids extracted with aqueous bicarbonate. The fatty acids were then purified by TLC, methylated, acetylated, and analyzed by capillary GLC. This provides a measure of the amount and composition of the lipid A component of the lipopolysaccharide of the gram-negative bacteria (Parker et al., 1982).

The water-soluble fraction was subjected to further hydrolysis ($6M$ HCl, reflux, 4.5 hr), the hydrolysate carefully neutralized and centrifuged, and the supernatant purified by TLC. The peracylated aldonitrile of muramic acid was assayed by GLC using a modification of the procedure of Fazio et al. (1979), further developed by Findlay et al. (1983).

Scanning Electron Microscopy

Scanning electron microscopy (SEM), after fixation in glutaraldehyde and successive dehydrations as described by Morrison et al. (1977) and the elemental analysis by a Tracor-Northern energy dispersive X-ray analysis, was performed by W. Miller III of the Biology Department, Florida State University.

Fatty Acid Designation

Fatty acids are designated as the number of carbon atoms: the number of double bonds with the position of the unsaturation nearest the w end of the molecule (the end opposite the carbonyl). The prefixes a, i, and delta indicate anteiso- or iso-branching or the presence of a cyclopropane ring in the chain, respectively. The prefix OH indicates a hydroxy fatty acid, and the number indicates the position.

RESULTS WITH DISCUSSION

Samples

Samples were collected from two cores, positions 6b and 7, at a site in Fort Polk, Louisiana by J. McNabb and M. R. Scalf from the R. S. Kerr Environmental Research Laboratory at Ada, OK, and S. Hutchins of Rice University. The apparatus and the precautions to prevent surface contamination are described by Wilson et al. elsewhere in this volume (Chapter 26). The samples were frozen in dry ice and shipped to the laboratory for analysis. SEM of the samples showed a fine clay with no obvious microbiota (Figure 14.2). The composition averaged 18,800 units of aluminum (K_a line at 1,476 meV), 25,300 units of silicon (K_a line at 1.741 meV), 934 units of potassium (K_a line at 3.315 meV), 357 units of calcium (K_a line at 3.650 meV), 1350 units of titanium (K_a line at 4.517 meV), 2590 units of iron (K_a line at 6.396 meV), and 161 units of nickel (K_a line at 7.507 meV).

Samples were taken from bore 7 at 4–5 ft and 12–13 ft in the unsaturated zone and 16.5 ft in the saturated zone, and from bore 6b at 13–14 ft and 18–19 ft in the unsaturated zone and 25–26 ft in the saturated zone.

Biomass

Four estimates of the microbial biomass were utilized in this study. Muramic acid is an amino sugar with a lactoyl ether that is unique to the prokaryotic cell wall (King and White, 1977; Fazio et al., 1979). Muramic acid concentration in the unsaturated zone is an order of magnitude less than that found in the saturated zone and at least a hundredfold less than that found in surface estuarine sediments (Table 14.1).

The conversion figures, developed by Millar and Casida (1970) and Findlay et al. (1983), of 88 (7.2) μmol/g dry wt (X $\pm$ S.D.) for gram-positive and 29 (1.9) μmol/g dry wt for gram-negative bacteria give indications of about 10^7 bacteria per gram dry wt if all were gram-positive and 14×10^8 per

Figure 14.2. Scanning electron micrographic structures of the ground water substrate. Top, 24-fold magnification; middle, 260-fold magnification; bottom, 2600-fold magnification.

Table 14.1   Estimation of the Microbial Biomass from the Fort Polk Ground Water Sediments

Measure	Unsaturated Zone	Saturated Zone	Estuarine Sediment[a]	Estimated Number of Microbes in Unsaturated Zone
Muramic acid (nmol/g dry wt)				
	2.06 (1.24)	11.3	22 (18)	2.3×10^7 $(+)^b$
				7.0×10^7 $(-)$
Lipopolysaccharide lipid A (pmol/g dry wt)				
3 OH 14:0	148 (68)	353 (175)	2260 (348)	9×10^{6c}
3 OH 16:0	28 (18)	168 (149)	662 (99)	
3 OH 18:0	3.5 (4.0)	<0.1	79 (158)	
12 OH 18:0	3.8 (4.3)	67 (83)	360 (245)	
Total lipid acyl and OH fatty acids (nmol/g dry wt)				
	7.13 (4.4)	14.7 (1.8)	77.6 (6.4)	1.4×10^{6d}
Total lipid glycerol phosphate (nmol/g dry wt)				
	0.98 (0.18)	5.5 (x.x)	26 (7)	1.9×10^{6e}

[a] Taken from surface of an estuarine embayment (29° 54'N, 84° 27.5'W).
[b] Calculated using 29×10^{-6} mol/g if all gram-negative, 88×10^{-6} mol/g dry wt if all gram-positive and 10^{12} cells/g wt.
[c] Calculated using 15×10^{-6} mol/g dry wt for *E. coli* 3 OH 14:0.
[d] Calculated using 1×10^{-4} mol/g dry wt.
[e] Calculated using 5×10^{-5} mol/g dry wt.

gram dry wt if all were gram-negative. This assumes that the bacteria were like those recovered from the surface substrates and were grown in relatively rich culture media.

Gram-negative bacteria contain the lipopolysaccharides, (LPS) that contain a lipid A component. The lipid A consists of a polysaccharide backbone that contains covalently linked fatty acids, particularly hydroxy fatty acids (Wilkinson, 1977). The LPS can be quantitatively recovered from environmental sediments (Saddler and Wardlaw, 1980; Parker et al., 1982) and assayed by the hydroxy fatty acids. *E. coli* in our hands contains 14 μmol 3 OH 14:0/g dry wt, which suggests that there are about 9×10^6 gram-negative organisms/g dry wt if all the organisms were like *E. coli* and grown in rich media. This methodology also gives some insight into the community structure since some microorganisms contain "fingerprints" of hydroxy fatty acids. Some examples are the *Pseudomonads,* which contain both shorter and longer hydroxy fatty acids (Wilkinson, 1977; Parker et al., 1982); the *Bacteroides,* which contain amide-linked branched hydroxy fatty acids

(Mayberry, 1980a,b); and the *Legionella*, which contain the dihydroxy fatty acids (Mayberry, 1981). A "fingerprint" of the ground water microbial LPS is shown in Figure 14.3.

The total lipid phosphate has proven a good measure of the biomass of many laboratory grown strains of eubacteria since the content of many of these strains averages about 50 μmol lipid phosphate/g dry wt (White et al., 1979a). The phospholipids are a good measure of the "living" membrane biomass since they have a relatively rapid turnover in monocultures (White and Tucker, 1969) and in living or dead organisms in sediments (White et al., 1979c). In the bacterial phospholipids there are 2 mol of fatty acids per molecule of phosphate on the average. The total fatty acids (acyl and hydroxyl) of the ground water microbiota, both saturated and unsaturated, indicates that there are about 8×10^7 organisms/g dry wt if the organisms are like the laboratory cultures that have been examined (Table 14.1). The estuarine surface microbiota contains at least an order of magnitude more organisms than does the ground water microbiota.

The lipid phosphate itself has been widely used in studies from this laboratory as a measure of "membrane" biomass (White et al., 1979c). The colorimetric analysis of the lipid phosphate is sensitive to only 0.05 μmol. For the ground water microbiota a new methodology based on the sensitivity of the glycerol assay by GLC was developed (Gehron and White, 1982). Water-soluble glycerol phosphate esters are quantitatively liberated from diacyl phospholipids by mild alkaline methanolysis (King et al., 1977). These esters can be hydrolyzed in 1 M methanolic HCl containing 0.1 volumes of chloroform to yield ethanolamine, choline, serine, glycerol, and so on, and cyclic glycerol phosphate. The cyclic glycerol phosphate resists acid hydrolysis—6 M HCl at reflux for 72 hr yielded only 6% of the phosphate from glycerol phosphate. Cyclic glycerol phosphate can be quantitatively cleaved by concentrated hydrofluoric acid at 0°C for 72 hr to yield glycerol and phosphate (Gehron and White, 1983). If the aqueous portion of the mild alkaline methanolysis is subjected to acid methanolysis, the glycerol phosphate esters are cleaved to cyclic glycerol phosphate and the choline, ethanolamine, serine, or glycerol, and so on that was a part of the ester. The glycerol released from the glycerol phosphoryl glycerol or bis glycerol phosphoryl glycerol, which are the esters released by mild alkaline methanolysis from the phosphatidyl glycerol or cardiolipin, can be estimated by GLC. This is the acid labile glycerol. The total glycerol released from the aqueous portion of the mild alkaline methanolysis after hydrofluoric acid hydrolysis from the unsaturated portion of the sediments was 1.6 (0.33) nmol/g dry wt, of which 26.2 (2.9)% was labile to mild acid (Table 14.1). This indicates about 0.98 (0.18) nmol/g dry wt of phospholipid. Since 5×10^{-4} mol of phospholipid/g dry wt are found in most bacteria (White et al., 1979a), the unsaturated ground water sediment contains about 2×10^7 cells/g dry wt (Table 14.1). The relatively high level of acid labile glycerol indicates a high concentration of phosphatidyl glycerol and cardiolipin, two lipids prominent in bacteria (Kates, 1964).

Community Structure

Examination of the phospholipid ester-linked acyl fatty acids shows a remarkable difference between the estuarine surface sediments from a sand bar (29° 54.0'N, 84° 27.5'W) and the ocean floor at 4620 m at 40°24'N, 63° 7.36'W (Table 14.2). Looking at the proportions of the acyl fatty acids relative to the palmitic acid (16:0), the short branched and short fatty acids characteristic of the gram-positive *cocci* and some other organisms (Bobbie and White, 1980) show little differences in proportions between the marine sediments (Table 14.2). Bacteria containing the anaerobic desaturase pathway that results in the formation of cis-vaccenic acid form a smaller proportion of the ground water microbiota than of the surface or deep sea microbiota, as estimated either by the proportion of the palmitic acid (16:0) or the

Table 14.2 Ester-Linked Fatty Acids from the Phospholipids of Saturated and Unsaturated Zones of the Fort Polk Ground Water Sediments, Deep Sea Sediments, and Surface Estuarine Sediments

Fatty Acid	Unsaturated Ground Water	Saturated Ground Water (pmol/g dry wt)[a]	Deep Sea Sediments	Estuarine Surface Sediments
i + a 15:0	132 (102)	274 (264)	1400 (800)	3470 (1000)
15:0	211 (160)	229 (235)	700 (300)	15,470 (750)
16:0	1770 (1650)	579 (368)	6400 (3800)	38,440 (1800)
delta 17:0 + 19:0	<0.1	<0.1	350 (130)	210 (140)
18:1w7	137 (138)	168 (180)	2700 (480)	4450 (1640)
18:1w9	1160 (1110)	413 (420)	4200 (1300)	2740 (950)
18:2w6	905 (931)	1213 (1549)	1270 (800)	1480 (840)
24:0	30 (30)	1.0 (1.3)	300 (80)	290 (40)
20:4w6	<0.1	<0.1	450 (140)	580 (170)
20:5w3	<0.1	<0.1	250 (50)	220 (120)
22:6w3	<0.1	<0.1	700 (50)	200 (110)
Total polyenoic >20	<0.1	<0.1	900 (350)	3370 (1510)
Community Structure				
a + i 15:0/15:0	1.1 (0.56)	1.5 (0.14)	1.9 (0.3)	0.25 (0.07)
a + i 15:0/16:0	0.16 (0.1)	0.24 (0.05)	0.13 (0.06)	0.11 (0.04)
18:1w7/16:0	0.11 (0.04)	0.1 (0.03)	0.33 (0.1)	0.16 (0.05)
18:1w7/18:1w9	0.17 (0.07)	0.26 (0.12)	0.69 (0.29)	1.81 (0.12)
18:2w6/16:0	0.49 (0.1)	0.37 (0.01)	0.15 (0.6)	0.04 (0.01)

[a] Values given as X (S.D.), for the samples ($n = 4$) from the unsaturated and saturated ($n = 2$) strata of the Fort Polk ground water aquifer, the deep sea (-4620 m) at 40°24' N, 63°7.36' W ($n = 6$), and from an estuarine sand bar at 29°54.0' N, 84°27.5' W ($n = 6$).

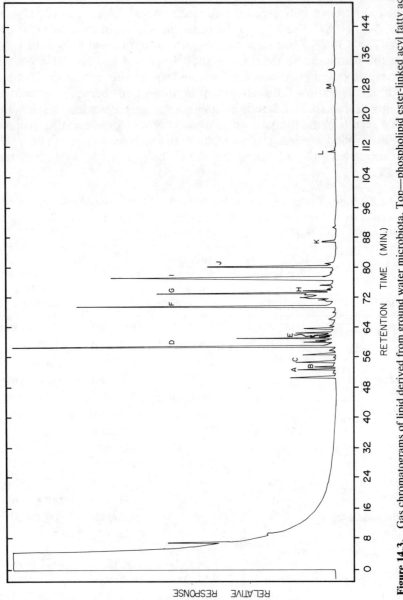

Figure 14.3. Gas chromatograms of lipid derived from ground water microbiota. Top—phospholipid ester-linked acyl fatty acid methyl esters: A = i 15:0, B = a 15:0, C = 15:0, D = 16:0, E = 16:1w7, F = 18:0, G = 18:1w9, H = 18:1w7, I = 19:0 (internal standard, 160 pmol), J = 18:2w6, K = 20:0, L = 22:0, M = 24:0.

RELATIVE RESPONSE

RETENTION TIME (MIN.)

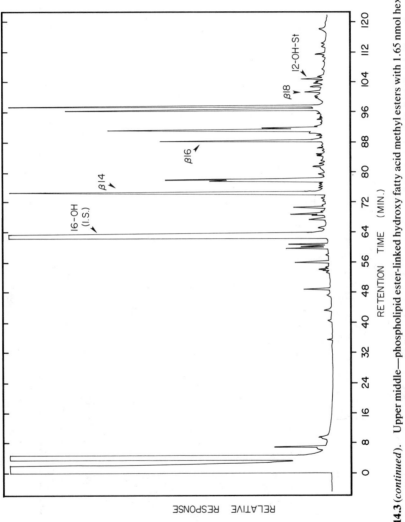

Figure 14.3 (*continued*). Upper middle—phospholipid ester-linked hydroxy fatty acid methyl esters with 1.65 nmol hexadecanol-heptafluorobutyrate (16 : 0H) as internal standard 3 OH 14 : 0, 3 OH 16 : 0, 3 OH 18 : 0, and 12 OH 18 : 0 indicated.

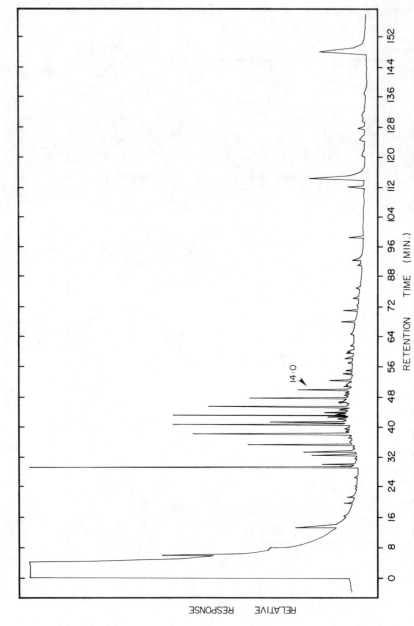

Figure 14.3 (*continued*). Lower middle—plasmalogen derived fatty aldehyde pentafluorobenzyloximes with the 14 : 0 oxime (18.9 pmol) as the internal standard.

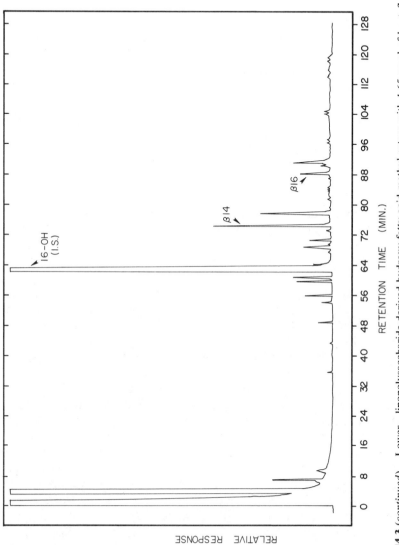

Figure 14.3 (*continued*). Lower—lipopolysaccharide derived hydroxy fatty acid methyl esters with 1.65 nmol of heptafluorobutaryl hexadecanol internal standard and the 3 OH 14 : 0 and 3 OH 16 : 0 indicated. The separations were performed on a Silar 10C coated 50-m glass capillary column at a helium flow rate of 1 mL/min, with splitless injection (0.5 min venting time) of a 2 μL hexane injection that was linearly programmed from 42 to 179°C at 1°/min with an isothermal period to the end of the chromatogram.

321

oleic acid (18 : 1w9) isomer. The ground water microbiota contains a higher proportion of linoleic acid (18 : 2w6) than the surface or deep marine sediments. Linoleic acid is found typically in gliding bacteria and some microeukaryotes. One striking feature of the acyl fatty acid analysis is the absence of polyenoic fatty acids, which indicates the absence of microeukaryotes such as the algae, fungi, protozoa or the micrometazoa common in surface sediments. This finding helps confirm the observations of Ghiorse and Balkwill in this volume (Chapter 20) of the absence of microbes with eukaryotic morphology or staining characteristics. Ghiorse and Balkwill, using transmission electron microscopy, show single cells and colonies of very small gram-positive and gram-negative *cocci* and *cocci*-bacillary prokaryotic forms surrounded by a ruthenium red-staining glycocalyx that binds the organisms to clay particles. These organisms do not show the morphological diversity seen in similar studies of the surface soils or sediments. The uniqueness of the aquifer-bound microbiota is reflected in the acyl fatty acids ester linked to the phospholipids, as illustrated in Figure 14.3. The "fingerprint" is remarkably different from the same type of chromatographic analysis of the estuarine detrital microbiota (Bobbie and White, 1980), the marine microfouling film (Nickels et al., 1981a), or the marine sedimentary microbiota (Nickels et al., 1981b).

The analysis of the hydroxy fatty acids of the lipid A of the LPS of the gram-negative bacteria provides information about the community structure. Comparison of the total muramic acid and the lipid A content strongly suggests that there is a high proportion of gram-positive bacteria in the ground water microbiota. This agrees with the electron micrographic observations of Ghiorse and Balkwill (Chapter 20). There is a progressive enrichment of gram-positive bacteria in marine sediments as the depth increases (Moriarty, 1980). The "fingerprint" of the covalently bound hydroxy fatty acids shows a difference between the saturated and unsaturated aquifer microbiota. The saturated zone contains gram-negative bacteria with a high content of 12 OH 18 : 0. The "fingerprint" of the covalently linked hydroxy fatty acids is illustrated in Figure 14.3.

In the studies of H. L. Fredrickson of this laboratory (1981), it has been possible to expand the observation that the sulfate reducing bacteria contain very distinctive branched unsaturated hydroxy fatty acids (Boon et al., 1977) by showing that these hydroxy fatty acids are ester-linked and a part of the extractable phospholipids. He has further shown that these signature lipid components rise or fall in parallel with the formation of $H_2{}^{35}S$ from ${}^{35}SO_4$, which is a measure of the sulfate reducing microbial activity. There are parallel increases in the signature lipids and the microbial activity when strong selection is applied in a chemostat or sediments are enriched in sulfate. The activities are depressed when the specific inhibitor molybdate is present (Sorensen et al., 1981). In addition, there are different "fingerprints" in the phospholipid ester-linked hydroxy fatty acids when the substrate is shifted from acetate to lactate. The parallelism in activity and "signature" lipids has existed for every sediment tested.

Table 14.3   Ester-Linked Hydroxy Fatty Acids from the Phospholipids of Saturated and Unsaturated Zones of the Fort Polk Ground Water Sediments and Surface Estuarine Sediments

Fatty Acid	Unsaturated Ground Water	Saturated Ground Water	Estuarine Surface Sediments
		(pmol/g dry wt)[a]	
3 OH 14:0	589 (295)	203 (109)	29 (65)
3 OH 16:0	290 (69)	81 (88)	240 (71)
3 OH 18:0	40 (50)	150 (211)	116 (79)
12 OH 18:0	67 (15)	340 (480)	31 (49)

[a] Values given as X (S.D.), for the samples ($n = 4$) from the unsaturated and saturated ($n = 2$) strata of the Fort Polk ground water aquifer, and from an estuarine sand bar 29°54.0' N, 84°27.5' W ($n = 6$).

Examination of the ground water sediment shows a large amount and variety of phospholipid ester-linked hydroxy fatty acids (Table 14.3), which are more numerous in the saturated zone than in the unsaturated zone. Many of the hydroxy fatty acids remain to be identified, but the pattern is clearly different from the signature hydroxy lipids of the sulfate reducer *Desulfovibrio vulgaris* (Figure 14.3).

The plasmalogens with the 1-alk-ene ether bound hydrocarbon chain are confined to the anaerobic fermenters in the microbial world (Goldfine and Hagen, 1972). The plasmalogens yield fatty aldehydes on hydrolysis. We have used the stability to mild alkali and lability to mild acid to demonstrate that there are 10 times more anaerobes in the aerobic sediment than in the anaerobic sediment (White et al., 1979b, 1979c). The anaerobes clearly live in association with heterotrophic bacteria, whose metabolism generates the anaerobic niches in the mixed microcolonies. In the ground water microbiota there is clearly a high proportion of fatty aldehydes released from the phospholipids of the ground water microbiota. The "fingerprints" are clearly different from the known fatty aldehydes released from *Clostridium butyricum* (Khuller and Goldfine, 1974).

Metabolic Activities

There are two general methods that could be applied to ground water systems to determine the activity of this unique microbial assembly.

Dynamic methods involve the application of test components to the microbiota to determine activities. The activity measurements will be important in relating the biodegradative activity of the microbes determined within the column microcosms described by Wilson et al. (Chapter 26) to that in the field. One particularly attractive method relies on the use of ^{13}C and the biochemical methodology described above. If a substrate such as ^{13}C-acetate can be placed in the aquifer and then sampled after a time, the activity of various components of the community can be analyzed by comparing the enrichment of ^{13}C in the "signature" lipids we have defined. The ^{13}C is particularly valuable because it is not radioactive and will not contaminate the field. It has specific activities at least 10^6 times that of the radioactive ^{14}C and can thus be used at much lower concentrations. The ^{13}C can be detected efficiently by combined capillary GLC-mass spectrometry operating in the selective ion mode. Such methods have not as yet been applied to the ground water microbiota but have proved useful in studies of detritus (Findlay et al., 1983).

The other method relies on the analysis of the products of unbalanced growth to give a nutritional history of the microbiota. Two compounds in the prokaryotes have been identified as useful in this analysis. Poly-β-hydroxy butyrate (PHB) accumulates during conditions of unbalanced growth when carbon and energy sources are adequate but some essential nutrient prevents growth (Nickels et al., 1979). The method involves an extraction with perchlorate and chloroform (Nickels et al., 1979) modified by a subsequent acid hydrolysis followed by derivatization and analysis of the 3 OH 4 : 0 by GLC; the method is sufficiently sensitive for the detection of PHB in the sparse microbiota of the ground water aquifer (Findlay and White, 1983). The unsaturated ground water sediments yielded 0.13 (0.04) nmol 3 OH 4 : 0/g dry wt. Surface estuarine sediments yielded 2.45 nmol 3 OH 4 : 0/g dry wt. The yield of PHB/bacteria estimated as muramic acid (MA) was 60 pmol PHB/nmol MA in the aquifer and 2.8 pmol PHB/nmol MA in the surface sediments. This indicates conditions of unbalanced growth, which drive PHB to accumulate to a 20-fold greater concentration per cell in the ground water microbiota than is found in the surface microbiota. The second component is the uronic acid-rich extracellular polysaccharide polymers, which can now be assayed quantitatively (Fazio et al., 1982). These polymers can be seen in the ruthenium red-staining material surrounding the organisms in electron micrographs (Ghiorse and Balkwill, Chapter 20). The formation of these polymers is stimulated by conditions of unbalanced growth in marine sediments (Uhlinger and White, 1983).

CONCLUSIONS

From these initial experiments based on one site and two samplings, a number of conclusions can be drawn:

1. There is a microbiota in the ground water sediments.
2. This microbiota is sparse compared to the surface microbiota by four measures of microbial biomass.

3. This microbiota is clearly different by a number of criteria from the surface microbiota. There is a large proportion of gram-positive bacteria. The microbiota is greatly enriched in sulfate reducing and fermentative bacteria that contain different lipid "fingerprints" compared to the microbiota found at the surface or deep sea.

4. There is an absence of microeukaryotes such as algae, fungi, protozoa, and micrometazoa in the ground water microbiota.

5. The microbiota of the ground water system exists under conditions of unbalanced growth in which the cells are metabolizing but not dividing. This is reflected in the 20-fold higher content of PHB/cell and the large amount of uronic acid-rich extracellular polysaccharide glycocalyx seen in the electron microscope. These microbes not only constitute a different community but exist in different metabolic status.

6. Methods exist or can be readily adapted by which the activity of this unique bacterial community can be correlated to its distinctive biodegradative activities. This will be essential in the management decisions necessary to protect this vital resource.

7. In the period between the conference and this publication, ground water aquifer sediments from 410 m below the surface were analyzed with these same methods and quite similar microbiota was detected (White et al., 1983b).

ACKNOWLEDGMENTS

This work was supported by contract N00014-75-C-0201 from the Department of the Navy, Office of Naval Research, Ocean Science and Technology Detachment, NSTL, MS; contract 80-7321-01 of the U.S. Environmental Protection Agency, administered by Gulf Breeze Environmental Research Laboratory, Gulf Breeze, FL; contract R-809994 of the U.S. Environmental Protection Agency, administered by the Robert S. Kerr Environmental Research Laboratory of the U.S. EPA; contract 31-109-38-4502 from the Department of Energy, Argonne National Laboratories, Argonne IL; grant 04-7-158-4406 from the National Oceanic and Atmospheric Administration, Office of Sea Grant, Department of Commerce; and grants OCE 76-19671 and DEB 78-18401 from the Biological Oceanography Program of the National Science Foundation.

REFERENCES

Berk, S. G., Mitchell, R., Bobbie R. J., Nickels, J. S., and White, D. C. (1981). Microfouling on metal surfaces exposed to seawater. *Int. Biodeterioration Bull.* **17**:29–37.

Block, K. (1969). Enzymatic synthesis of monounsaturated fatty acids. *Acc. Chem. Res.* **2**:193–202.

Bobbie, R. J. and White, D. C. (1980). Characterization of benthic microbial community structure by high resolution gas chromatography of fatty acid methyl esters. *Appl. Environ. Microbiol.* **39**:1212–1222.

Bobbie, R. J., Nickels, J. S., Smith, G. A., Fazio S. D., Findlay, R. H., Davis, W. M., and White D. C. (1981). Effect of light on the biomass and community structure of the estuarine detrital microbiota. *Appl. Environ. Microbiol.* **42**:150–158.

Boon, J. J., deLeeuw, J. W., v. d. Hoek, G. J., and Vosjan, J. H. (1977). Significance and taxonomic value of iso and anteiso monoenoic fatty acids and branched β-hydroxy acids in *Desulfovibrio desulfuricans*. *J. Bacteriol.* **129**:1183–1191.

Davis, W. M. and White, D. C. (1980). Fluorometric determination of adenosine nucleotide derivatives as measures of the microfouling, detrital and sedimentary microbial biomass and physiological status. *Appl. Environ. Microbiol.* **40**:539–548.

Fazio, S. D., Mayberry, W. R., and White, D. C. (1979). Muramic acid assay in sediments. *Appl. Environ. Microbiol.* **38**:349–350.

Fazio, S. D., Uhlinger, D. J., Parker, J. H., and White, D. C. (1982). Estimations of uronic acids as quantitative measures of extracellular polysaccharide and cell wall polymers from environmental samples. *Appl. Environ. Microbiol.* **43**:1151–1159.

Findlay, R. H., Moriarty, D. J. W., and White, D. C. (1983). Improved method of determining muramic acid from environmental samples. *Geomicrobiol. J.* **3**:135–155.

Findlay, R. H. and White, D. C. (1983). Polymeric β-hydroxy alkanoates from environmental samples and *Bacillus megaterium*. *Appl. Environ. Microbiol.* **44**:71–78.

Fredrickson, H. L. (1981). Lipid characterization of sedimentary sulfate reducing communities. *Abstr. Am. Soc. Microbiol.,* p. 206.

Gehron, M. J. and White, D. C. (1982). Quantitative determination of the nutritional status of detrital microbiota and the grazing fauna by triglyceride glycerol analysis. *J. Exp. Mar. Biol. Ecol.* **64**:145–158.

Gehron, M. J. and White, D. C. (1983). Sensitive measurements of phospholipid glycerol in environmental samples. *J. Microbiol. Methods* **1**:53–61.

Goldfine, H. and Hagen, P. O. (1972). Bacterial plasmalogens. In: F. Snyder (Ed.), *Ether Lipids: Chemistry and Biology.* Academic Press, New York, pp. 329–350.

Johns, R. B. and Perry, C. J. (1977). Lipids of the marine bacterium *Flexibacter polymorphus*. *Arch. Microbiol.* **114**:267–271.

Kates, M. (1964). Bacterial Lipids. *Adv. Lipid Res.* **2**:17–90.

Kenyon, C. N., Rippka, R., and Stanier, R. Y. (1972). Fatty acid composition and physiological properties of some filamentous blue-green algae. *Arch. Microbiol.* **83**:216–236.

Khuller, G. K. and Goldfine, H. (1974). Phospholipids of *Clostridium butyricum*. V. Effects of growth temperature on the fatty acid, alk-l-enyl ether group, and phospholipid composition. *J. Lipid Res.* **15**:500–507.

King, J. D. and White, D. C. (1977). Muramic acid as a measure of microbial biomass in estuarine and marine samples. *Appl. Environ. Microbiol.* **33**:777–783.

King, J. D., White, D. C., and Taylor, C. W. (1977). Use of lipid composition and metabolism to examine structure and activity of estuarine detrital microflora. *Appl. Environ. Microbiol.* **33**:1177–1183.

Knivett, V. A. and Cullen, J. (1965). Some factors affecting cyclopropane acid formation in *Escherichia coli*. *Biochem. J.* **96**:771–776.

Law, J. H. (1971). Biosynthesis of cyclopropane rings. *Acc. Chem. Res.* **4**:199–203.

Mayberry, W. R. (1980a). Hydroxy fatty acids in *Bacteroides* species: D(−) 3-hydroxy-15-methyl decanoate and its homologues. *J. Bacteriol.* **143**:582–587.

Mayberry, W. R. (1980b). Cellular distribution and linkage of the D(−) 3-hydroxy fatty acids in *Bacteroides sp. J. Bacteriol.* **144**:200–204.

Mayberry, W. R. (1981). Dihydroxy and monohydroxy fatty acids in *Legionella pneumophilia*. *J. Bacteriol.* **147**:373–381.

Millar, W. N. and Casida, L. E., Jr. (1970). Evidence for muramic acid in the soil. *Can. J. Microbiol.* **18**:299–304.

Moriarty, D. J. W. (1980). Measurement of bacterial biomass in sandy sediments. In: P. A. Trudinger, M. R. Walter, and R. J. Ralph (Eds.), *Biogeochemistry of Ancient and Modern Sediments*. Australian Academy of Science, Canberra, pp. 131–138.

Morrison, S. J., King, J. D., Bobbie, R. J., Bechtold, R. E., and White, D. C. (1977). Evidence for microfloral succession on allochthonous plant litter in Apalachicola Bay, Florida, U.S.A. *Mar. Biol.* **41**:229–240.

Morrison, S. J. and White, D. C. (1980). Effects of grazing by estuarine gammaridean amphipods on the microbiota of allochthonous detritus. *Appl. Environ. Microbiol.* **40**:659–671.

Nickels, J. S., Bobbie, R. J., Lott, D. F., Martz, R. F., Benson, P. H., and White, D. C. (1981a). Effect of manual brush cleaning on the biomass and community structure of the microfouling film formed on aluminum and titanium surfaces exposed to rapidly flowing seawater. *Appl. Environ. Microbiol.* **41**:1442–1453.

Nickels, J. S., Bobbie, R. J., Martz, R. F., Smith, G. A., White, D. C., and Richards, N. L. (1981b). Effect of silicate grain shape, structure, and location on the biomass and community structure of colonizing marine microbiota. *Appl. Environ. Microbiol.* **41**:1262–1268.

Nickels, J. S., King, J. D., and White, D. C. (1979). Poly-beta-hydroxy-butyrate metabolism as a measure of unbalanced growth of the estuarine detrital microbiota. *Appl. Environ. Microbiol.* **37**:459–465.

Nickels, J. S., Parker, J. H., Bobbie, R. J., Martz, R. F., Lott, D. F., Benson, P. H., and White, D. C. (1981c). Effect of cleaning with flow-driven brushes on the biomass and community composition of the marine microfouling film on aluminum and titanium surfaces. *Int. Biodeterioration Bull.* **17**:87–94.

Oremland, R. S. and Taylor, B. F. (1978). Sulfate reduction and methanogenesis in marine sediments. *Geochim. Cosmochim. Acta* **42**:209–214.

Parker, J. H., Smith, G. A., Fredrickson, H. L., Vestal, J. R., and White, D. C. (1982). Sensitive assay, based on hydroxy-fatty acids from lipopolysaccharide lipid A for gram-negative bacteria in sediments. *Appl. Environ. Microbiol.* **44:**1170–1177.

Perry, G. T., Volkman, J. K., and Johns, R. B. (1979). Fatty acids of bacterial origin in contemporary marine sediments. *Geochim. Cosmochim. Acta* **43:**1715–1725.

Saddler, J. N. and Wardlaw, A. C. (1980). Extraction, distribution, and biodegradation of bacterial lipopolysaccharides in estuarine sediments. *Antonie van Leeuwenhoek J. Micro Serol.* **46:**27–29.

Schultz, D. M. and Quinn, J. G. (1973). Fatty acid composition of organic detritus from *Spartina alterniflora. Estuarine Coastal Mar. Sci.* **1:**177–190.

Smith, G. A., Nickels, J. S., Bobbie, R. J., Richards, N. L., and White D. C. (1982a). Effects of oil and gas well drilling fluids on the biomass and community structure of the microbiota that colonize marine sands in running seawater. *Arch. Environ. Contam. Toxicol.* **11:**19–23.

Smith, G. A., Nickels, J. S., Davis, W. M., Martz, R. F., Findlay, R. H., and White D. C. (1982b). Perturbations of the biomass, metabolic activity, and community structure of the estuarine detrital microbiota: Resource partitioning by amphipod grazing. *J. Exp. Mar. Biol. Ecol.* **64:**145–158.

Sorenson, J., Christensen, D., and Jorgensen, B. B. (1981). Volatile fatty acids and hydrogen as substrates for sulfate-reducing bacteria in anaerobic marine sediments. *Appl. Environ. Microbiol.* **42:**5–11.

Uhlinger, D. J. and White, D. C. (1983). Relationship between the physiological status and formation of extracellular polysaccharide glycocalyx in *Pseudomonas atlantica. Appl. Environ. Microbiol.* **45:**64–70.

White, D. C. and Benson, P. H. (1984). Determination of the biomass, physiological status, community structure and extracellular plaque of the microfouling film. In: J. D. Costlow and R. C. Tipper (Eds.), *Symposium on Marine Biodeterioration and Fouling.* U.S. Naval Institute Press, Annapolis, MD, pp. 68–74.

White, D. C., Bobbie, R. J., Herron, J. S., King, J. S., and Morrison, S. J. (1979a). Biochemical measurements of microbial mass and activity from environmental samples. In: J. W. Costerton and R. R. Colwell (Eds.), *Native Aquatic Bacteria: Enumeration, Activity and Ecology.* ASTM STP 695, American Society for Testing and Materials, Philadelphia, pp. 69–81.

White, D. C., Bobbie, R. J., King, J. D., Nickels, J. S., and Amoe, P. (1979b). Lipid analysis of sediments for microbial biomass and community structure. In: C. D. Litchfield and P. L. Seyfreid (Eds.), *Methodology for Biomass Determinations and Microbial Activities in Sediments.* ASTM STP 673, American Society for Testing and Materials, Philadelphia, pp. 87–103.

White D. C., Bobbie R. J., Nickels, J. S., Fazio, S. D., and Davis, W. M. (1980a). Non-selective biochemical methods for the determination of fungal mass and community structure in estuarine detrital microflora. *Bot. Mar.* **23:**239–250.

White, D. C., Davis, W. M., Nickels, J. S., King, J. D., and Bobbie, R. J. (1979c). Determination of the sedimentary microbial biomass by extractible lipid phosphate. *Oecologia* **40:**51–62.

White, D. C., Findlay, R. H., Fazio, S. D., Bobbie, R. J., Nickels, J. S., Davis, W. M., Smith, G. A., and Martz, R. F. (1980b). Effects of bioturbation and predation by *Mellita quinquiesperforata* on the sedimentary microbial community structure. In: V. S. Kennedy (Ed.), *Estuarine Perspectives*. Academic Press, New York, pp. 163–171.

White, D. C., Livingston, R. J., Bobbie, R. J., and Nickels, J. S. (1979d). Effects of surface composition, water column chemistry, and time of exposure on the composition of the detrital microflora and associated macrofauna in Apalachicola Bay, Florida. In: R. J. Livingstone (Ed.), *Ecological Processes in Coastal and Marine Systems*. Plenum, New York, pp. 83–116.

White, D. C., Nickels, J. S., Gehron, M. J., Parker, J. H., Martz, R. F., and Richards, N. L. (1983a). Biochemical measures of coral metabolic activity, nutritional status and microbial infection with exposure to oil and gas well drilling fluids. In: I. W. Duedall (Ed.), *Wastes in the Ocean*. Wiley-Interscience, New York.

White, D. C., Smith, G. A., Gehron, M. J., Parker, J. H., Findlay, R. H., Martz, R. F., and Fredrickson, H. L. (1983b). The ground water aquifer microbiota: biomass, community structure and nutritional status. *Dev. Indust. Microbiol.* **24:**201–211.

White, D. C. and Tucker, A. N. (1969). Phospholipid metabolism during bacterial growth. *J. Lipid Res.* **10:**220–233.

Wilkinson, S. G. (1977). Composition and structure of bacterial lipopolysaccharides. In: I. Sutherland (Ed.), *Surface Carbohydrates of Prokaryotic Cells*. Academic Press, New York, pp. 97–175.

15

MICROCOSM FOR GROUND WATER RESEARCH

Göran Bengtsson*

National Center for Ground Water Research
University of Oklahoma
Norman, Oklahoma

Ground water has traditionally been a major source of domestic and industrial process water, and is likely to be increasingly used because of population growth coupled with limited supplies of high quality surface waters. The possibility of serious contamination of ground water has long been largely ignored, but recent abandonment of local aquifers necessitated by contamination problems has directed attention to the need for protection of clean ground water supplies. More information on transport and fate of contaminants in aquifers is now being requested by regulatory agencies that have been assigned ground water protection responsibilities. However, research aimed at providing such information is limited by the relatively large inputs of capital and time needed for field studies of saturated subsurface environments. Less expenive and more flexible approaches need to be addressed.

In response to these requirements, the National Center for Ground Water Research has undertaken a research program to develop laboratory microcosms suitable for investigations of the behavior of pollutants in saturated subsurface environments, and to utilize these systems for studying various processes governing pollutant fate in ground water. A main concern in this work was the design of microcosms of sufficient veracity and reproducibility to be useful for estimating process parameters of relevance for conceptual mathematical modeling.

* Present address: Laboratory of Ecological Chemistry, University of Lund, Ecology Building, Helgonavägen 5, S-223 62 Lund, Sweden.

A microcosm may be defined as any part of an ecosystem that may be subject to laboratory control due to its reduction in size or complexity. The feedback between processes in the physical model, or microcosm, and a mathematical model, which may be used to predict the behavior of a large physical system, is important. The microcosm can be a mere simulation of some greater physical system, or it can serve to estimate parameters that may be used in mathematical equations that predict the behavior of the greater physical system.

Microcosms have been used to address several levels of biological complexity—including cellular, tissue, organ, organismal, population, community, and ecosystem—to assess chemical transport, fate, and effects (Witherspoon et al., 1976). Three critical problems have been identified in connection with use of microcosms (Ausmus et al., 1980). First is the problem of realism, the difficulty of applying results obtained in simplified systems to complex natural environments. Second, is the problem of reproducibility, the lack of information on the comparability of data from microcosms of different size and structure. Last is the problem of replicability, the difficulty of estimating confidence limits on parameters measured in microcosms.

Wilson and Noonan (1984) have identified two broad classes of microcosms. Microcosms of the first class are designed to reveal effects on the biological community, whereas microcosms of the second class are used to define processes that affect the transport and fate of pollutants. The microcosm described in this paper was constructed primarily for studies pertaining to transport and fate of pollutants in ground water, but it also has utility for investigation of the response of the subsurface biological community to environmental perturbations.

MATERIALS AND METHODS

Design of the Microcosm

The basic microcosm unit comprised a glass column filled with aquifer matrix material, through which ground water amended with selected chemicals could be passed at a controlled rate by means of a dosing system constructed entirely from glass and Teflon (Figure 15.1).

The column was constructed from Kimax beaded process pipe components (obtained from ACE Glass Incorporated, Vineland, NJ 08360). It consisted of an appropriate length of pipe, an end plug, and an end cap, held together by Teflon-lined couplings. Internally threaded glass connectors fitted with Teflon bushings (ACE Glass Incorporated, codes 5027-2 and 5029-35, respectively), sealed into the end plug and end cap on the longitudinal axis of the column, served as influent and exit ports. An identical connector in the side of the pipe component, near the mid-point of the column, served

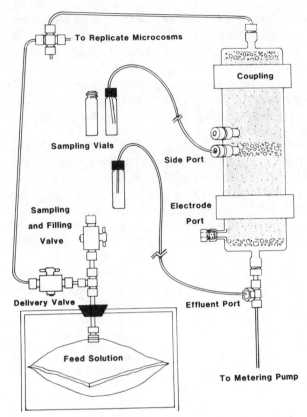

Figure 15.1. Microcosm designed for ground water research.

as a sampling port, while two others, one adjacent to the mid-column sampling port and the other in the side of the end cap, permitted insertion of microelectrodes into the aquifer material. Plugs of glass wool in the exit and side sampling ports prevented escape of solids from the column during operation. Column dimensions were either 8 × 40 cm (3 × 16 in.) or 4 × 20 cm (1.5 × 8 in.).

The primary component of the dosing system was a 2.6-L Teflon bag (Alltech Associates, Houston, TX) submerged in a water-filled plexiglass vat which was maintained under positive pressure by an external water reservoir. The outlet of the bag was attached to a Teflon union tee, with the other two arms of the tee being attached to Teflon plug valves (valves and tees were obtained from Daigger Scientific, Chicago, IL). One valve was used for filling the Teflon bag with feed solution and for sampling the contents of the bag. The other was attached to a 3-mm Teflon tube through which feed solution was delivered to the column, and served primarily to isolate the column from the dosing system during filling and sampling of feed solution. A

Teflon union cross in the feed solution delivery tube allowed three replicate columns to be dosed from a single Teflon bag. Or in some cases, the delivery valve outlet was attached directly to a union tee, to which were connected two 3-mm Teflon delivery tubes. Each delivery tube was connected to a union cross, thus permitting a single Teflon bag to service up to six columns.

The flow rate of feed solution through the column was controlled by a peristaltic pump placed downstream of the effluent port of the column to preclude contamination of the system by the tubing used in the pump. A Teflon union tee attached to the effluent end of the column upstream of the peristaltic pump permitted sampling of effluent before it had contacted any materials other than glass or Teflon.

Collection of Aquifer Matrix Material

A modification of procedures described by Dunlap et al. (1977) was used to acquire samples of saturated aquifer solids that were not contaminated by surface microorganisms. A continuous-flight auger was used to drill a hole to the desired depth, 2–3 ft below the ground water table at the field site near Pickett, OK. The auger was removed, and the saturated soil was sampled with an autoclaved thin-wall core barrel (9.5 cm ID × 45 cm length). The core barrel was provided at one end with a core retainer supplied with spring-loaded metal teeth to prevent the soil core from slipping out of the barrel during withdrawal from the hole (Figure 15.2). The barrel was filled to 75–80% of its capacity by a hydraulic ram.

The core was extruded at the drilling site to avoid dewatering the soil. Extrusion was accomplished inside a specially constructed wind shield to minimize possibilities for contamination. For extrusion, an autoclaved aluminium block was fitted into the barrel on top of the core and the core retainer removed. The core was hydraulically extruded 3–5 cm and then broken off to produce an aseptic face. The remainder of the core was extended past a device that pared away the outer 0.5 cm of the matrix (Figure

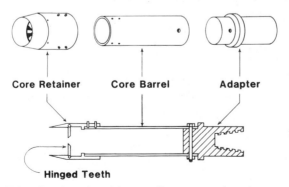

Core Retainer   Core Barrel   Adapter

Hinged Teeth

Figure 15.2.   Core barrel used for sampling uncontaminated saturated soil.

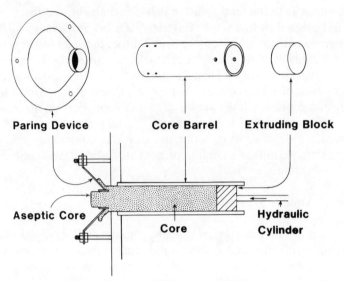

Figure 15.3.   Extruding device for uncontaminated saturated soil sample.

15.3). The autoclaved glass column was fitted into the device so that the pared sample slipped into the column without fractionation. After half of the column was filled, washed and autoclaved pea gravel (1.19–2.00 mm diameter, sieved by U.S. Standard sieve series) was added to fill the space adjacent to the sampling port. Pea gravel was also added on top and bottom of the column to ensure even distribution of the feeding solution into and out of the soil core. Since the core material was slightly compressed when extruded, some interstitial pore water was released and collected at the bottom of the column. The pores were refilled with water by gently tapping the column. The column was taken to the laboratory in an ice chest. In a hood equipped with ultraviolet germicidal lights, pea gravel and distilled autoclaved water were added on top of the column, and the end plug was attached. The bushings were plugged with short pieces of glass rods until the column was installed in the complete microcosm system.

Maintenance of Microcosm

A special Teflon well was constructed to acquire water from the same lens of saturated soil that was sampled by the core barrel. Water was pumped into a 4-L bottle at a flow rate of approximately 1 L/hr. The water was filter sterilized using autoclaved Millipore sanitary sterilizing equipment (293 mm diameter filter, 0.22 μm) under nitrogen pressure. The filtered water was collected in an autoclaved 2-L volumetric flask via a delivery tube. The water was purged for 1 day with nitrogen (pre-filtered through a 0.22-μm disposable Millex filter) to reduce the oxygen concentration to the same level as in the interstitial pore water. The water could then be dosed with selected chemicals. The bottle was sealed with a glass stopper so that no headspace was created, and then stirred for 3 days in darkness at room temperature to dissolve the chemicals. The solution was aseptically transferred to one of the flexible, autoclaved 2-L Teflon bags. One week was required to prepare a dosed feeding solution.

Flow through the columns was maintained at 2 cm/day, which corresponded to an effluent volume of 38 mL/day from the 8 cm columns and 9 mL/day from the 4 cm columns. With three large columns connected to the bag, it was necessary to renew the feeding solution every second week.

Samples for volatile constituents were drawn by connecting a sterile gastight syringe to the sample port. Samples for non-volatile constituents were collected in 4-mL glass vials connected to the sampling ports by 3-mm Teflon tubings. The tubings were inserted into the vials through perforated screw caps and Teflon-faced septa. Vials were maintained in place at all times during operation of the system and were kept tightly closed except during sampling. For sampling, the in-place vial was unscrewed, allowing 2–3 mL of water to flow slowly into it. The vial was then replaced by a clean, autoclaved vial, into which the following 4 mL, constituting the sample, were collected. The first 2–3 mL were discarded.

The volume of effluent was monitored daily. All ports were covered by autoclaved Al foil to prevent contamination by dust particles and subsequent growth of contaminating microorganisms. The Al foil was changed regularly. The columns were examined daily for leaks. The microcosm was incubated at 17°C, the *in situ* temperature of the aquifer matrix material.

Measurement of DOC and Colonies of Bacteria

A 4-mL sample was drawn from the column, acidified with 1 drop of concentrated HCl, and centrifuged. The supernatant was purged with N_2 for 5 min to remove CO_2 and a 20-μL sample was injected into a Beckman 915 carbon analyzer. Three replicates were analyzed for each sample.

A 1-mL sample was properly diluted in sterile peptone water (1 g/L) and spread onto nutrient agar plates. Numbers of colonies were counted after 1 week of incubation at room temperature. A 1-g soil sample was homogenized in peptone water and treated in the same way.

Water Chemistry

The concentration of major nutrients was lower in the well water supplied to the columns than in the interstitial pore water, collected by pressure filtration of saturated soil in nitrogen atmosphere or by pressing the water out of the soil in the core barrel (Table 15.1). The pH and oxygen concentration were higher in the well water, probably owing to exposure of the water to air. The

Table 15.1   Determination of Major Constituents in Interstitial Pore water of the Saturated Soil and in the Well Water Supplying the Microcosm

	pH	Conductivity (μS/cm)	o-P (mg/L)	SO_4^{-2} (mg/L)	NO_3^{-1} (mg/L)	NH_4^{+1} (mg/L)	O_2 (mg/L)
Interstitial pore water	5.7	120	0.22	70	6.1	0.2	4.0
Water supply well	7.3	150	0.01	12	4.1	<0.005	8.4

organic carbon content of the soil (0.02%) was low (Marvin Piwoni, U.S. EPA, Ada, OK—personal communication).

The Mann–Whitney test was used for statistical comparisons.

RESULTS

Bacterial densities of the aquifer material acquired for preparation of microcosms during this study were quite consistent, averaging about 6000 colony-forming units per gram of material when enumerated on a complex medium (Table 15.2). This density was about 1000 times lower than bacterial densities indicated by the same enumeration procedure to be present in topsoil from the field site. Obviously, precautions to keep the saturated aquifer material uncontaminated and the indigenous microflora intact during core acquisition and processing and microcosm maintenance operations were essential.

Since the soil cores generally were aerobic (cf. Table 15.1) with isolated patches of anaerobiosis, as indicated by precipitated iron sulfide, total exclusion of oxygen during the acquisition and processing of cores was not necessary, but the aquifer material was protected from excessive exposure to air

Table 15.2   Density of Bacteria in the Aquifer Material

Date of Acquisition	Colony Forming Units/ g Material[a]
6-15-81	6.4×10^3
6-17-81	6.2×10^3
7-21-81	5.4×10^3
7-28-81	5.9×10^3

[a] Enumerated using nutrient agar media.

and possible de-watering. The procedures employed for core acquisition and column loading appeared to effectively prevent bacterial contamination of the aquifer material used in microcosm construction. Although the leading face of the core adjacent to the core retainer often became contaminated, extrusion and breaking off of the first 3–5 cm of the core to expose a fresh face removed this contamination, as indicated by bacterial densities. Also, plate counts of bacteria in the soil from the center and edges of fresh core faces thus exposed were not significantly different, indicating no contamination from the core barrel.

To avoid contamination of the filtrate during maintenance of the columns, less than 2.5 L of well water was filtered at one time. It was necessary to pass the nitrogen gas used to purge excess oxygen from well water employed for preparation of feed solutions through a 0.22-μm filter to eliminate contamination from tubing connected to the gas supply, and great care was required to prevent contamination when adding chemicals to the water. It was also necessary to limit the quantity of dosing solution loaded into each 2.6-L Teflon bag to a maximum of 1.5 L, since these bags tended to develop leaks along the seams if loaded more fully, resulting in both loss and contamination of dosing solution. The filling and sampling valves were flooded with ethanol and then flamed between each use.

The systems were equilibrated initially by dosing with well water for 14 days; then dosing of six systems (three each large and small) with well water amended with a mixture of phenols was begun, while dosing of two each small and large systems with unamended well water was continued. The columns were monitored for numbers of bacteria and dissolved organic carbon (DOC), and the data were used to illustrate the replicability and reproducibility of the columns. The replicability was significantly higher for DOC than for plate counts (Tables 15.3 and 15.4, $p < 0.01$), and it was higher when the concentration of DOC increased (Table 15.3, $p < 0.01$). The variability was nearly twice as high for plate counts as for DOC, a difference partly explained by the difference in variability between the methods: the coefficient of variation of consecutive injections of the same sample on the carbon analyzer was less than 5% while variation on plate counts was between 5 and 10%.

Differences in variability of biological and chemical parameters is common in field monitoring of pollution and emphasizes the realism of the microcosm. Assuming that a 20% coefficient of variation can be maintained for monitoring fate of pollutants in a microcosm, the number of replicates required to detect a significant change in concentration can be calculated. For example, if a 50% difference in concentration is regarded as important and an 80% probability of detecting this difference in a one-tailed 5% test of significance is desired, at least two replicates would be required (see Snedecor and Cochran, 1967). A 10% difference would not be detected unless 50 replicates were used, assuming the same probability and significance level. Assuming a practical limit of 10 replicates, a 25% change in concentration of

Table 15.3   Variability of the Concentration of Dissolved Organic Carbon (DOC) in the Effluent from the Microcosms

	DOC mg/L[a]			
	Large Columns		Small Columns	
Microcosms	$\bar{X}$	C.V.	$\bar{X}$	C.V.
Controls[b]	3.56	0.58	3.38	0.14
	2.17	0.25	3.36	0.26
	2.54	0.12	3.84	0.34
	2.69	0.20	4.65	0.43
	2.71	0.50	1.87	0.47
	1.30	0.28	1.59	0.30
	1.90	0.35	2.28	0.58
		Av = 0.33		Av = 0.36
Treated[c]	8.58	0.33	10.19	0.26
	10.58	0.11	7.75	0.13
	6.80	0.16	7.48	0.16
	6.40	0.08	6.86	0.10
	4.71	0.23	4.42	0.35
	5.55	0.15	6.58	0.20
		Av = 0.18		Av = 0.20

[a] Means ($\bar{X}$) and coefficients of variation (C.V.) calculated from data for aqueous samples from side and exit ports at different sampling dates.
[b] $N = 4$.
[c] Dosed with ground water amended with mixture of phenols, total concentration of about 10 mg/L; $N = 6$.

the pollutant might be detectable. Efforts to reduce the variability would be worthwhile if a higher resolution were required. If the coefficient of variation could be reduced to 10%, the resolution would be doubled. It may be possible to improve the sampling technique to some extent in order to reduce variability, but the dynamic nature of the microcosm and natural variations in organic matter, indigenous microflora, and so on, in soils is more likely to determine the lower limit on variability among replicate microcosms.

The reproducibility of the microcosm was demonstrated by comparison of the smaller and the larger columns. Coefficients of variations for DOC were not significantly different for the two column sizes, whereas the mean values for DOC were slightly, though insignificantly, higher for the smaller columns than for the larger columns, indicating that either sorption was less efficient or the channeling more severe in the smaller columns. A significantly

Table 15.4   Variability in the Density of Bacteria in the Effluent from the Microcosms

| | Colony-Forming Units ($\times 10^5$) per ml Column Effluent[a] | | | |
| | Large Columns | | Small Columns | |
Microcosms	$\overline{X}$	C.V.	$\overline{X}$	C.V.
Controls[b]	0.19	0.84	0.45	0.51
	0.45	0.47	0.51	0.27
	1.10	0.45	1.58	0.77
	0.31	0.64	0.27	0.41
	0.43	0.58	0.32	0.47
	0.23	0.48	0.27	0.52
		Av = 0.58		Av = 0.49
Treated[c]	9.20	0.64	15.70	0.53
	5.25	0.47	14.08	0.40
	6.55	0.90	4.50	0.66
	6.22	0.64	4.72	0.49
	6.87	0.48	5.76	0.07
	5.04	0.85	7.38	0.33
		Av = 0.66		Av = 0.41

[a] Means ($\overline{X}$) and coefficients of variation (C.V.) calculated from data for aqueous samples from side and exit ports at different sampling dates.

[b] $N = 4$.

[c] Dosed with ground water amended with mixture of phenols, total concentration of about 10 mg/L; $N = 6$.

($p < 0.05$) greater coefficient of variation was found for bacterial numbers in effluents from the larger columns than from the smaller columns dosed with phenols, but the mean number of colony-forming units was not different in the smaller and larger columns.

To test whether channeling would explain the slight difference in DOC from smaller and larger columns, the elution volume (corresponding to breakthrough of conductivity) was compared with the pore volume of the columns (Figure 15.4). The pore volume and the elution volume for the larger columns coincided, but there was a deviation for the smaller columns, indicating a channeling effect. The effect of channeling on the data reported here was not dramatic, so the smaller columns may still be useful for monitoring and screening purposes. The size of the smaller columns seemed to be close to the lowest practical size limit.

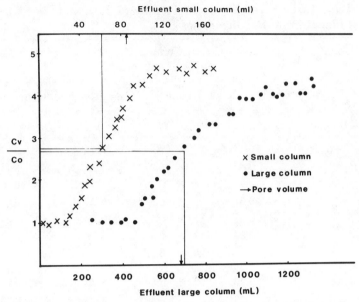

Figure 15.4. Dispersion of nutrients, mainly chloride, passing through microcosm columns. The breakthrough curves of two small columns (average shown) and one large column was achieved by determining the conductivity of effluent samples after the small columns were dosed with NaCl solution and the large column with renovated sewage water. C_0 is conductivity at volume 0 and C_v at volume v. The pore volume is indicated by arrows.

The realism of the microcosm was not evaluated, but its extensive variability can be taken as an indication of the similarity between the microcosm and a natural system. A static microcosm with a homogenous matrix would probably improve replicability to some extent, but at the expense of reduced reproducibility and realism. The major advantages of the microcosm described in this paper are its reproducibility and its use of authentic aquifer material with indigenous microflora to evaluate processes in the saturated subsurface environment.

ACKNOWLEDGMENTS

John T. Wilson, U.S. EPA, Ada, OK, made suggestions and improvements of vital importance for the design and performance of the microcosm. Montie Fraser, Marion R. Scalf, John T. Wilson, and Lynn Wood of Ada, OK advised and assisted in drilling and sampling of saturated soil and construction of the well. James F. McNabb advised in monitoring of microbial activity and density. Cynthia Hammons, Teresa King, and Marita Bengtsson maintained the microcosms and did plate counts of bacteria. Kenneth Grider performed the analyses of DOC.

REFERENCES

Ausmus, B. B., Eddlemon, G. K., Draggan, S. J., Giddings, J. M., Jackson, D. R., Luxmoore, R. J., O'Neill, E. G., O'Neil, R. V., Ross-Todd, M., and Van Voris, P. (1980). *Microcosms as Potential Screening Tools for Evaluating Transport and Effects of Toxic Substances.* EPA-600/3-80-042, U.S. Environmental Protection Agency, Athens, GA.

Dunlap, W. J., McNabb, J. F., Scalf, M. R., and Cosby, R. L. (1977). Sampling for organic chemicals and microorganisms in the subsurface. EPA-600/2-77-176, U.S. Environmental Protection Agency, Ada, OK.

Snedecor, G. W. and Cochran, W. G. (1967). *Statistical Methods.* Iowa State University Press, Ames, IA.

Wilson, J. T. and Noonan, M. J. (1984). Microbial activity in model aquifer systems. In: G. Bitton and C. P. Gerba (Eds.), *Ground Water Pollution Microbiology.* Wiley-Interscience, New York, pp. 117–133.

Witherspoon, J. P., Bondietti, E. A., Draggan, S., Taub, F. P., Pearson, N., and Trabalka, J. R. (1976). State-of-the-art and proposed testing for environmental transport of toxic substances. EPA-560/5-76-001, Oak Ridge National Laboratory, Oak Ridge, TN.

PART THREE

SUBSURFACE CHARACTERIZATION IN RELATION TO GROUND WATER POLLUTION

16

OVERVIEW OF SUBSURFACE CHARACTERIZATION RESEARCH

Philip B. Bedient

National Center for Ground Water Research
Rice University
Houston, Texas

Subsurface characterization of soil and aquifer systems has taken a prominent role in research efforts related to pollution of ground water. The disciplines that must be interrelated in the characterization process include hydrogeology, mineralogy, geochemistry, microbiology, and contaminant transport. All of these processes tend to control the propagation and attenuation of inorganic and organic contaminants in ground water. Simple descriptions of bulk aquifer properties, useful in the past for ground water supply and development, are being rapidly replaced by more sophisticated methods. However, given the thousands of potentially available chemical compounds which may interact in hundreds of different soil or geologic strata, it is important that research be directed to help develop predictive relationships between the subsurface and pollutant attenuation and migration.

Hydrogeologic properties of an aquifer include the measurement of porosity, hydraulic conductivity, grain size, and other topographic patterns. Variations of these parameters in the vertical and horizontal dimensions are of great importance to the ultimate migration of pollutants from a waste source. Geologic history available from maps or existing well data is useful in a regional context, but the need for numerous wells and borings in the vicinity of a waste source is never completely satisfied.

An understanding of subsurface mineralogy may be obtained through field study and laboratory experiment. The primary techniques used to characterize solid phases include X-ray diffraction, infrared spectroscopy, electron beam microscopy, and other chemical analyses. The characterization of the subsurface necessarily implies relationships between solid and liquid phases and associated geochemical reactions. Water in contact with sediment material may result in the leaching out of more soluble species, precipitation of saturated species, absorption of inorganic ions or certain organics, or the formation of clays or colloidal oxides via chemical reaction.

Geochemical processes determine to a large extent the persistence and mobility of contaminants in ground water. Redox potential and the distribution coefficient K_d are presented as two particularly important geochemical parameters in Young's chapter (17). In particular, the organic content of soil and octanol–water partition coefficient have been related to K_d by Karickhoff et al. (1979), as a useful predictive tool for evaluating migration of retarded organics compared to conservative tracers. Field techniques for the measurement of K_d by radial injection dual-tracer tests has recently been presented in Pickens et al. (1981). Much more work is needed to evaluate these parameters at actual field sites where heterogeneities exist. Roberts et al. report in Part Four (Chapter 23), on a large-scale tracer study, designed to evaluate retardation of organics injected into the aquifer.

Recently, attention has been directed to the characterization and adaption of subsurface microbial populations capable of degrading organics in the field and the laboratory. Ghiorse and Balkwill (Chapter 20) and Wilson et al. (Chapter 26) address these problems in related chapters. Studies are underway to determine where microbial degradation may be occurring at ground water waste sites and under what conditions. Some of the major hurdles to be overcome include the development of sterile sampling methods and the proper measurement of dissolved oxygen in the ground water. Both anaerobic and aerobic degradation of organics can occur given the right conditions. As microbial measurement techniques are perfected in the subsurface, the process of aquifer decontamination using microbes may become an increasingly important area of research.

One of the most perplexing problems in subsurface characterization relates to pollutant transport and the ability to predict the rate of migration of a contaminant plume. Transport processes of advection, dispersion, adsorption, and decay may all produce a complex and sometimes unexplainable outcome in the field. While work has been completed on laboratory column studies, much research needs to be developed at actual field sites where tracers are applied and observed through time. The two-well and three-well injection-production tracer tests offer a real possibility in understanding some of these complex transport mechanisms.

Our own research at Rice University for the National Center for Ground Water Research has involved a major effort on subsurface characterization at an abandoned creosote waste site in Texas. A number of wells and boreholes installed adjacent to and down gradient of the waste pits have yielded high levels of polynuclear aromatic hydrocarbons such as naphthalene. Conservative chloride has migrated about 300 ft (90 m) down gradient. Organic contaminants have been significantly attenuated compared to Cl^-. Mechanisms of adsorption and microbial decay under aerobic conditions have occurred at the creosote site, based on detailed field and laboratory analyses by Rice University and John Wilson at EPA RSKERL. More work is planned for the future, including a multiwell tracer test using organic wastes at the site. Adsorption can be determined through such a controlled injection-production scheme.

The papers contained in this symposium present detailed descriptions of subsurface characterization in relation to ground water pollution. It is fitting that each paper was prepared by a researcher in a different field of scientific investigation; all of these subjects must be included in the overall interdisciplinary process if ground water migration and attenuation is to be understood and eventually controlled.

REFERENCES

Karickhoff, S. W., Brown, D. S., and Scott, T. A. (1979). Sorption of hydrophobic pollutants on natural sediments. *Water Res.* **13**:241.

Pickens, J. F., Jackson, R. E., and Inch, K. J. (1981). Measurement of distribution coefficient using a radical injection dual tracer test. *Water Resour. Res.* **17**:529–544.

17

SUBSURFACE CHARACTERIZATION IN RELATION TO GROUND WATER POLLUTION

C. P. Young

Groundwater, Environmental Protection
Water Research Centre, Medmenham Laboratory
United Kingdom

The increased awareness of the vulnerability of many aquifer systems to pollution, and continued growth of ground water as a resource for direct public supply, for conjunctive use with surface waters and for regulation of rivers, have provided the impetus toward more detailed examinations of the subsurface environment. The simple descriptions of aquifer systems in the past—often defined only in terms of bulk hydraulic conductivity, broad classifications of lithologic, and general hydraulic gradients—are no longer sufficient to define the physical, geochemical, and microbial systems which control the propagation and attenuation of contaminants. Interest in the capacity of unsaturated zones to attenuate pollutants has given rise to more detailed investigation of the pore size distributions and of unsaturated hydraulic conductivities, with measurements of physicochemical properties such as self-diffusion coefficients of specific solutes through porous media and estimations of partition coefficients for particular pollutants with respect to different aquifer materials. The relationship between fissure and matrix flow has received much attention. The age of ground waters has been estimated from their content of radioisotopes such as tritium (^{3}H) and ^{14}C and from the ratio of stable isotopes such as ^{16}O and ^{18}O. The measurement of inert gases He and Rn has recently been investigated as a tool for dating certain ground waters. The importance of rock-water interactions and the chemical equilibrium status of the ground water has encouraged not only more complete analyses of the fluid phase, with measurements of the redox-potential, but also detailed identification of the minerals forming the aquifer matrix. The possibility that microbial transformation may modify significantly ground water chemistry within the aquifer itself has led to studies of the distribution of bacterial populations and of substrates they may colonize in aquifers.

PHYSICAL CHARACTERIZATION

The problem of the mechanisms of movement of solvents and solutes through unsaturated media involves both physical and geochemical considerations. In the case of the Upper Cretaceous Chalk of northern Europe, the physical problems are compounded by the structure of the rock, the matrix of which has an interconnected porosity typically in the range 0.35–0.50 but with mean pore diameters of less than 1 micron, so that its hydraulic conductivity is only about 10^{-6} cm/sec. The rock mass is characteristically traversed by an intersecting plexus of nearly vertical and horizontal fractures ranging from less than 1 mm in width up to 10 mm or more (Ward et al., 1968; Foster and Milton, 1974). Starting with a hypothesis of piston displacement of natural recharge through the matrix to explain the measured distribution of thermonuclear tritium (Smith et al., 1970), a considerable debate has ensued in the United Kingdom, with theoretical consideration being given to both piston displacements and diffusive interchange between fissure and intergranular waters (Foster, 1975; Young et al., 1976; Oakes, 1977; Young et al., 1979; Foster and Smith-Carington, 1980; Oakes, 1981; Barker and Foster, 1981), with contributions from Ballif (1978) and Brossier et al. (1980) in France on the distribution of Chalk pore dimensions and solute movement. The development of automatic soil moisture neutron probes and tensiometers, capable of operation to depths of 7–10 m (Kitching et al., 1981), has enabled Wellings and Bell (1980) to characterize the hydraulic properties of the Upper Chalk of part of southern England and to show that matrix flow is dominant, with fissure flow occuring only during times of intense recharge.

The role of fissure flow in saturated porous media in determining the quality of mobile ground water has led to studies of the Triassic Sandstone aquifer by Brereton and Skinner (1974) and of contaminated Chalk by Headworth et al. (1980). Further development of the induced-polarization technique for the remote sensing of active fracture patterns described by Finch and Griffiths (1977) may provide a valuable tool for characterizing fracture patterns.

Characterization of ground water flow regimes in karstic limestone aquifers in the United Kingdom has been based on extensive tracer studies

(Atkinson, 1971, 1977; Atkinson and Smart, 1981). Atkinson and Smart made use of a two-well pulsed tracer technique developed by Ivanovitch and Smith (1978) in a study of the interrelationship between the storage and movement of ground water in solution-enlarged fractures and that in the porous matrix of the Chalk aquifer in southern England; they showed that although velocities in the fractures were an order of magnitude greater than in the matrix, storage in the matrix was some 10 times that in the fractures.

GEOCHEMICAL ASPECTS

The role of rock–water interactions and of the equilibria of solutions (Langmuir, 1972) in determining the persistence and mobility of contaminants in ground water has been recognized. In particular, the redox potential (Edmunds, 1977; Champ et al., 1979; UNESCO, 1980; Edmunds and Lloyd, 1981) and the distribution coefficient (Jackson et al., 1977) have been stressed as important controls, with field techniques for the measurement of the latter parameter by radial injection dual-tracer tests having been recently described by Pickens et al. (1981).

The recognition of the potential of rock–water interactions has given rise to detailed characterization of both the mineralogy and porewater composition of the Chalk aquifer in the United Kingdom (Edmunds et al., 1973; Morgan-Jones, 1977; Spears, 1979), leading to hydrogeochemical studies of ground water catchments such as those described by Ineson and Downing (1963) and Young and Morgan-Jones (1980).

Studies of the geochemistry of ground waters in the Triassic Sandstone aquifer have been reported by Edmunds and Morgan-Jones (1976), who found trace element mobility to be limited by the reactivity of hydrated ferric oxides present in the formation. Significant changes in the direction of ground water flow from the recharge zone into the confined zone in terms of pH, dissolved oxygen content, cation exchange capacity, and redox potential in Jurassic limestones in southwestern England give rise to a progression from a calcium bicarbonate water in the recharge zone to a sodium bicarbonate-sodium chloride water within the confined zone and have been described by Morgan-Jones and Eggboro (1981).

A field study of the influence of superficial glacial deposits on the quality of ground waters in the Triassic Sandstone aquifer of Yorkshire by Spears and Reeves (1975) indicated that pyrite oxidation and carbonate solution were the major influences. Subsequently Spears (1976), using the same glacial sediments, published the results of laboratory studies of rock–water interactions and was able to demonstrate that such work is capable of providing meaningful information in studies of ground water chemistry.

Soil and rock monoliths and *in situ* lysimeters have played an important role in the investigation of contaminant mobility in the United Kingdom's principal research program into the effects of waste disposal on ground water quality (Department of the Environment, 1978). In particular, a series of 50 m³ *in situ* lysimeters were constructed in the Lower Greensand, a poorly cemented, fine sand and sandy clay aquifer of Lower Cretaceous age; they have been employed to assess the attenuation of persistent organic compounds, inorganic anions, and heavy metals under controlled experimental conditions. Reporting the studies, Ross (1980) concluded that biodegradation was the primary mechanism for the attenuation of both inorganic anions and the organic species. Selective, sequential chemical extraction of heavy metals from the sediments suggested that their rate of movement was controlled by the stability of the geochemical phases in which they were retained, with uptake by sesquioxides and clay minerals being particularly important.

Determinations of the age of ground waters by measurements of stable and radioisotopic concentrations have been used to assess the residence time of waters. Studies of ^{14}C, ^{18}O, and deuterium in the confined Chalk of the London Basin (Smith et al., 1976) indicated an age in excess of 25,000 yr in the central part of the basin. Previously, Mather et al. (1973) had reported ages of 24,000 yr for ground waters contained in the confined portion of the Lower Greensand aquifer of the London Basin, again based on ^{14}C measurements. Comparable ages have been deduced for confined ground waters in the Jurassic Limestone aquifer of Lincolnshire (Downing et al., 1977), and in the confined Triassic Sandstone aquifer of Nottinghamshire (Andrews and Lee, 1979). In the latter case the authors made use of the dissolved argon and krypton contents of the confined ground waters to estimate the mean temperature of the water at the time of its recharge and showed that the values obtained were consistent with infiltration during the late Pleistocene.

MICROBIAL DISTRIBUTIONS

The presence in ground waters of pathogenic organisms derived from waste disposals, cess-pits, and leaking sewage reticulations has long been recognized (Mallman and Mack, 1961). The examination of aquifers to determine whether they contain viable microbial populations, capable of significantly modifying ground water quality, is, however, a recent development. Reviews of the probable situation by McNabb and Dunlap (1975) and McNabb (1977) concluded that most aquifer systems would encompass micro-environments favorable to bacterial colonization, but suggested that the lack of detailed information was, in part, due to the problems of ensuring sterile sampling in the subsurface environment. Recently, Vogel et al. (1981) have reported slow denitrification from a confined aquifer under the Kalahari desert, with evidence from stable isotope ratios comparable to those reported for microbial denitrification.

However, Lind (1975) reported the successful sampling of unconsolidated, heterogeneous alluvial aquifers in Denmark and found evidence of biological nitrate reduction. More recently, Whitelaw and Rees (1980) have recorded the presence of nitrate-reducing and ammonium-oxidizing bacteria at two sites in the unsaturated zone of the Chalk aquifer in southern England. Samples were obtained to a depth of 10 m beneath an area of permanent, unfertilized grassland; oxidizing forms in these samples exceeded the numbers of reducing forms by at least one order of magnitude, and generally by three orders of magnitude. At the second site, beneath fertilized arable land, sampling was continued to 50 m depth and populations of ammonium-oxidizers in the range 10^7–10^8/g dry wt of Chalk were recorded at several depths. Nitrate-reducing bacteria were reported from a restricted vertical interval, at between 30 and 45 m depth, but were in numbers comparable to the oxidizing forms.

The same two sites, plus a fertilized grassland site and an organically fertilized arable site, both on Chalk, were examined to determine the distribution of carbohydrates in the unsaturated zone (Whitelaw and Edwards, 1980). Carbohydrate concentrations were measured on whole-rock samples and in interstitial water. Whole-rock values of about 100 mg/kg dry Chalk were recorded from beneath the grassland sites, with whole-rock levels of nearer 50 mg/kg dry wt from the two arable sites. Interstitial water concentrations in the range of 5–10 mg/L were recorded at both arable and grassland sites. Analyses of the composition of carbohydrates in the whole-rock samples showed the material to be principally Mannose (77%) followed by Glucose (17%), Galactose (3%), and Xylose (2%), and the authors concluded that sufficient carbohydrates to support bacterial activity were present at all depths penetrated in the Chalk.

ACKNOWLEDGMENTS

This paper is published with the permission of the Director, Environmental Protection, Medmenham Laboratory, Water Research Centre, United Kingdom.

REFERENCES

Andrews, J. N. and Lee, D. J. (1979). Inert gases in groundwater from the Bunter Sandstone of England as indicators of age and palaeoclimatic trends. *J. Hydrol.* **41:**233–252.

Atkinson, T. C. (1971). The dangers of pollution of limestone aquifers. *Proc. Univ. Bristol Spelaeological Soc.* **12:**281–290.

Atkinson, T. C. (1977). Diffuse flow and conduit flow in limestone terrain in the Mendip Hills, Somerset (Great Britain). *J. Hydrol.* **35:**93–110.

Atkinson, T. C. and Smart, P. L. (1981). Artificial tracers in hydrogeology. In: *A Survey of Hydrogeology,* 1980, Royal Society of London, pp. 173–190.

Ballif, J. L. (1978). Porosite de la craie: Appreciation de la taille et de la reparatition des pores. *Ann. Agron.* **29**:123–131.

Barker, J. A. and Foster, S. S. D. (1981). A diffusion exchange model for solute movement in fissured porous rock. *Q. J. Eng. Geol. London* **14**(1):17–24.

Brereton, N. R. and Skinner, A. C. (1974). Groundwater flow characteristics in the Triassic Sandstone in the Fylde area of Lancashire. *Water Serv.* **78**(942):275–279.

Brossier, G., Kerbaul, A., Landreau, A., and Morfaux, P., avec contributions par D. B. Oakes et C. P. Young (1980). Impact des pratiques agricoles sur la mineralisation des eaux interstitielles des terrains sousjacents. Consequence sur la qualite de l'eau de la nappe. Bureau de Recherches Géologiqus et Minières, Départment Hydrogéologie, B.P. 6009-45060, Orleans, France. Report 79 SGN 768 HYD.

Champ, D. R., Gulens, J., and Jackson, R. E. (1979). Oxidation–reduction sequences in ground water flow systems. *Can. J. Earth Sci.* **16**(1):12–23.

Department of the Environment (1978). Cooperative programme of research on the behaviour of hazardous wastes in landfill sites: Final report of the Policy Review Committee. Her Majesty's Stationery Office, London.

Downing, R. A., Smith, D. B., Pearson, F. J., Monkhouse, R. A., and Otlet, R. L. (1977). The age of groundwater in the Lincolnshire Limestone, England, and its relevance to the flow mechanism. *J. Hydrol.* **33**:201–216.

Edmunds, W. M. (1977). Groundwater geochemistry—controls and processes. In: *Proceedings of a Conference on Groundwater Quality—Measurement, Prediction and Protection,* Reading, Sept. 1976, Water Research Centre, U.K., pp. 115–147.

Edmunds, W. M. and Lloyd, J. W. (1981). Application of geochemistry to British hydrogeological studies. In: *A Survey of British Hydrogeology 1980.* Royal Society of London, pp. 125–139.

Edmunds, W. M., Lovelock, P. E. R., and Gray, D. A. (1973). Interstitial water chemistry and aquifer properties in the Upper and Middle Chalk of Berkshire, England. *J. Hydrol.* **19**:21–23.

Edmunds, W. M. and Morgan-Jones, M. (1976). Geochemistry of groundwaters in British triassic sandstones: the Wolverhampton–East Shropshire area. *Q. J. Eng. Geol.* **9**:73–101.

Finch, J. W. and Griffiths, D. H. (1977). The detection of contaminated groundwater in the Nottinghamshire Trias by geophysical techniques. In: *Proceedings of a Conference on Groundwater Quality*—Measurement, Prediction and Protection, Reading, Sept. 1976, Water Research Centre, U.K., pp. 364–372.

Foster, S. S. D. (1975). The Chalk of groundwater tritium anomaly—a possible explanation. *J. Hydrol.* **25**:159–165.

Foster, S. S. D. and Milton, V. A. (1974). The permeability and storage of an unconfined Chalk aquifer. *Hydrol. Sci. Bull.* **19**(4):485–500.

Foster, S. S. D. and Smith-Carington, A. K. (1980). The interpretation of tritium in the Chalk unsaturated zone. *J. Hydrol.* **46**:343–364.

Headworth, H. G., Puri, S., and Rampling, B. H. (1980). Contamination of a Chalk aquifer by mine drainage at Tilmanstone, East Kent, U.K. *Q. J. Eng. Geol.* **13**:105–117.

Ineson, J. and Downing, R. A. (1963). Changes in the chemistry of groundwaters of the Chalk passing beneath argillaceous strata. *Bull. Geol. Survey Great Britain* **20:**176–192.

Ivanovitch, M. and Smith, D. B. (1978). Determination of aquifer parameters by a two-well pulsed method using radioactive tracers. *J. Hydrol.* **36:**35–45.

Jackson, R. E., Merritt, W. F., Champ, D. R., Gulens, J., and Inch, K. J. (1977). The distribution coefficient as a geochemical measure of the mobility of contaminants in a groundwater flow system. In: *The Use of Nuclear Techniques in Water Pollution Studies,* IAEA, Vienna.

Kitching, R., Bell, J. P., Edworthy, K. J., and Tate, T. K. (1981). New instrumentation in hydrogeology. In: *A Survey of British Hydrogeology 1980,* Royal Society of London, pp. 113–124.

Langmuir, D. (1972). Controls on the amounts of pollutants in subsurface waters. *Earth Mineral Sci.,* **42:**9–13.

Lind, A. M. (1975). Nitrate reduction in the subsoil. *Proceedings of International Association on Water Pollution Research Conference—Nitrogen as a Water Pollutant,* Copenhagen, 18–20 August 1975. **1:**14 pp.

Mallman, W. L. and Mack, W. N. (1961). Biological contamination of groundwater. In: *Groundwater Contamination, Proceedings of 1961 Symposium,* Cincinnati USDHEW, Robert A. Taft Sanitary Engineering Center, Technical Report W 61-5, 35–43.

Mather, J. D., Gray, D. A., Allen, R. A., and Smith, D. B. (1973). Groundwater recharge in the Lower Greensand of the London Basin—results of tritium and ^{14}C determinations. *Q. J. Eng. Geol.* **6:**141–152.

McNabb, J. F. and Dunlap, W. J. (1975). Sub-surface biological activity in relation to groundwater pollution. *Groundwater,* **13**(1):33–44.

McNabb, J. F. (1977). Current developments in assessing the role of sub-surface biological activity in groundwater pollution. In: *Proceedings of a Conference on Groundwater Quality—Measurement, Prediction and Protection,* Reading, Sept. 1976, Water Research Centre, U.K., pp. 95–114.

Morgan-Jones, M. (1977). Mineralogy of the non-carbonate material from the Chalk of Berkshire and Oxfordshire, England. *Clay Mine.* **12:**331–344.

Morgan-Jones, M. and Eggboro, M. D. (1981). The hydrogeochemistry of the Jurassic Limestones in Gloucestershire, England. *Q. J. Eng. Geol.* **14**(1):25–40.

Oakes, D. B. (1977). The movement of water and solutes through the unsaturated zone of the Chalk United Kingdom. Paper presented at Third International Hydrology Symposium, Colorado State University, Fort Collins, USA.

Oakes, D. B. (1981). Nitrate pollution of groundwater resources—mechanisms and modelling. IIASA Task Force meeting—The Management and Control of Non-point Nitrate Pollution of Municipal Water Supply Sources, Laxenburg, Austria, Feb. 10–12.

Pickens, J. F., Jackson, R. E., and Inch, K. J. (1981). Measurement of distribution coefficients using a radial injection dual-tracer test. *Water Resour. Res.* **17**(3):529–544.

Ross, C. A. M. (1980). Experimental assessment of pollutant migration in the unsaturated zone of the Lower Greensand. *Q. J. Eng. Geol.* **13**(3):177–188.

Smith, D. B., Downing, R. A., Monkhouse, R. A., Otlet, R. L., and Pearson, F. J. (1976). The age of groundwater in the Chalk of the London Basin. *Water Resour. Res.* **12**(3):392–404.

Smith, D. B., Wearn, P. L., Richards, H. G., and Rowe, D. C. (1970). Water movement in the unsaturated zone of high and low permeability strata using natural tritium. In: *Proceedings of the International Symposium on Isotope Hydrology,* IAEA, Vienna, pp. 73–87.

Spears, D. A. (1976). Information on groundwater composition obtained from a laboratory study of sediment–water interaction. *Q. J. Eng. Geol.* **9**:25–36.

Spears, D. A. (1979). Pore water composition in the unsaturated zone of the Chalk, with particular reference to nitrates. *Q. J. Eng. Geol.* **12**:97–105.

Spears, D. A. and Reeves, M. J. (1975). The influence of superficial deposits on groundwater quality in the Vale of York. *Q. J. Eng. Geol.* **8**:255–269.

UNESCO (1980). R. E. Jackson (Ed.), Aquifer contamination and protection. Report of Project 8-3, International Hydrological Programme. UNESCO, Paris.

Vogel, J. C., Talma, A. S., and Heaton, T. H. E. (1981). Gaseous nitrogen as evidence for denitrification in groundwater. *J. Hydrol.* **50**(1/3):191–200.

Ward, W. H., Burland, J. B., and Gallois, R. W. (1968). Geotechnical assessment of a site at Mundford, Norfolk, for a large proton accelerator. *Geotechnique* **18**:339–431.

Wellings, S. R. and Bell, J. P. (1980). Movement of water and nitrate in the unsaturated zone of the Upper Chalk, near Winchester, Hants., England. *J. Hydrol.* **48**:119–136.

Whitelaw, K. and Edwards, R. A., 1980. Carbohydrates in the unsaturated zone of the Chalk, England. *Chem. Geol.* **29**:281–291.

Whitelaw, K. and Rees, J. F. (1980). Nitrate-reducing and ammonium-oxidizing bacteria in the vadose zone Chalk aquifer of England. *J. Geomicrobiol.* **2**(2):179–187.

Young, C. P., Oakes, D. B., and Wilkinson, W. B. (1976). Prediction of future nitrate concentrations in groundwater. *Ground Water* **14**:426–438.

Young, C. P., Oakes, D. B., and Wilkinson, W. B. (1983). The impact of agricultural practices on the nitrate content of groundwater in the principal United Kingdom aquifers. In: *Environmental Management of Agricultural Watersheds* (Golubev, G., Ed.). IIASA Collaborative Proceedings Series CP-83-51, International Institute for Applied Systems Analysis, Laxenburg, Austria, pp. 156–197.

Young, C. P. and Morgan-Jones, M. (1980). A hydrogeochemical survey of the Chalk groundwater of the Banstead area, Surrey, with particular reference to nitrate. *J. Inst. Water Eng. Sci.* **34**:13–236.

18

STRATEGY FOR SUBSURFACE CHARACTERIZATION RESEARCH

A. W. Hounslow

Department of Geology
Oklahoma State University
Stillwater, Oklahoma

Since its inception in 1965, the Ground Water Research Branch of the U.S. Environmental Protection Agency's Robert S. Kerr Environmental Research Laboratory has concentrated its research in three of the seven areas of its research strategy (Figure 18.1), methods development, transport and fate, and technical assistance. Most of this research has been directed toward organic compounds and biological contaminants. Although technical assistance is not research in the same sense, it is an extremely important aspect of the laboratory mission in that it directs the attention of the research personnel to real-life situations and counteracts a common tendency of research laboratories to concentrate on interesting but esoteric problems.

Over the last few years the research thrust has been toward the examination of the attenuation and degradation of selected synthetic organic contaminants through the use of soil columns and simulated aquifers.

It has become increasingly apparent that the organic and inorganic constituents of the soil and subsurface materials play an important part in the retardation of the organic pollutants, and thus allow a greater time period for both biological and abiotic degradation to occur. It is a tenet of ground water geochemistry that the composition of the ground water is primarily the result of the mineralogy and chemistry of the rocks through which it passes.

The ultimate aim of subsurface study is to develop definitive information concerning the behavior of contaminants in subsurface environments in or-

U.S E.P.A
GROUND WATER RESEARCH STRATEGY

Categories

Methods Development

Contaminant Transport and Fate

Subsurface Characterization

Specific Sources of Contamination

Aquifer Rehabilitation

Information Transfer

Technical Assistance

Figure 18.1.   The U.S. Environmental Protection Agency ground water research strategy.

der to provide a valid scientific basis for protection of ground water quality. The three basic aspects of this type of study are:

1.   The prediction of the impact on ground water quality of contaminants in the environment.
2.   The optimization of existing conditions so as to enhance the attenuating biotic or abiotic processes affecting contaminants in subsurface environments.
3.   The development of cost-effective methods of rehabilitating contaminated aquifers.

The second and third aspects of such a study must to a great extent await the results of the first part.

The successful prediction of the transport of contaminants in the subsurface requires an understanding of the interactions between the pollutant and the soil or other subsurface geological materials.

Movement of water through the saturated zone has been extensively studied by hydrologists and is reasonably well understood. Movement of water through the unsaturated zone has been extensively studied by soil physicists and is partially understood. Although useful mathematical models describing both situations already exist, further research is needed on unsaturated flow and the effects of inhomogeneity of strata and soil.

Poor understanding of the interactions between pollutants and subsurface materials limits our ability to predict the behavior of pollutants in the subsurface. However, sorption of a number of nonpolar organic compounds to organic matter has been used successfully to predict their movement through soil and aquifers. The same approach can be applied, although with less confidence, to more polar compounds.

With the exception of certain agricultural chemicals, sorption of organic chemicals to other subsurface components—including clay and amorphous oxides of iron, manganese, and aluminum—is poorly understood. Again, with the exception of certain agricultural chemicals, our understanding of biodegradation in soil is, at best, qualitative. Our understanding of biodegradation in the deeper subsurface is essentially nonexistent.

A major aim of ground water research is to determine the transport and fate characteristics of introduced organic or inorganic compounds as well as pathogens, primarily by answering the following questions: How far or how fast do the pollutants (and any degradation products) move? How long do they remain in the system?

At the present time these questions are answered primarily by experiments using laboratory physical models with the geological material of particular immediate interest. These experiments are tedious, time consuming, and expensive. These physical models may be soil columns or simulated aquifers, and in accord with present usage will be called subsurface microcosms.

The impracticability of this approach is apparent when one considers that there are 5000 different geological materials and soils and 40,000 organic chemicals. If 25 tests can be done per year, then 8 million yr would be required to complete the project.

An alternative is to understand the nature of the subsurface environment in terms of its solid components (including its biological composition), of the composition of the liquid component, and of the extent of movement of this fluid phase through the subsurface strata.

If one could then group and classify these parameters, a manageable number of experiments could be designed to confirm the concepts developed and allow a prediction of how a particular compound will behave at a particular geographic location.

It cannot be overemphasized that such a conceptual scheme must be the result of field observation, laboratory characterization, and laboratory experimentation. All phases will be supportive of each other, and none should be allowed to overshadow the others.

CHARACTERIZATION OF THE SUBSURFACE

The composition and properties of subsurface solids are very important when considering the degree to which sorption of organic and inorganic contaminants occurs.

The quality of ground water reflects the mineralogic composition of the aquifer. During the slow movement of water through an aquifer, its composition will gradually change, and this change will reflect the composition of the rocks through which the water has passed. These reactions generate geochemical parameters such as Eh, pH, and ionic strength, which in turn may determine the types of microbial processes that may occur.

A study of this type is multidisciplinary in nature but is primarily concerned with hydrogeology, mineralogy, geochemistry, and microbiology. Some of these interactions are shown diagramatically in Figure 18.2.

GROUND-WATER GEOCHEMISTRY

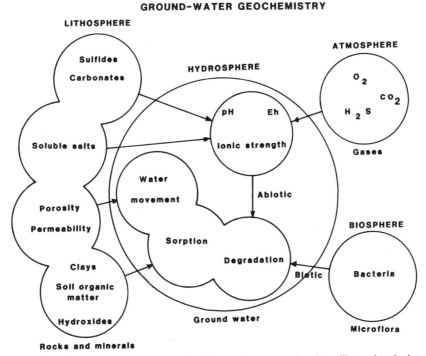

Figure 18.2. Diagrammatic representation of ground water geochemistry illustrating the interactions between the lithosphere, hydrosphere, atmosphere, and biosphere.

Hydrogeologic Parameters

A study of the hydrogeology of an area should include determining the number and type of aquifers present and the rate and direction of movement of the waters in them. These determinations, of course, control the speed of migration and the size and location of the area that may be affected by a pollutant.

Pollutant movement occurs in both the unsaturated and saturated zones; thus, the hydraulic properties of the unsaturated zone are of great importance. Parameters such as porosity, permeability, and heterogeneity, together with the shape and size of voids, are likely to be of considerable importance in controlling the movement of pollutants through the unsaturated zone to the saturated zone.

Mineralogical Parameters

Of primary importance are those phases with sorptive properties, specifically clays and zeolites. Although not minerals, but included here, are the various constituents of organic matter and the hydrous oxides of iron, manganese, and aluminum usually present in parts of the regolith.

Next in importance are those minerals that govern the pH or other chemical properties of water in contact with them. These include carbonates, sulfides, and soluble salts. These may also react with certain pollutants.

Geochemical Parameters

The mobility of many elements and compounds and the efficiency of many microbial reactions are controlled by chemical parameters or environmental factors such as pH, temperature, and redox potential. Of particular importance here are geochemical boundaries such as Eh–pH boundaries.

Biological Parameters

Biodegradation may be the major factor affecting the ultimate fate of many pollutants in the subsurface. Estimates of subsurface microbial populations and their degradative capabilities under various environmental conditions are needed. Microbial populations depend on having an adequate supply of organic carbon, mineral nutrients, and electron acceptors in addition to a suitable environment.

DISCUSSION

The movement of a pollutant in the subsurface can be studied in three more or less distinct stages, each stage providing an improved description of pollutant behavior. The first stage results in a mathematical description of the water movement, the second in a mathematical description of mass transport processes, and the third in a mathematical description of biotic and abiotic degradation processes.

On each of these stages is superimposed the overall strategy already mentioned, namely, methods development, subsurface characterization, and transport and fate. A further overlay, necessary but distinct, is the classification of the pollutants themselves. Whether they be pathogens, elements or ions, or synthetic organic compounds, they must be classified and grouped so that a limited number of these materials selected for study reflect the behavior of a much larger population of potential pollutants.

Rate of Ground Water Movement—Hydrogeology

A hydrogeologic study of an area includes the determination of the number, type, and placement of aquifers present, as well as the geologic characterization of the saturated and unsaturated zones. Subsequently, for each aquifer the water quality, the direction, and rate of movement of the ground water is determined, and the area subdivided into recharge and discharge zones. All these data are important in a study of the extent and movement of pollution plumes.

The main problem encountered in obtaining any of the above parameters is that their acquisition classically requires drilling, casing, and developing wells, measuring depths to water, collecting and analyzing water samples, and carrying out aquifer tests—an expensive and time consuming project.

With some care, many of the above data can also be obtained by other less direct means. For example, because many streams contain a significant portion of water derived from ground water runoff, estimates of ground water recharge can be obtained by stream hydrograph separations with the use of computerized techniques. Further, a great deal of hydrologic data can be obtained from published reports, topographic and geologic maps, aerial photographs, satellite imagery, and geophysical investigations.

It is important, then, to develop and refine techniques for acquiring hydrogeologic information with minimal involvement in extensive and expensive drilling programs.

A large portion of the existing data and most of the hydrogeologic techniques depend primarily on the assumption that the aquifer is homogeneous, an assumption generally far from correct. Considerable study and research must be done on the spatial variations of hydraulic properties in both the saturated and unsaturated zones.

Mass Transport Processes

Once ground water velocity has been determined, the degree of actual and possible contamination may be ascertained, assuming no biotic degradation, if one considers:

1. Extent and rate of volatilization.
2. Extent of dispersion and diffusion.
3. Nature and extent of subsurface attenuation.
4. Effects of large amounts of pollutants.

The geochemical implications are illustrated diagrammatically in Figure 18.3.

Hydrogeology

Although unreactive solutes move at a rate equal to the average linear velocity of the ground water, they tend to spread out because of a phenomenon called dispersion. This process is not well understood, especially when applied to heterogeneous aquifers, and further study is obviously necessary.

Characterization of Aquifer Materials

This in actuality includes the mineralogy of the subsurface materials, but because some of the subsurface materials consist of organic matter and hydrous metal oxides of variable crystallinity, a broader study is implied.

An understanding of the surface processes operating in a particular case of ground water contamination may be obtained by field observations and

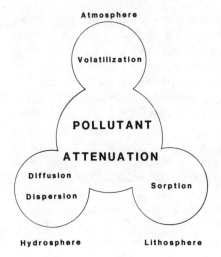

Figure 18.3. Factors leading to pollutant attenuation not involving biodegradation.

laboratory simulations. Instrumental and chemical analyses of the various sorbent-sorbate phases may indicate the nature of these associations.

The primary techniques used to characterize solid phases (Figure 18.4) include X-ray diffraction, electromagnetic absorption techniques such as infrared spectroscopy, microscopy (including both optical and electron beam), and chemical analysis. All these techniques can be applied before and after selective dissolution or to the various separates derived from mechanical separation methods.

X-Ray Diffraction. X-ray diffraction is used to identify, and sometimes quantify, the crystalline phases present, as well as to determine any changes in the lattice spacings of clays. Before a subsurface sample can be analyzed, a sample must be obtained, usually by coring, and the recovered sample prepared for analysis by crushing and grinding. The majority of identifications relies on X-ray diffraction as a major tool supplemented when neces-

Figure 18.4. The various facets of materials analysis that may be applied to subsurface characterization.

sary by microscopy or chemistry. It should be noted that X-ray diffraction techniques generally are applicable only to materials with a definite crystalline structure, and that detection limits are of the order of 1–2% without prior concentration.

The minerals primarily responsible for the sorption of contaminants are the clay minerals. Their identification by X-ray diffraction relies on their separation from other components and their alignment such that their interlayer spacing can be measured.

Several basic problems occur with clays: some clays are poorly crystalline and give anamalous X-ray patterns, and some are amorphous and may not give an identifiable X-ray pattern. In addition, a soil organic matter–clay–mineral complex may show greater sorptive properties than either substance alone.

Infrared Spectroscopy. Whereas X-ray diffraction reflects the total crystal structure, infrared spectroscopy looks at individual bonds. Therefore, amorphous materials can readily be studied by infrared spectroscopy, but unlike X-ray diffraction, there is no simple relationship between an infrared peak and the bond from which it originates. At the present time infrared spectroscopy is the best tool for studying the bonding between clays, soil organic matter, amorphous hydroxides, and introduced organic compounds. The bonding between clays and organics is strongly dependent on whether the organic material is polar or non-polar, charged or uncharged, and whether the clay is expanding or non-expanding. Adsorption can take place within the structure or around the outside of the clay particle. Aromatic rings may be aligned parallel to or perpendicular to the layer structure of the clay. Some complexes require a bridging inorganic ion, whereas some complexes occur more readily if the clay occurs as a clay–soil organic matter complex.

Many soil pollution studies ignore the fact that soil organic matter is made up of a variety of constituents, each having different properties. Humic acids are soluble in alkalis but insoluble in acids, whereas fulvic acids are soluble in both acids and alkalis. As well as enhancing adsorption, fulvic acids also solubilize some materials.

Infrared spectroscopy has been used to identify some of these humus and non-humus materials as well as to detect changes in their structures after various treatments.

Further soil constituents that are amorphous but also very important with respect to their adsorptive properties are the colloidal metal hydroxides. These also can be detected by infrared spectroscopy.

Microscopy. Textural relationships among the phases present can be readily determined by optical or electron beam examination of the impregnated sedimentary materials.

The physical nature of clays and humic materials can change if the chemical environment in which they occur changes. For example, humic acids may occur as either a sphere, sheet-like material, or bundle of fibers depending on the pH. In addition, the movement of water and adsorption of pollutants may be strongly dependent on the shape and size of the voids in the subsurface materials.

The optical microscope generally lacks the resolution to solve these problems, so use must be made of electron microscopy. The resolution of electron microscopes has increased so much in recent years that some large molecules can now be resolved. Also, the chemical composition of the samples can be determined by means of electron microscopes.

None of the techniques described above are entirely adequate to solve the problems posed, and new methods for elucidating mineral–colloid–organic interactions are needed.

Geochemistry.   Ground water geochemistry is concerned with the distribution and mobility of elements and compounds at or near the earth's surface, with an environment resulting from the interaction of the hydrosphere, atmosphere, lithosphere, and biosphere.

When water is in contact with minerals (or amorphous materials) in rock, sediment, or soil, five basic processes may take place:

1. The more soluble constituents are leached out and move into the ground water system. Under normal conditions this leaching determines the ground water quality.
2. Some solution of the less soluble materials, such as quartz, may take place, but the proportion dissolved is so small that it has little effect on the overall abundance of this phase in the rock.
3. Insoluble materials, such as clays and colloidal oxides and hydroxides, may be formed by the chemical alteration of existing minerals. These secondary materials frequently have high sorption capacities and greatly influence the movement of pollutants in the subsurface.
4. If the water is saturated with a particular component, or if the geochemical environment changes, new minerals may be precipitated in the pore spaces of the existing rocks, soil, or sediment.
5. Inorganic ions or dissolved organic matter may be adsorbed onto an existing solid phase. If this is an ion-exchange reaction, other ions will be released to the liquid phase. The solid phases with the highest sorption capabilities are the clays, amorphous sesquioxides, and naturally occurring organic matter. The presence of inert mineral grains such as quartz is also important, because the amorphous sesquioxides frequently occur as coatings on such minerals, a phenomenon that allows them to exert chemical influence far out of proportion to their total concentration.

Geochemical Parameters.   The geochemical parameters such as pH, redox potential, temperature, and ionic strength influence most of the reactions that may occur in the subsurface, and thus must be accurately determined if predictive techniques are to be evolved. Of particular importance in subsurface studies is the detection of geochemical boundaries, that is, boundaries on either side of which geochemical parameters differ drastically. Examples are oxidizing–reducing, acid–neutral, and fresh water–saline water boundaries.

Although these parameters are readily measured in surface waters, their measurement in ground waters presents innumerable difficulties. The redox potential of a solution may be obtained directly by means of millivolt probes, or indirectly either by chemically determining appropriate ion activity ratios or by measuring the concentrations of certain dissolved gases. Because it is virtually impossible to drill and sample a well without affecting the redox potential of the water sample, the development of *in situ* techniques is highly recommended. Some redox sample–activity ratios that change slowly on exposure to air have been reported.

An alternative to actual Eh measurements (for waters having a nearly neutral pH), would be to define three key hydrogeochemical environments based on the presence or absence in ground water of dissolved oxygen and hydrogen sulfide. In the case when neither gas is present, the resulting water will frequently contain more than 1 mg/L of dissolved ferrous iron.

Biotic Kinetic Processes

This section emphasizes synthetic organic contaminants and introduced pathogens such as viruses. The transformation of inorganic pollutants will not be discussed primarily because much more is known about inorganic contaminants than about organic contaminants.

A primary environmental goal is the ability to estimate the residence time of a contaminant in the subsurface. We need to know its mobility, its likelihood of degradation, its degradation products, and whether either contaminant or degradation product will reach the ground water. The mobility of a compound can sometimes be predicted based on studies of the sorption of the contaminant by the soil or subsurface material.

The degradation of organic chemicals in the environment is a topic about which our ignorance is abysmal. However, any chemical reactions, whether they be biotic or abiotic, will be highly dependent on the geochemical parameters, including pH, Eh, and temperature. It must also be recognized that the presence of clays, amorphous sesquioxides, and soil organic matter in the subsurface will at most reduce the mobility of the organic contaminants and possibly retain them in an environment wherein they may be degraded.

Microbiology

The characterization of the subsurface in terms of its biological properties can be divided into five more or less distinct parts:

1. Identify those subsurface habitats that harbor an indigenous microflora and determine the nature and vertical extent of the microbes.

2. Determine the effect of interactions between subsurface organisms. Included here would be the effects of predation and competition, and synergistic relationships within the microbial consortia, especially as it relates to the degradation of pollutants.

3. Determine the concentrations of various pollutants at waste-disposal sites that can either modify existing microbial communities or induce a community where one did not exist before.

4. Determine the ability of both indigenous and induced microfloras to degrade trace organic contaminants.

5. Determine the limits imposed by geochemical parameters on biologic activity.

Subsurface habitats harboring a microflora can be identified by a reconnaissance of subsurface materials based on microscopic techniques. Exclusive use of techniques that require growing the organisms should be avoided because of the difficulty of culturing all the organisms from any natural habitat. Because organic matter leached from the surface soil is the most likely source of food for an indigenous microflora in the deeper subsurface, subsurface materials should be examined under soils with wide ranges in organic matter content and in rate of aquifer recharge. Because water below the water table is often depleted of oxygen, which profoundly affects microbial activity, material from both the saturated and unsaturated zones should be examined.

The impact of pollutants on the structure and composition of the subsurface microflora may best be determined in microcosms, where the effects of variations in concentrations of organic pollutants (with respect to both toxicity and availability as a substrate), of inorganic nutrients, and of other anions can readily be studied.

The vulnerability of organic contaminants to transformation by microbes could be screened in simple batch biodegradation assays conducted with authentic subsurface materials protected from microbial contamination.

However, quantitative descriptions of the fate of an organic pollutant in the subsurface requires an understanding of the kinetics of its degradation. At present most descriptions of degradation in natural systems assume first-order kinetics (i.e., the rate of degradation is described as a half-life). This characterization will likely prove inadequate over the wide range of concentrations of pollutants in ground water and can be improved by determining the concentration at which degradation becomes saturated (e.g., where the rate of degradation no longer increases with concentration of pollutant).

These kinetic parameters can be determined only by cumbersome and expensive experiments or field studies at a limited number of unique sites. The rate of disappearance of the pollutant could be monitored with time when present at varying concentrations, or the rate of production of $^{14}CO_2$ or other metabolic product could be monitored from a radiolabeled pollutant. Such monitoring will require either the development of methods to efficiently extract the compounds of interest from subsurface materials, or the use of custom-synthesized radiolabeled compounds.

Correlation analysis might identify inexpensive tests that could predict the kinetic parameters within acceptable limits, if the materials used to determine the kinetics of degradation are also extensively characterized with respect to physical, chemical, and biochemical properties. Use of such relationships would greatly lower the cost of site-specific and compound-specific prediction of fate, and would provide particularly useful and cost-efficient tools to the regulatory agencies.

Laboratory Experimentation

Many of the coefficients required for mathematical models will be obtained by the experimental application of organic chemicals to subsurface microcosms.

This ostensibly simple task is beset with difficulties, starting with the selection and collection of the geological material and the selection of the organic pollutant to be used. Also, microcosm work is both time consuming and tedious, and because it requires constant attention when running, only a limited number of experiments are feasible. Other factors that must be considered are open versus closed systems, the oxygen budget, new nutrient inputs, and saturated versus unsaturated conditions.

Selection of Geological Material.   The only time that this may become a problem is when a soil *(sensu stricto)* is to be tested. There are two basic geochemical approaches to preparing soil columns. The first involves collecting soils of a limited number but wide variety and analyzing them for the appropriate parameters. The second involves the preparation of synthetic soils using appropriate and measured amounts of clays, silt, sand, and organic matter. The first is more realistic in that, with care, at least some of the soil structure may be retained. In the second case, the composition (texture) is well documented, but the structure is completely unrelated to that of most soils; considerable caution must therefore be used in the interpretation of the results.

Organic Chemical Selection.   The second major input variable in subsurface microcosm studies is the choice of the organic compounds to be investigated. There are essentially two major approaches to this problem.

1. Select for study individual compounds of obvious and immediate concern because of high probability for adverse health and/or ecological effects, implication in reported cases of ground water pollution, or high potential for release in significant quantities to the subsurface environment because of large production volumes, manufacturing and transport practices, uses, and/ or probable methods of ultimate disposal.

2. Choose compounds on a systematic basis by grouping the broad range of organic compounds into a limited number of divisions or classes and selecting representative compounds from each class for investigation. The key to this approach is the development of a manageable classification system effectively encompassing the vast array of organic compounds that may be encountered. Several possibilities are presented below:

(a) Classify organic compounds on the basis of easily measured physical–chemical properties, such as water solubility, octanol/water partition coefficient, or vapor pressure.

(b) Group organic chemicals into divisions according to major structural features, such as carbon skeleton and type or number and position of attachment of functional groups. A significant problem in this classification system, however, is the difficulty encountered in categorizing compounds of multiple functionality and complex structure.

(c) Classify organic chemicals according to their behavior in specially devised empirical separation schemes. One possibility would be the use of the dissolved organic carbon fractionation analysis, which separates organic compounds into six classes on the basis of sorption by macroreticular resins. In its present form this scheme provides too few classes, which are consequently far too broad, but it might be modified and extended to provide a grouping of the broad range of organics into a reasonable number of specific classes. Another possibility might be the use of empirical separation schemes based on a group of standard geological materials rather than synthetic resins.

Of the two major approaches for selecting compounds for study, the first has the advantage of providing information of immediate value for solution of current problems. Furthermore, the study of individual compounds of critical environmental concern will undoubtedly continue to be necessary for some time to come. However, if carried to the extreme, this approach could require the individual study of a large majority of known organic compounds. As previously noted, this is completely impractical. Obviously, the major effort must be devoted to the second, systematic approach in order to develop the database and tools required for a broadly applicable predictive capability for behavior of organic pollutants in subsurface environments. This will require the establishment of a classification scheme for organic chemicals that permits the effective grouping of the vast array of potential organic pollutants into a manageable number of classes based on properties or characteristics related to subsurface transport and fate.

ACKNOWLEDGMENTS

This chapter is the result of numerous discussions and many hours of meetings with RSKERL personnel, particularly those in the Ground Water Research Branch. It is impractical to list all those involved; however, the nucleus group included Bill Dunlap, James McNabb, Marv Piwoni, Dick Scalf and John Wilson. An earlier version of this paper was reviewed by Virgil Freed, Dick Jackson, Don Langmuir, Jay Lehr, Perry McCarty, David Miller, and Wayne Pettyjohn. I would like to extend my thanks to these reviewers for their comments and suggestions, many of which are incorporated in this paper. In writing this paper I have attempted to maximize areas of agreement; however, the prejudices that remain are mine, for which I take full responsibility. Finally, I must extend my thanks to Jack Keeley for his help and continuous encouragement during the course of this sometimes seemingly hopeless task.

REFERENCES

Forstner, U. and Wittman, G. T. W. (1979). *Metal Pollution in the Aquatic Environment*. Springer-Verlag, New York.

Freeze, R. A. and Cherry, J. A. (1979). *Groundwater*. Prentice-Hall, Englewood Cliffs, NJ.

Grim, R. E. (1968). *Clay Mineralogy*. 2nd ed. McGraw-Hill, New York.

Jenne, E. A. (1968). Controls on Mg, Fe, Co, Ni, Cu, and Zn concentrations in soils and water: The significant role of hydrous Mn and Fe oxides. In: *Trace Organics in Water*. ACS Advances in Chemistry Series **73**:337–387.

Schnitzer, M. and Khan, S. U. (1978). *Soil Organic Matter*. Elsevier, New York.

Theng, B. K. G. (1979). *Formation and Properties of Clay–Polymer Complexes, Developments in Soil Science,* Vol. 9. Elsevier, New York.

19

INTERACTION BETWEEN ORGANIC MOLECULES AND MINERAL SURFACES*

M. M. Mortland

Department of Crop and Soil Sciences
Michigan State University
East Lansing, Michigan

The subject of mineral–organic interactions is of interest in many natural systems as well as in an industrial context. Most of the research in this area has been performed utilizing clays as the mineral portion of the complex. There are several reasons for this:

1. Clays have large surface areas per unit weight as compared with the silt and sand fractions of soils and sediments, and thus have much greater adsorptive properties than do the coarser mineral materials. Even in sandy soils and sediments where, in terms of amount present, they comprise a very small percentage of the total mineral material, they may still provide the greater proportion of the adsorptive capacity.

2. Clays are more easily studied since they do have large adsorption capacities and reactivities. In addition, their small size makes them more amenable to investigation by spectroscopic methods (infrared, ultraviolet–visible, and electron spin resonance).

3. Various kinds of clay minerals may be obtained in reasonably pure form from supply houses.

4. The surface properties of silt and sand-sized silicate minerals are probably not greatly different from those of non-expanding clay minerals.

* Published as Michigan Agricultural Experiment Station Journal Article No. 10233.

Clays are ubiquitous in nature, occurring in almost all kinds of sedimentary deposits. They are the products of weathering of primary minerals during soil formation and may end up in sedimentary deposits as the result of various kinds of erosional processes. In both soils and lake and sea beds, clays may be intimately complexed with natural organic materials. In fact, it has been suggested that clays may have played a role in the synthesis of macromolecules during prebiotic times (Bernal, 1951) and as catalysts in the formation of petroleum from biological materials (Almon and Johns, 1975).

The clay minerals have platy layer structures; details of the various kinds may be found in Grim (1968) and in Dixon and Weed (1977). An important property of clays is that they possess cation exchange capacity. That is, they have negative electrical charge arising from isomorphous substitution of lower valent cations within the silicate structure, which leads to a net negative charge compensated for by cations at the mineral surface. These sites are permanently charged and exist regardless of the pH. Cation exchange sites may occur at mineral edges because of unsatisfied negative charge. Also, amorphous aluminosilicates, which may occur in clays, have cation exchange capabilities, but these properties are usually pH dependent, increasing with alkalinity. Cation exchange capacities of clay minerals range from 5 to 200 milliequivalents per 100 g, depending upon structure. A second important property clays possess is a large surface area. Clay minerals that have the ability to swell through the solvation of the regions between the individual mineral layers have surface areas near 800 m^2/g. Those that do not swell have surface areas ranging from 5 to perhaps near 40 m^2/g, depending upon particle size. These two properties—high cation exchange capacity and large surface area per unit weight—give clays chemical and physical effects in soils and sediments in great disproportion to the amounts which may be present, and in many sandy deposits clays are chiefly responsible for whatever adsorptive properties the sediments may have. Thus, in soil and ground water, the interaction between pollutants and mineral surfaces most often involves the clay material present. Since clay minerals themselves differ in properties, it is often important to know how much and what kind of clay minerals are present in soils and sediments if one is going to make any kind of prediction regarding behavior.

MECHANISMS OF ADSORPTION OF ORGANIC SPECIES ON CLAYS

There are a variety of interactions possible between organic molecules and mineral surfaces. These interactions vary considerably in energy and thus have different effects on the overall stability of the organic–mineral complex. In many cases several kinds of interaction may be operating at once, particularly in the case of large molecules or polymers such as proteins.

The mechanisms of mineral–organic interaction have been studied by a

number of laboratory techniques. Among these are use of adsorption isotherms, calorimetry, X-ray diffraction, UV-visible spectroscopy, electron
spin resonance (ESR) spectroscopy, and infrared (IR) spectroscopy. All
these methods may give information of one kind or another on organic–
mineral interaction. Probably the technique that has contributed most to our
knowledge in this area is infrared spectroscopy (Mortland, 1970; Theng,
1974). With the development of high quality double beam instruments coupled with the breakthroughs in clay sample preparation, spectra of organic
molecules adsorbed on mineral surfaces can be directly obtained. Since
molecular vibrations often are affected by the environment in which they
exist, unequivocal conclusions regarding organic–mineral interactions can
often be reached.

Cationic Adsorption

The first organic–clay interaction to be recognized was cation exchange. As
suggested above, clays possess a net negative electrical charge that, in nature, is most often neutralized by exchangeable metal cations on the mineral
surface. If the organic species is cationic (i.e., a protonated amine), it may be
adsorbed on the mineral surface by simple ion exchange:

$$RNH_3^+ + M^+ - Clay \rightarrow RNH_3^+ - Clay + M^+ \qquad (19.1)$$

where RNH_3^+ is a protonated amine and M^+ is some exchangeable metal ion.
Many organic species are cationic because of protonation of amine groups,
as in the case of amino acids and alkyl amines. Obviously the cationic
property of such molecules is dependent upon pH. On the other hand, there
are organic species that possess positive charge regardless of the pH, in
particular the quaternary ammonium cations. These materials are very
strong bases. Examples of such a species that might occur in nature as a
result of application to soils are the two herbicides diquat [6,7-dihydrodi-
pyrido(1,2-a : 2',1'-c)-pyrazidium dibromide] and paraquat (1,1'-dimethyl-
4,4'-dipyridium dichloride).

Organic cations that have been exchanged onto the mineral surfaces are
usually more strongly adsorbed than the metal counterion they have replaced because they are usually larger in size and may have other kinds of
interaction with the mineral surface (e.g., hydrogen bonding or physical
forces).

Protonation

In addition to adsorption of organic compounds by ion exchange of organic
with inorganic cations, organic molecules that are neutral but have the basicities to become protonated may become cationic after adsorption at the
mineral surface. This reaction requires that the mineral surface function as a

proton donor (Brönsted acid). The sources for such protons are (1) exchangeable protons at exchange sites on the mineral, (2) water associated with exchangeable metal ions, and (3) proton transfer to a newly adsorbed organic base from an organic cation already occupying an exchange site. It is obvious that the existence of an organic base in molecular or cationic form on the mineral depends upon the basic strength of the molecule and the proton-supplying power of the mineral surface.

The reaction of an organic base with exchangeable protons is quite straightforward.

$$RNH_2 + H^+ - Clay \rightarrow RNH_3^+ - Clay \qquad (19.2)$$

This reaction goes strongly to the right since it is merely the neutralization of an acid with a base and large energies of reaction are involved (several kilocalories). On the other hand, increase in pH of the system may consume protons and return the organic base to its molecular form and possibly result in its desorption from the mineral surface.

Organic base reaction with protons supplied by hydrated exchangeable metal cations is in accordance with the following reaction.

$$M(H_2O)_x^{+n} - Clay + B \rightleftarrows [M(H_2O)_{x-1}OH]^{+n-1}BH^+ - Clay \qquad (19.3)$$

Here M is an exchangeable metal cation and B is the base in question. Ordinary water is not a strong enough acid to protonate many bases. However, when water is associated with metal cations in spheres of hydration, hydrolysis of the complex produces more or less protons depending upon the kind of metal ion involved. The more electronegative the cation, the greater polarizing effects it has on its associated water molecules and the greater will be its ability to donate protons. Thus solutions containing Al^{3+} are relatively acidic while those with K^+ are much less so. Obviously, then, the acidity of a mineral surface will depend on the kind of cations occupying the exchange sites and on the consequent ability to protonate bases.

In addition to the kind of metal exchange ion, the water content of clay affects its Brönsted acidity (Mortland and Raman, 1968). As the amount of adsorbed water decreases, proton donating ability increases. An example is work by the author in which calcium-smectite (montmorillonite) was equilibrated over water–ammonia solutions where the partial pressure of water (P/Po) vapor was 0.2 and 0.98. At the lower partial pressure 80 me of NH_4^+ per 100 g of clay were formed on the mineral surface as determined by infrared spectroscopy. At the higher vapor pressure only 16 meq of NH_4^+ were formed. These results suggest that as absorbed water decreases, the remaining water molecules become more acidic, probably as a result of the cationic polarization forces becoming more localized and thus increasing the dissociation of the fewer remaining water molecules.

With the advent of application of infrared spectroscopy to organic–clay interactions, it has been possible to study interaction at low water contents since the technique permits the scientist to obtain spectra of molecules on the surface of the clay and, in the case of organic bases, to determine whether they are protonated.

Not only does acidity of clay surfaces increase with decreasing amounts of adsorbed water, but also the strength of that proton-donating ability may become quite extraordinary. Molecules such as urea and acetamide, which are very weak bases, may become protonated on H^+-smectite surfaces (Mortland, 1966; Tahoun and Mortland, 1966). In homogenous solutions a pH near 0.5 is required to protonate acetamide. Thus a pH measurement of a clay in suspension does not truly represent the acid nature of the clay in drier environments; this suggests that conclusions regarding clay surface chemistry obtained in suspensions may not be directly extended to other situations.

Another mechanism by which organic molecules may become protonated at mineral surfaces is the donation of a proton from another organic base already occupying an exchange site. The general reaction is

$$AH^+ - Clay + B \rightleftarrows BH^+ - Clay + A \qquad (19.4)$$

Here AH^+ represents some protonated species on the exchange sites of the mineral, and B represents the organic base to which a proton is donated. The degree to which this reaction will go to right or left depends upon the relative basicities and concentrations of the two species involved. This kind of proton transfer has been observed for a number of systems by Russell et al. (1968) and Raman and Mortland (1969).

Another kind of interaction that has been observed on clay surfaces by infrared spectroscopy is the formation of hemisalts by organic bases. This interaction comprises the sharing of a proton by two molecules of organic base to form a symmetrical hydrogen bond in which a proton is shared on an equal basis and does not belong to either molecule. This kind of complex occurs when the number of base molecules on the mineral surface exceeds the number of available protons and has been observed for such molecules are urea, pyridine, aliphatic amines, and amides (Mortland, 1970).

Anionic Adsorption

Normally, anionic species are repulsed by the negative electrical charge of clay mineral surfaces. However, in special cases organic anions may be observed on clays, as for example in dry environments where water is not a factor. Also, when polyvalent cations are on the exchange sites of the clay, one charge may be freed to interact with organic anions to form essentially the organic salt of that particular cation. Such interactions have been noted for benzoate anion on clays by Yariv et al. (1966). In such environments as ground waters, it is not likely that organic anions will be adsorbed by a clay surface unless they possess other properties that favor adsorption—such as amino groups in amino acids or large molecular weights in polymers, where entropy change might favor adsorption.

Adsorption via Coordination or Ion–Dipole Interaction

Much of the older literature regarding the adsorption of polar but nonionic organic molecules by mineral surfaces has attributed a major function to oxygen and hydroxyl groups of the silicate surface. This type of interaction has been said to be one of hydrogen bonding between those groups and functional groups of polar organic molecules. This kind of interaction is still believed to operate, but the most energetic reaction polar molecules have with mineral surfaces is that of coordination or ion–dipole interaction with exchangeable metal cations. Through the application of infrared spectroscopy it is possible to view the condition of the adsorbed molecule directly, and from the spectra to draw relatively unambiguous conclusions regarding the nature of the interaction.

Studies of a large number of polar organic molecules adsorbed on mineral surfaces prove that the nature of the exchangeable metal cation plays a decisive role in the adsorption process. Not only do these cations determine the acidity of the surface by their interaction with and hydrolysis of water, and thus the possibility of protonating organic bases, but they also serve as adsorption sites for polar nonionic organic molecules by coordination or ion–dipole type reactions. The metal cations have some affinity for electrons and thus interact with organic molecules capable of donating electrons. Transition metal cations, in particular, have unfilled d orbitals and thus will interact strongly with electron-supplying groups to form classical types of coordination complexes. Alkaline earth and alkali metal cations, though not having unfilled d orbitals, interact strongly with polar molecules through ion–dipole interactions. Thus these cations on mineral exchange sites will react to adsorb polar molecules. The energies involved in these kinds of reactions may be quite large; for example, the solvation of some metal cations by water involves hundreds of kilocalories. This reaction is the main reason minerals such as vermiculite and smectite (montmorillonite) will swell with polar molecules like water and minerals of similar structure such as talc and pyrophyllite do not. The former possess exchangeable metal cations between the silicate sheets while the latter are electrically neutral and do not have cation exchange capacity.

It is important to realize that in nature polar organic molecules and water molecules compete for solvating or coordinating with exchangeable metal cations on mineral surfaces. Such a reaction is as follows:

$$M(H_2O)_x^{n+} - Clay + ROH_y \rightleftarrows M(ROH)_y^{n+} - Clay + H_2O_x \quad (19.5)$$

Here M is the exchangeable cation and ROH is some alcohol. The alcohol can completely dehydrate the mineral, or on the other hand, water can displace all the alcohol according to mass action requirements. Factors affecting the direction of this reaction are the relative affinity of the metal cation for the organic and the water molecules, and the relative concentrations of the two. Some organic molecules possess functional groups that are better electron donors than others and thus may compete better with water for interaction with the cation. An example would be a comparison between ketones and alcohols, which compete with water for ligand positions on exchangeable metal cations much better than do ketones. The consequences of this competition are extremely important because, depending upon the environment, polar organic molecules may or may not be adsorbed by mineral surfaces. In ground water, for example, competition of an organic species with water would be at a maximum and its adsorption through this mechanism consequently more difficult. However, many clay–organic complexes of this kind have been reported and summarized by Theng, 1974.

Adsorption Through Hydrogen Bonding

This is a very important mechanism of interaction in many clay–organic complexes. It is less energetic than electrostatic interactions but becomes significant particularly in large molecules and polymers where additive bonds of this type in addition to a large molecular weight may result in relatively stable complexes.

One kind of hydrogen bonding involves the linking of a polar organic molecule to an exchangeable metal cation through a water molecule in the primary hydration shell. This is sometimes called a water bridge. An example would be a ketone interacting with a hydrated metal cation on the exchange sites:

$$
\overset{\displaystyle H}{\underset{}{\vert}} \qquad\qquad \overset{\displaystyle R}{\underset{}{\vert}}
$$

$$
M^{+n}\overset{H}{\overset{\vert}{O}}-H \;\cdot\cdot\;\; O{=}\overset{R}{\underset{R}{C}}
$$

Such an interaction would come about when the organic is unable to replace the directly coordinated water molecule and hydrogen bonds with that molecule. Thus cations having high solvation energies for water (i.e., polyvalent cations) retain their sphere of water of hydration in spite of the importunities of the competing organic molecule. The resulting hydrogen bond has been observed by infrared spectroscopy for a number of organic molecules (Mortland, 1970; Theng, 1974). This kind of hydrogen bonding is probably very common in nature, where exchangeable cations are likely to retain primary hydration shells in relatively moist or wet environments.

Another kind of hydrogen bonding observed on mineral surfaces occurs between an organic cation on the exchange sites and another organic molecule, as for example:

$$R-\overset{\overset{\displaystyle R}{|}}{\underset{\underset{\displaystyle R}{|}}{N}}{}^{+}-H \cdots O=\overset{\overset{\displaystyle R}{|}}{\underset{\underset{\displaystyle R}{|}}{C}}$$

Here the amine is the neutralizing cation on the mineral surface. This kind of reaction has also been verified by infrared spectroscopy (Doner and Mortland, 1969a). Obviously the energy of the bond would depend upon the relative basicity of the neutral organic molecule.

Much of the earlier literature ascribes important hydrogen bonding properties to surface oxygens and hydroxyls of mineral surfaces. Thus interaction of such molecules as those of water, alcohols, and amines with mineral surface oxygens or hydroxyls was suggested to be the primary mode of action. However, with the development of infrared studies on these systems, the exchange cation effects on adsorption have been shown to be overriding through the interactions discussed earlier. For such molecules as those of water or urea, for example, intermolecular hydrogen bonding is more energetic than hydrogen bonding with mineral surfaces (Mortland, 1970). This is not to say such hydrogen bonding does not occur. It does take place but is quite weak. It can be important in the stabilization of large organic molecules such as polymers, where such additive bonds promote stability. Their role in adsorption of organics will be at a maximum when the exchange cations are of low solvation energy. The author believes that hydrogen bonding through the water bridge described earlier is a much more important type of hydrogen bonding in organic–clay interactions.

Another type of interaction that may contribute to organic adsorption on minerals consists of van der Waal's or physical forces. These are extremely weak energetically and decrease rapidly with distance between interacting species. They may, however, become significant in mineral–organic complexes where the organic is of large molecular weight because these interactions are additive. Greenland (1965) suggests an increment of 400 cal/mol for each CH_2 segment of n-alkylammonium ions. When these large cations are adsorbed on mineral surfaces, the principal van der Waal's interactions occur between the adsorbed molecules rather than between adsorbed molecules and the mineral surface.

Entropy change may be a factor in the adsorption of large molecules such as polymers. In homogenous solutions they will have a particular physical conformation that represents the most stable condition, thermodynamically. Upon adsorption at a surface there will be not only a decrease in entropy for the polymer due to adsorption, but also usually a change in conformation. In addition, as polymer is adsorbed at mineral surfaces, water molecules may be desorbed, resulting in an increase in their entropy. Also, for every molecule of polymer adsorbed, many molecules of water may be desorbed into the bulk phase, which may result in an overall increase in entropy for the whole reaction. This has been observed for glycine polymer on smectite by Greenland et al. (1965).

The preceding discussion of mechanisms of organic–clay interaction should not be interpreted to mean that these reactions operate independently of each other. As a matter of fact, more than one kind of interaction may be taking place, particularly for large molecules. For example, it is entirely possible to have an organic cation that may adsorb on a mineral by ion exchange and have alcoholic groups elsewhere in the structure that may hydrogen bond with the mineral surface, other organic species, or water. Obviously, such many faceted interactions may lead to great stability of the mineral–organic complex.

INTERACTION OF ORGANIC POLLUTANTS WITH CLAYS

Organic pollutants in soils and ground water may encompass a large variety of chemical species. Usually they are the result of municipal, industrial, or agricultural activities. Since these compounds are of such great variability, their chemistries and thus their interactions with mineral surfaces will vary widely. Such intrinsic properties as molecular structure, functional groups, solubility, and molecular weight are responsible for such variation. Whether or not an organic species will adsorb on clay depends upon the nature of the clay and its exchange cations as well as upon the structure and properties of the organic compound. The following discussion attempts to cover the possible interaction of several classes of organic pollutants with mineral surfaces. At this point it is necessary to point out that natural organic matter in soils and sediments is a very important adsorbent for most organic molecules, and in some cases organic matter is preferred over mineral surfaces.

Pesticides

These organic compounds are quite varied in their structures and thus also in their chemical properties. There are also extreme variations in their rates of decomposition in soils and sediments. Some are very soluble in water and others almost insoluble. Usually the mobility of a pesticide in soils and sediments involves the solution of the compound into water from which it may be adsorbed by minerals or organic matter or move with the water flow. In general it may be stated that adsorption will retard movement more or less depending upon the residence time of the molecule on the adsorbing surface. If adsorption energies are low, desorption may occur rapidly; conversely, if adsorption energies are high, residence times may be so long that the molecule may be considered to be immobilized.

The herbicides diquat and paraquat were mentioned earlier in the discussion of adsorption of organic cations by ion exchange. These compounds are quaternary ammonium compounds and remain cationic regardless of pH. They are both divalent and are strongly preferred to metal exchange cations on clay surfaces. The consequence of this strong adsorption is the removal of these compounds from the environment, where they can exert biological activity. In addition, it prevents any great degree of mobility in soils and sediments. It was shown by Scott and Weber (1967) that paraquat was rendered unavailable to plant roots when it was adsorbed within the interlamellar regions of a swelling clay such as smectite (montmorillonite). Weber and Coble (1968) showed that diquat was unavailable for microbial degradation when it was adsorbed on the internal surfaces of smectite, but was degradable when it was on the external surfaces of the mineral kaolinite.

As suggested above, one consequence of strong adsorption by minerals or organic matter is the reduction of their biological activity and thus more of the compound must be applied to soils to achieve the goal of control of the target species. The material bound to the colloid fraction of soil, while perhaps not active in that environment, may undergo erosion, transport, and deposition in another environment where it might exert biological activity. Thus a reservoir of strongly adsorbed pesticide may constitute a potential hazard at a future time and place.

Many pesticides are organic bases that have the potential for being adsorbed through protonation at mineral surfaces. As suggested earlier, the likelihood of such a reaction will depend upon the basicity of the molecule and the surface acidity of the mineral. Examples of pesticides that may undergo this kind of adsorption process are all the triazines, which have furnished a large number of pesticides. Weber (1970) has discussed the chemistry of these chemicals and their reactions with and adsorption by colloids in soils and sediments. In addition to adsorption on minerals via protonation, these chemicals possess structural properties that may lead to interaction with exchangeable metal cations through coordination. Transition metal ions in particular are capable of forming these complexes. In general nitrogen atoms within the organic structures possess paired electrons for donation to unfilled d orbitals of the transition metal ion. An example is the work of Russell et al. (1968) in which 3-aminotrizole, a herbicide, was shown to form these kinds of relatively stable coordination complexes with clays possessing Cu(II) or Ni(II) on the cation exchange sites.

Other kinds of pesticides may complex with exchange cations on minerals but through less energetic bonds. For example, the herbicide ethyl dipropyl thiolcarbamate (EPTC) has been shown by infrared spectroscopy to complex with exchange cations on clays (Mortland and Meggitt, 1966). These complexes are stable against atmospheric moisture, but when placed in water, the EPTC is quantitatively displaced from the mineral surface by the reaction described previously in Eq. (19.5). Thus such a compound would probably not complex at all on mineral surfaces in a ground water environment where water would compete so favorably for solvation of the exchange ions. In such a situation the EPTC would likely stay in the solution phase and move with the ground water.

Other kinds of pesticides may have very little affinity for mineral surfaces. In general these are chemicals that have limited solubility in water but may be relatively more soluble in some organic solvents. They thus have organophilic properties and are on the other hand hydrophobic. Since mineral surfaces in the natural state are generally hydrophilic, such species are thereto not attracted. An example would be the insecticide DDT, which has very low water solubility. Work by Shin (1970) strongly suggests that certain fractions of soil organic matter are the main sites of DDT adsorption and that the mineral fraction adsorbs very little of the compound. Thus in the ground water situation where little adsorption is likely, DDT would be expected to move with ground water. The qualifying point should be made that the solubility in water is so small (1–2 parts per billion) that solution concentrations would be very low.

Aromatic Molecules

This group of organic compounds is extremely varied in structures and therefore properties. Halogenated varieties constitute some of the most persistent and non-refractory pollutants found in the environment. Examples of these are polychlorobiphenyls (PCB), polybromobiphenyls (PBB), and various chlorobenzenes. Benzene itself has been shown to be a carcinogen and is found as a pollutant in some situations, primarily in association with industrial activities. Many of the nitrogen derivatives of benzene are also toxic and in some cases considered to be carcinogenic (e.g., benzidine).

The reactions of this group of chemicals with mineral surfaces are extremely varied. Their properties depend upon the structure and in particular the functional groups they possess. Thus benzene itself has low solubility in water while its derivatives such as analine (aminobenzene) and phenol (hydroxybenzene) are much more soluble because of the polar functional groups on the benzene ring. Since analine is a base, it may be adsorbed in its cationic form by mineral surfaces through ion exchange or be protonated at an acidic surface after adsorption through the reactions described earlier in Eq. (19.1–19.3). Also, since the nitrogen of the amino group possesses a lone pair of electrons, it may be adsorbed via coordination with electron accepting ions (e.g., transition metal ions) on the exchange sites of the mineral. This molecule can be considered to some degree representative of many amino-bearing aromatics regarding interaction with mineral surfaces.

The halogenated aromatic molecules are generally characterized by low solubilities in water and relatively little adsorption on mineral surfaces from water solution. Organic matter in soils and sediments is a much more effective adsorbent for these materials (Filinow et al., 1976). Thus in ground water there are not likely to be actual concentrations of PCBs or PBBs beyond a few parts per billion. However, it is possible for these compounds to exist as extremely small particulates and thus be carried in the solid state by moving water. Chlorobenzenes are not adsorbed from water to any degree on mineral surfaces containing metal exchange ions. However, if the mineral surface is made hydrophobic, or conversely, organophilic by adsorbing long chain alkylammonium cations on the exchange sites, chlorobenzenes will be adsorbed (McBride et al., 1977). While such artificial modifications of the surface properties of minerals are possible in the laboratory, they may not be possible in most natural environments. Naturally occurring clay–organic complexes, as in soils and sediments, will have more adsorptive interaction with hydrophobic organic species than would the inorganic clay component by itself.

Benzene itself is a relatively non-polar molecule with very limited solubility in water. Again evidence shows little benzene will be adsorbed on ordinary mineral surfaces from a water solution. However, if the surfaces of clays are made hydrophobic and thus organophilic by exchanging of alkylammonium cations, benzene may be adsorbed in large quantities (McBride et al., 1977). Although benzene has no functional groups in its structure, its π electron cloud is capable of interacting with some electron accepting transition metal cations such as Cu(II), Ag(I), Fe(III), and VO(II) when they occupy exchange sites on some clay mineral surfaces (Doner and Mortland, 1969b; Pinnavaia et al., 1974). In the cases of Ag(I) and Cu(II) smectites, coordination complexes with benzene through the π electrons may be formed when hydrating water molecules are removed. In addition, complete electron donation may occur in the Cu(II), Fe(III), and VO(II) smectites, resulting in radical cation formation in the benzene and reduction in valence of the metal cation. A number of alkylbenzenes and other aromatic molecules such as toluene, anisole, napthalene, and anthracene undergo similar reactions. The radical cation thus formed is very reactive and may lead to dimerization or polymerization reactions (Mortland and Halloran, 1976; Fenn et al., 1973).

Enzymes, Viruses, and Bacteria

Microorganisms tend to become concentrated at solid surfaces of particles rather than to be uniformly dispersed. Such interactions with clays or organic matter may have important consequences for their survival. Just as adsorption of natural organic matter on clays protects it to some degree against decomposition, so also bacteria–clay agglomerates may give some protection to the organism and even perhaps provide some nutrition via the exchangeable cations, which might be of the kind required for some physiological functions within the organism. After all, in soils these same exchangeable cations and adsorbed organic matter provide a reservoir for many nutrients required by plants for their growth. Many studies have demonstrated that extensive microbial growth takes place in nature on surfaces of particles and inside loose aggregates of solid particles (Rice et al., 1975). Thus, only a few free microorganisms will be found in soil solutions and ground water. The environment provided by the microorganism–clay complex may permit growth that would not be possible in the absence of such an interaction. The nature of the physical–chemical attachment of microorganisms to sand, silt, and clays has been studied by Stotzky (1966a, 1966b), Lammers (1967), and Boyd et al. (1969). Stotzky showed that cation exchange capacity (CEC) of the solid phase had an important effect on complexation of the organisms and particles. This suggests an electrostatic interaction between the organism and the mineral surface, probably resulting from the neutralization of negative charge sites on the mineral by positive sites on the organism. This kind of bond is very strong and tends to explain reports of great difficulty being experienced in separating the organism from mineral phases.

Adsorption of viruses by particulate matter, including minerals, is an important phenomenon in soils and sediments. For example Berg (1973) showed that poliovirus was strongly adsorbed on silt particles and that very little of it could be subsequently eluted. This kind of interaction is not surprising since viruses are protein-like molecules, and extensive studies of protein adsorption by clays indicate strong interactions (Theng, 1974). A number of different interactions are involved when these kinds of large molecules are adsorbed by clays. For example, when the pH is below the isoelectric point of the molecule, positive charges are created via protonation of some amino groups, so the molecule will be adsorbed by ion-exchange type reactions. This is a very strong type of binding and can be reversed only by an increase in pH above the isoelectric point, and even then the protein type molecules only partially desorb. Obviously the other kinds of mineral–organic reactions discussed earlier will also be operating. Hydrogen bonding and ion–dipole-coordination interactions will be important for these large molecules, which possess a variety of functional groups. Enzymes (protein catalysts) have been immobilized on minerals and their properties studied extensively (Zaborsky, 1974). It has been demonstrated that some adsorbed enzymes still show catalytic activity, the degree of which depends upon the enzyme, the nature of the mineral, and the general environment of the system.

Viral adsorption on mineral particles is consistent with the above discussion on bacterial–mineral interactions. As a result, concentration of viruses will be much greater on particulate surfaces than in soil or ground water (National Academy of Science, 1977). The resulting interference with disinfection suggests that this association between virus and particulate produces a resistant system that is not easily dissociated.

ALTERATION OF ORGANICS AT MINERAL SURFACES

When some organic molecules are adsorbed at mineral surfaces, the possibility exists for catalytic alteration. In the past, clays have been employed as catalysts in the cracking of petroleum into gasoline. Such reactions involve high pressures and temperatures as well as activation of the clay by acid treatment. One theory on petroleum formation itself involves clays as natural catalysts (Almon and Johns, 1975). In recent years it has become known that clays may act as catalysts for some reactions at ambient temperatures and pressures. Catalytic reactions usually involve either Brönsted or Lewis acids, and mineral surfaces may function in either case under appropriate conditions. Brönsted acidity involves the donation of protons; their generation at clay surfaces was discussed in an earlier section. Lewis acidity involves electron acceptance which can be accomplished at mineral edges where ions such as Al^{3+} may be exposed, or on the exchange complex by appropriate ions, in particular the transition metal cations with unfilled d orbitals.

An example of clays catalyzing a reaction via their Brönsted acidity is the hydrolysis of esters demonstrated by McAuliffe and Coleman (1955). Other demonstrated reactions are the conversion of atrazine to hydroxy-atrazine by Russell et al. (1968) and the hydrolysis of nitriles to amides by Sanchez et al. (1972). As suggested earlier, the ability of a mineral surface to act as a Brönsted acid will depend upon the nature of the exchangeable cations and the water content.

Clays have been shown by Solomon (1968) to catalyze organic reactions through Lewis acidity developed at the edges of the mineral. Transition metal cations on exchange sites were shown by Mortland (1966) to act as Lewis catalysts in the decomposition of urea to ammonium ions. Aromatic molecules such as benzene will complex via π electrons with clay minerals containing Cu(II), Fe(III), or VO(II) ions on exchange sites and under rigorous desiccating conditions form a radical cation (Doner and Mortland, 1969b). These radical cations will react with neutral molecules to produce parapolyphenyl polymers (Mortland and Halloran, 1976). Such catalytic effects require low water contents since the presence of water will often result in those molecules occupying ligand sites on the transition metal cation and thus preventing direct interaction of the organic matter with the electron accepting ion.

CONCLUSIONS

Organic molecules may be adsorbed by minerals through a variety of interactions of varying energy or not adsorbed at all. Whether an organic species is adsorbed depends upon its structure and chemistry as well as the nature of the surfaces. Water content may be an important factor affecting adsorption of some molecules. The consequences of adsorption include a reduction of the rate of movement of the particular organic species through soil or sediments. Extremely strong adsorption may render the molecule essentially immobile while weak adsorption results in a short residence time on the surface and merely a delay in its movement with soil or ground water. Biological agents such as bacteria and viruses interact with minerals to the extent that their concentration in the water phase may be very low compared with that in the particulate phase. In addition, such complexation may protect the biological agent and provide an environment beneficial to its preservation and growth. In some cases, catalytic alteration of an organic species may occur after adsorption on a mineral surface.

REFERENCES

Almon, W. R. and Johns, W. D. (1975). Petroleum forming reactions: Clay catalyzed fatty acid. *Proceedings of the International Clay Conference* (Mexico), pp. 451–464.

Berg, G. (1973). Removal of viruses from sewage, effluents, and water. *Bull. WHO* **49:**451–460.

Bernal, J. D. (1951). *The Physical Basis of Life*. Routledge and Kegan Paul, London.

Boyd, J. W., Yoshid, T., Vereen, L. E., Cada, R. L., and Morrison, S. M. (1969). Bacterial response to the soil environment. *Sanitary Engineering Papers*. Colorado State University, Ft. Collins, No. 5, pp. 1–22.

Dixon, J. B. and Weed, S. B. (1977). *Minerals in Soil Environments*. Soil Science Society of America, Madison, WI.

Doner, H. E. and Mortland, M. M. (1969a). Intermolecular interaction in montmorillonites: NH–CO systems. *Clays and Clay Minerals* **17:**265–270.

Doner, H. E. and Mortland, M. M. (1969b). Benzene complexes with Cu(II) montmorillonite. *Science* **166:**1406–1407.

Fenn, D. B., Mortland, M. M., and Pinnavaia, T. J. (1973). The chemisorption of anisole in Cu-montmorillonite. *Clays and Clay Minerals* **21:**315–322.

Filinow, A. B., Jacobs, L. W., and Mortland, M. M. (1976). Fate of polybrominated biphenyls (PBB's) in soils: Retention of hexabromobiphenyl in four Michigan soils. *J. Agric. Food Chem.* **24:**1201–1204.

Greenland, D. J. (1965). Interaction between clays and organic compounds in soils. I. Mechanisms of interaction between clays and defined organic compounds. *Soils Fertilizers* **28:**415–425.

Greenland, D. J., Laby, R. H., and Quirk, J. P. (1965). Adsorption of amino acids and peptides by montmorillonite and illite. Part 1. Cation exchange and proton transfer. *Trans. Faraday Soc.* **61:**2013–2023.

Grim, R. E. (1968). *Clay Mineralogy*. McGraw-Hill, New York.

Lammers, W. T. (1967). Separation of suspended and colloidal particles from natural water. *Environ. Sci. Technol.* **1:**52–57.

McAuliffe, C. and Coleman, N. T. (1955). H-ion catalysis by acid clays and exchange resins. *Soil Sci. Soc. Am. Proc.* **19:**156–160.

McBride, M. B., Pinnavaia, T. J., and Mortland, M. M. (1977). Adsorption of aromatic molecules by clays in aqueous suspension. In: I. H. Suffet (Ed.), *Fate of*

Pollutants in the Air and Water Environments, Part 1, Vol. 8, Wiley, New York, pp. 145–154.

Mortland, M. M. (1966). Urea complexes with montmorillonite: An infrared absorption study. *Clay Minerals* **6**:143–156.

Mortland, M. M. (1970). Clay–organic complexes and interactions. *Adv. Agron.* **22**:75–114.

Mortland, M. M. and Halloran, L. J. (1976). Polymerization of aromatic molecules on smectite. *Soil Sci. Soc. Am. Proc.* **40**:367–370.

Mortland, M. M. and Meggitt, W. F. (1966). Interaction of ethyl-*N*,*N*-di-*n*-propyl-thiolcarbamate (EPTC) with montmorillonite. *J. Agric. Food Chem.* **14**:126–129.

Mortland, M. M. and Raman, K. V. (1968). Surface acidity of smectites in relation to hydration, exchangeable cation, and structure. *Clays and Clay Minerals* **16**:393–398.

National Academy of Science (1977). *Drinking Water and Health.* Washington, D.C., pp. 179–183.

Pinnavaia, T. J., Hall, P. L., Cady, S., and Mortland, M. M. (1974). Aromatic radical cation formation on the intracrystal surfaces of transition metal layer lattice silicates. *J. Phys. Chem.* **78**:994–999.

Raman, K. V. and Mortland, M. M. (1969). Proton transfer reactions at clay mineral surfaces. *Soil Sci. Soc. Am. Proc.* **33**:313–317.

Rice, C. W., Uydess, I. L., Hempfling, W. P., and Vishniac, W. V. (1975). Isolation of microorganisms from soil of the Antarctic "Dry Valleys." *Abstr. Ann. Meet. Am. Soc. Microbiol.,* New York, p. 134.

Russell, J. D., Cruz, M. I., and White, J. L. (1968). The adsorption of 3-amino-triazole by montmorillonite. *J. Agric. Food Chem.* **16**:21–24.

Sanchez, A., Hidalgo, A., and Serratosa, J. M. (1972). Adsorption des nitriles dans la montmorillonite. *Proceedings of the International Clay Conference* (Madrid), pp. 617–626.

Scott, D. C. and Weber, J. B. (1967). Herbicide phytotoxicity as influenced by adsorption. *Soil Sci.* **104**:151–158.

Shin, Y. O. (1970). Adsorption of DDT by soils and biological materials. Ph.D. Thesis, Michigan State University, E. Lansing, MI.

Solomon, D. H. (1968). Clay minerals as electron acceptors and/or electron donors in organic reactions. *Clays and Clay Minerals* **16**:31–39.

Stotzky, G. (1966a). Influence of clay minerals on microorganisms. II. Effect of various clay species, homoionic clays, and other particles on bacteria. *Can. J. Microbiol.* **12**:831–848.

Stotzky, G. (1966b). Influence of clay minerals on microorganisms. III. Effect of particle size, cation exchange capacity and surface area on bacteria. *Can. J. Microbiol.* **12**:1235–1246.

Tahoun, S. and Mortland, M. M. (1966). Complexes of montmorillonite with primary, secondary, and tertiary amides: I. Protonation of amides on the surface of montmorillonite. *Soil Sci.* **102**:248–254.

Theng, B. K. G. (1974). *The Chemistry of Clay–Organic Reactions*. Adam Hilger Ltd., London.

Weber, J. B. (1970). Mechanisms of adsorption of s-triazines by clay colloids and factors affecting plant availability. *Residue Rev.* **132**:93–130.

Weber, J. B. and Coble, H. D. (1968). Microbial decomposition of diquat adsorbed on montmorillonite and kaolinite clay. *J. Agric. Food Chem.* **16**:475–478.

Yariv, S., Russell, J. D., and Farmer, V. C. (1966). Infrared study of the adsorption of benzoic acid and nitrobenzene in montmorillonite. *Isr. J. Chem.* **4**:201–213.

Zaborsky, O. (1974). *Immobilized Enzymes*. CRC Press, Cleveland, OH.

20

MICROBIOLOGICAL CHARACTERIZATION OF SUBSURFACE ENVIRONMENTS

W. C. Ghiorse

Cornell University
Ithaca, New York

D. L. Balkwill

University of New Hampshire
Durham, New Hampshire

Subsurface environments are important as reservoirs of ground water—95% of the U.S. fresh water supply is ground water (Josephson, 1980)—and as regions where surface water can be purified as it enters the ground water. In contrast to the numerous studies on microorganisms in surface soil environments, investigations that characterize the indigenous microorganisms in subsurface environments are few (Dunlap and McNabb, 1973). In surface soil, microorganisms are thought to control most of the biological activity (Alexander, 1977). Furthermore, current evidence suggests that in aquatic and terrestrial environments microorganisms are the chief agents of biodegradation of environmentally important molecules (Alexander, 1981). In subsurface zones, however, although microorganisms are presumed to be present and active (Dunlap and McNabb, 1973; Ehrlich, 1981), evidence for their presence in sparse and evidence for their activity is nonexistent.

Most past investigations of subsurface microorganisms have been concerned with sulfate reducing bacteria, methane bacteria, and hydrocarbon-utilizing bacteria in oil- and gas-bearing deposits (Dunlap and McNabb, 1973) or with the movement of disease-indicator bacteria and their viruses injected into subsurface environments (Allen, 1980; Bitton and Gerba, 1984). Few previous investigations have dealt with indigenous microflora of nonpetro-liferous subsurface environments.

The aim of the present work is to characterize indigenous microorganisms of ground water-bearing subsurface environments with respect to their identity, abundance, and metabolic activities. In the past, identification and estimates of microbial abundance in subsurface material have relied mainly on cultural techniques. However, cultural techniques are not ideal for identification because the types of microorganisms selected by cultivation on laboratory growth media may not truly represent the indigenous microflora. Thus, new methods are needed to aid identification of microorganisms in subsurface environments. In addition, cultural techniques are capable of detecting only a fraction—1–10% in soil (Alexander, 1977)—of the microorganisms in a given environment; therefore, they can be expected to give only an underestimate of the abundance of subsurface microorganisms. Clearly, new methods capable of detecting a larger proportion of the microbial biomass are also needed.

In this report, several promising methods for the study of subsurface microorganisms are reviewed. In addition, results from our preliminary studies of Oklahoma subsurface material are presented.

REVIEW OF AVAILABLE METHODS

A major problem in the characterization of subsurface microorganisms is procurement of samples that are uncontaminated by surface microorganisms. Fortunately, researchers at the Robert S. Kerr EPA Laboratory, Ada, Oklahoma have developed procedures using specially designed equipment for obtaining minimally contaminated subsurface samples (Dunlap et al., 1977). These procedures have been used to obtain samples for our preliminary studies (reported below).

Methods for electron microscopic (EM) examination of subsurface material can provide a wealth of information that will help to characterize indigenous microorganisms. Descriptive properties such as sizes and shapes of cells, structures of cell walls (e.g., gram-positive vs. gram-negative), and presence of spores or cysts can be determined using EM techniques. It is also possible that unusually small (Bae et al., 1972; Balkwill and Casida, 1973) or previously undescribed forms will be detected. Furthermore, the presence of morphologically unusual bacteria, such as budding or stalked bacteria, or morphologically undistinguished bacteria that grow under limited nutrient conditions may be used as indicators of environmental quality (Hirsch, 1974; Staley, 1971). In some instances, the presence of internal structures visible only by EM (e.g., cross walls in dividing bacteria or intracytoplasmic membranes found in nitrifying or methane-oxidizing bacteria) may indicate the presence of actively growing or metabolically specialized organisms. In these instances, additional information on viable cells or the levels of chemical species such as ammonium, nitrite, nitrate, methane, and oxygen in the water could help to establish that the bacteria were, in fact, present and active. Other information that may be derived from EM studies of subsurface material includes the structures of microenvironments and the distributions and population densities of microorganisms within the microenvironments.

A number of methods have been published in which soil microorganisms can be examined directly by EM techniques (Bae and Casida, 1973; Bae et al., 1972; Balkwill and Casida, 1973; Balkwill et al., 1975, 1977; Gray, 1967; Waid, 1973). Those involving release and concentration of microorganisms from soil particles prior to thin-sectioning and examination by transmission EM (TEM) are considered most promising for this work because they have been used successfully with a variety of soil types of different textures and different microbial population densities (Bae and Casida, 1973; Bae et al., 1972; Balkwill and Casida, 1973; Balkwill et al., 1975, 1977). This success suggests that they will also be applicable to a variety of types of subsurface samples as well.

Scanning EM (SEM) has also been employed for directly viewing soil samples (Gray, 1967). However, the success of this method may depend largely on the density of microbial populations in the samples. Because it is expected that subsurface materials have low population densities, SEM techniques should have limited application.

Floatation of gel-like surface films from soil particles (Waid, 1973) is another technique that holds some promise for examining subsurface microorganisms by EM. The soil films are picked up on Formvar-coated EM grids and examined directly by TEM after shadow casting or negative staining. This method has been used to reveal the diversity of bacteria inhabiting certain soils and probably can be used for routine examination of soil microorganisms (Waid, 1973; unpublished data, W. C. Ghiorse and P. Hirsch). This method may not be optimal for saturated subsurface material of high water content; however, the method would be applicable to drier material from unsaturated zones. For samples with low population densities, it should be possible to concentrate the microorganisms in the films by centrifugation and examine them by thin-sectioning techniques in TEM.

A very effective method for determining the identity of some of the microorganisms in a natural environment is the classical *Aufwuchs* method (Cholodny, 1930). With this method, glass slides or other substrates can be inserted, in our case, into samples of subsurface material. After days or weeks of incubation during which growth of microorganisms can occur, the substrates are removed and examined microscopically, generally by phase contrast epifluorescence light microscopy. A modification of this technique, in which specially constructed, flat glass capillaries are employed (Perfilev and Gabe, 1969), may provide an even better method of examining microorganisms in subsurface material. The glass capillaries mimic capillaries in the soil providing micro-environments for fastidious microorganisms. Thus, populations that normally would not be seen can grow and be observed. *Aufwuchs* methods have also been adapted for EM observations of aquatic environments (Hirsch and Pankratz, 1972); however, EM-*Aufwuchs* methods for soil microorganisms are more problematic because films on EM grids are fragile and are often disrupted in soil (unpublished observations, W. C. Ghiorse). On the other hand, EM-capillary methods hold more promise because microorganisms growing in the capillaries can be fixed and embedded for thin-sectioning.

Estimates of biomass are of primary concern for microecological investigations of subsurface environments. Without such estimates information on community structure, metabolic activities, and nutritional status will have little meaning. Because it is probable that the majority of microorganisms in subsurface material are bacteria (Dunlap and McNabb, 1973), methods for estimating bacterial biomass are emphasized here. Thus, in the following discussion the terms "microbial biomass" and "bacterial biomass" are interchangeable.

Of the existing methods for determining microbial biomass, direct counting of cells by epifluorescence microscopy is judged to be the most reliable. Methods developed for epifluorescence enumeration of soil microorganisms (Trolldenier, 1973) can be applied to subsurface material (see below). The results of these cell counts can then be converted to biomass by applying appropriate factors (Dale, 1974).

Chemical measurements of microbial cell components as estimates of biomass are less direct, but may be more sensitive than cell counts. Thus, these methods can be used in parallel with epifluorescence microscopy for estimating biomass, especially when low population densities are experienced. The chemical methods involve the extraction and measurement of substances that are unique to bacterial cell walls (e.g., muramic acid) or that are found in high concentrations in bacterial cells (e.g., phospholipids).

Generally, the greater sensitivity of the chemical methods is due to increased sample sizes. Furthermore, in the case of muramic acid analyses, sensitivity can be extended by linking the analysis to a light-producing luciferase measuring system (Moriarty, 1977). Perhaps the simplest method for estimating bacterial biomass is the determination of chloroform-extractable lipid phosphate (White et al., 1979). In this method amounts of lipid phosphates in a sample can be taken as a measure of bacterial biomass since bacteria have a higher proportion of phospholipids than other microorganisms (D. C. White, personal communication).

Measurement of adenosine triphosphate (ATP) is another commonly employed method for estimating biomass; however, ATP measurements are considered to be less reliable than muramic acid analyses for estimating biomass in subsurface material because the rapid turnover of ATP in microbial cells (Karl, 1980) may cause wide fluctuations in the amount of ATP per cell.

The foregoing methods for biomass estimation do not generally distinguish between viable and non-viable cells, nor can they be expected to distinguish metabolizing from non-metabolizing cells. In order to determine the viability and metabolic activities of microorganisms in natural environments, several indirect methods can be applied depending on the type of environment (see Rosswall, 1973 for a partial listing). For subsurface environments, measurements of ^{3}H-thymidine incorporated into DNA (Tobin and Anthony, 1978) or determination of the ratio of guanosine triphosphate (GTP) to ATP (Karl, 1978) seem most appropriate. With both of these methods, it is possible to determine the proportion of actively growing bacteria in a given environment. However, these methods may not be applicable to the study of subsurface material where activities are expected to be low.

Other promising methods for determining microbial activities in subsurface populations utilize epifluorescence microscopy in conjunction with either autoradiography (Meyer-Reil, 1978) or tetrazolium dye reduction (Zimmermann et al., 1978). With the epifluorescence-autoradiography technique, subsurface samples can be incubated with radiolabeled compounds such as ^{3}H-glucose added, after which the labeled sample is spread on a microscope slide and coated with a photographic emulsion. Uptake of radioactive compounds by microbial cells can be determined in an epifluorescence-phase contrast microscope after a second incubation period by counting dense silver grains associated with fluorescent cells. This method has the advantage that it can reveal both total biomass and the proportion thereof that takes up the radioactive compounds in the same preparation (Meyer-Reil, 1978).

The epifluorescence-dye reduction technique is based on the ability of respiring microorganisms to reduce a tetrazolium dye to an insoluble colored formazan that is deposited inside cells. Subsurface material can be incubated in the presence of the dye, then spread on a slide, stained with acridine orange, and examined by epifluorescence-phase contrast microscopy. The ratio of formazan-containing and fluorescent cells to fluorescent cells without formazan can be taken as a measure of the proportion of respiring to non-respiring bacteria in the sample (Zimmermann et al., 1978).

PRELIMINARY STUDIES OF SUBSURFACE MATERIAL

Materials and Methods

Aseptically procured subsurface material was obtained from two locations near Ada, Oklahoma (Pickett and Lula) using equipment and procedures adapted from those described by Dunlap et al. (1977). The subsurface samples were packed into sterile containers, and identical samples were shipped as soon as possible to Ithaca, New York and Durham, New Hampshire. Samples were kept cool during shipping and usually arrived within 24 hr after sampling. Upon arrival in Ithaca, subsamples were removed aseptically for enumeration of microorganisms by epifluorescence microscopy. Additional subsamples were removed for determination of muramic acid content. The remaining sample material was stored at 4°C. In Durham, subsamples were removed for plate counting and for electron microscopy.

The material from both locations was sandy in texture and orange-brown in color. It contained from 10 to 25% water and <0.1% organic carbon. The average temperature of subsurface zones in these locations was 17°C. The zones were assumed to be aerobic.

Total numbers of microorganisms were estimated using an epifluorescence counting procedure that was a modification of the procedure described by Trolldenier (1973). In the modified procedure 2.5 g of material was transferred aseptically to a clean, sterile 125-mL Erlenmeyer flask containing 22.5 mL of filter-sterilized 0.1% sodium pyrophosphate ($Na_4P_2O_7 \cdot 10H_2O$) adjusted to pH 7.0 with HCl. The flask and contents were then shaken on a rotary shaker for 30 min at 160 rpm. After shaking, the suspension was allowed to settle for 2 min, then 9.0 mL was transferred with a pipet to a 25-mL scintillation vial and 0.1 mL of 50% glutaraldehyde plus 1.0 mL of molten 1% Noble agar were added. The diluted and fixed samples were stored until smears were prepared for counting by epifluorescence microscopy.

For epifluorescence cell counts, diluted samples were shaken vigorously by hand, then 5–10 μL samples were removed with a mechanical pipetting device and smeared evenly over a circular area of approximately 1 cm^2 (1.1 cm diameter) marked on a glass microscope slide. The smear was allowed to dry at room temperature and then was stained with filter-sterilized 0.01% acridine orange solution for 2.0 min. Excess acridine orange was washed away with tap water and the wet smear covered with a coverglass and sealed with clear nail polish or paraffin.

Sealed preparations were examined and cells counted using 40× and 100× oil immersion objectives on a Zeiss Standard 18 microscope equipped for epifluorescence and phase contrast viewing. The epifluorescence system included a mercury HB050 light source, a 466300 vertical fluorescence illuminator and a 487709 exciter-barrier filter and reflector combination. Photomicrographs of appropriate fields were recorded on Kodak Ektachrome 160T film using a Zeiss MC63/35 mm camera system. Color slides were copied on Kodak Tri X film for black and white reproductions.

Cells that fluoresced green against orange-red or orange-brown soil material were enumerated in 50 microscopic fields that were in a statistically representative peripheral zone of each smear (Trolldenier, 1973). Numbers of microorganisms per gram of oven-dried (105°C) material were calculated from mean numbers of cells per microscopic field using appropriate conversion factors. The limit of detection in this method was approximately 10^5 cells per gram.

Viable cell counts were obtained by aseptically blending 10 g of material in 100 mL of 0.1% sodium pyrophosphate for 1 min in a Waring blender. Appropriate dilutions of the blended suspension were made immediately and plated in triplicate on three different growth media containing 1.5% agar. The growth media were (1) a nutritionally rich medium containing 10 g glucose, 5 g peptone, 5 g tripticase, and 10 g yeast extract per liter of a mineral salts solution (PYG); (2) a 20-fold dilution of PYG (5% PYG); and (3) a soil extract medium (10% SEA) prepared by autoclaving 100 g of surface soil from the sampling site or from Durham in 100 ml of distilled water for 1 hr at 15 psi, then filtering the soil water mixture. The 10% SEA medium was prepared by diluting the extract 10-fold with distilled water. All plates were incubated at 22–23°C until the numbers of colony forming units (CFU) per gram of oven-dried material was constant.

Muramic acid contents were determined using an enzyme-linked method in conjunction with a luciferase assay described by Moriarty (1975; 1977). Reagent 2 of Moriarty's luciferase assay (1977) was modified by substituting a sonicated suspension of decyl aldehyde in H_2O for dodecyl aldehyde saturated in ethanol and decreasing the pH of the phosphate buffer to 7.0. Freshly purchased bacterial luciferase (Analytical Luminescence Laboratory, Inc.) was reconstituted to 1 mg/mL in $0.1M$ phosphate buffer and frozen in small quantities until used in the assay. The use of fresh reagents was found to be essential to obtain the highest sensitivity with this method. Light produced in the luciferase assay was measured in a Beckman LS-230 liquid scintillation counter.

For electron microscopy of subsurface microorganisms, a release-and-concentration method was employed (Balkwill et al., 1975, 1977). In this method 10 g of material was blended with 100 mL of 0.1% sodium pyrophosphate for no more than 30 sec at a time with 30 sec rest intervals. After blending, the entire volume was transferred to a 100-mL graduated cylinder and allowed to settle for 5 min. The top 50 mL was then transferred to a centrifuge bottle and centrifuged at $650g$ for 5 min in a Sorvall RC-2B centrifuge. The supernatant was decanted and saved, and the pellet was resuspended in another 50 mL of pyrophosphate and the centrifugation repeated. This centrifugation and washing procedure was repeated four additional times, each time saving the supernatant. After the fifth round, the pellet was discarded, and the five supernatants were combined and centrifuged at $23,000g$ for 20 min. The pellet derived from this process was shown to contain approximately 85% of the viable bacteria in the original 50 mL sample. It was previously determined (D. L. Balkwill, 1977, Ph.D. Thesis, Penn. State Univ.) that the majority of the bacterial cells along with humic substances and fine clay particles were deposited in a brown-black layer that formed on the surface of the final pellet. Thus, the brown-black surface layer was carefully removed with a spatula, fixed with glutaraldehyde-OsO_4, and embedded in plastic as previously described (Balkwill et al., 1977).

Thin sections of the embedded material were obtained with glass knives on an LKB Ultramicrotome. The sections were double stained with uranyl acetate and lead citrate and examined in a JOEL TEM-100S electron microscope.

RESULTS AND DISCUSSION

Examination of subsurface samples by phase contrast microscopy revealed numerous particles on the order of 1 μm or less in size (Figures 20.1A and C). Some of the particles fluoresced green when stained with acridine orange (compare Figure 20.1A with 20.1B and 20.1C with 20.1D) indicating that they contained double-stranded deoxyribonucleic acid (DNA) (Daley and Hobbie, 1975). Since bacteria are known to contain significant amounts of DNA in their nucleoids, it can be deduced from these results that green

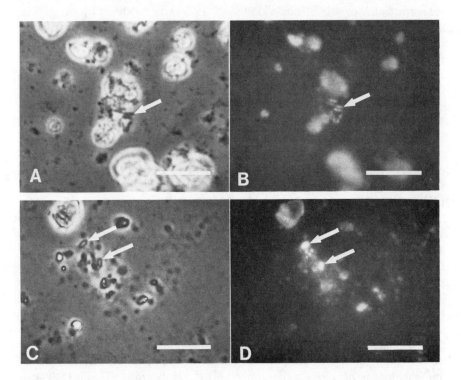

Figure 20.1. Identical fields containing subsurface material photographed under phase contrast (left) and epifluorescence microscopy (right). Subsurface samples were obtained from 18 ft deep at the Pickett sample site (A and B) and 16 ft deep at the Lula sample site (C and D). Arrows indicate bacteria-sized particles that fluoresced green when stained with acridine orange. Bars = 10 μm.

fluorescent particles 1 μm or less in size were probably bacterial cells. Thus, for epifluorescence cell counts, particles that met these two criteria were counted as bacterial cells. If doubt existed about the identity of a green fluorescent particle, it was examined more closely to determine if its morphology was bacterial. If doubt still existed, the particle was not counted. For the most part, bacterial cells were easily distinguished from non-living forms and were therefore easily enumerated in subsurface samples. Significantly, neither green fluorescent forms larger than bacteria nor forms that were internally complex were observed, suggesting that few, if any, eukaryotic microorganisms inhabited the subsurface material. Occasionally, large fluorescent particles were observed; however, invariably the fluorescence could be attributed to the presence of two or more bacterial cells (Figure 20.1). In these cases, the number of cells in an aggregate was estimated and included in the count.

An important problem that may be encountered in studies of this nature is the possibility that microbial numbers will change during storage. Normally, our subsurface samples were shipped on ice and stored at 4°C. As a rule they were processed as soon as possible after arrival. However, since immediate processing may not always be possible in the future, it was imperative to determine the effect of storage on microbial cell counts. For this, a series of epifluorescence counts was performed every 2 weeks on samples from the Pickett site that were stored at 4°C. During 6 weeks of storage nine separate counts were made on four different occasions. The variation between the averages of the biweekly counts was within one standard deviation of the mean of all nine counts (Table 20.1), suggesting that the number of cells in the sample remained constant for the entire 6 weeks of storage. The mean of all nine determinations was 5.2 ($\pm$3.1) $\times$ 10^6 cells per gram of oven-dried material (Table 20.1). This number can be taken as a measure of the bacterial population density in subsurface material from the Pickett site.

Table 20.1   Effect of Storage on Epifluorescence Cell Counts of Subsurface Microorganisms[a]

Weeks at 4°C	Number of Determinations	Cells/g[b] $\times$ 10^6	
		Average[c]	Mean ($\pm$S.D.)[d]
0	1	5.7	
2	1	1.3	5.2 ($\pm$3.1)
4	4	4.4	
6	3	7.6	

[a] Subsurface material from 18-ft depth at the Pickett sample site.

[b] Oven-dried material.

[c] For each biweekly sampling.

[d] For all 9 determinations.

The results of the epifluorescence counts raise the question of whether the cells that were counted were viable and potentially active. Classically, this question has been answered by enumerating viable cells using plate counts. Although our investigation was initiated to find more reliable ways of estimating active biomass, plate counting methods were included for comparison. Normally, in plate counting methods, rich organic media are employed; however, because subsurface samples were poor in organic material, it was decided to compare viable cell counts derived from three different media: a rich organic medium (PYG), a dilution of the same medium (5% PYG), and a soil extract medium (10% SEA). The results of this comparison showed clearly that PYG underestimated the number of viable cells in material from the Pickett site by at least a factor of 10^3 (Table 20.2). Diluted media, on the other hand, yielded numbers of CFU that were approximately a factor of 2 below the mean value of epifluorescence counts (compare Tables 20.1 and 20.2). If these results are compared to viable versus total counts in surface soils (Alexander, 1977), our viable counts on diluted media are reasonably high. Because more than one bacterial cell may constitute a CFU, these results would suggest that 50% or more of the bacteria counted by epifluorescence in these samples were capable of growth.

Examination of thin-sections of subsurface material by EM (Figure 20.2) supported light microscopic observations that few, if any, eukaryotes were present in these samples. To date, only bacterial forms have been observed by EM in samples from both sample sites. In samples from the Lula site, both gram-positive (Figure 20.2A, B, and D) and gram-negative (Figure 20.2C) bacterial cell wall types were observed, but gram-positive types appeared to be more abundant. Significantly, the bacterial cells seen in EM preparations were generally smaller (<0.8 μm in diameter) than cells cultivated in the laboratory. The observation of small bacterial forms is similar to the findings of Bae et al. (1972) and Balkwill and Casida (1973) that small forms of bacteria were abundant in surface soil samples examined by TEM. Apparently many of the bacteria in soil environments exist as cells smaller

Table 20.2 Effect of Nutrient-Rich and Nutrient-Poor Growth Media on Viable-Cell Counts of Subsurface Microorganisms[a]

Medium	Mean CFU ($\pm$S.D.)/g[b]
PYG (rich)	6.3 ($\pm$1) $\times$ 10^2
5% PYG (poor)	3.1 ($\pm$1.0) $\times$ 10^6
10% SEA (poor)	2.5 ($\pm$1.0) $\times$ 10^6

[a] Subsurface material from 18-ft depth at the Pickett sample site.
[b] Oven-dried material.

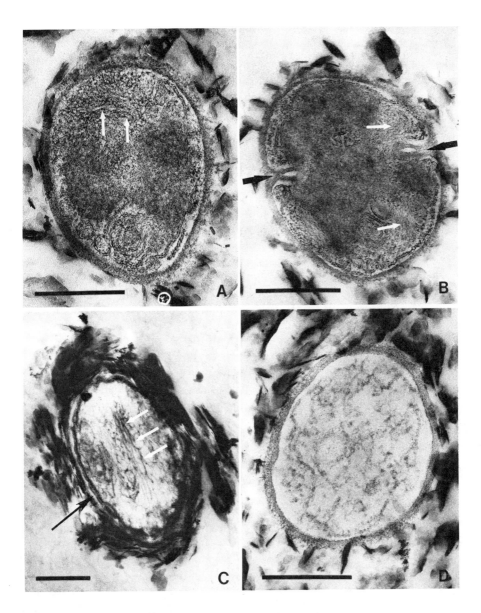

Figure 20.2. Electron photomicrographs of thin-sectioned subsurface material from the Lula sample site. Prokaryotic (bacterial) forms, such as those depicted, were the only life forms observed in these samples. White arrows indicate fibrous bacterial DNA that would bind acridine orange and fluoresce green in the epifluorescence microscope. Note electron-dense subsurface material at the periphery of these bacteria. (A) Coccoid cell possessing a typical gram-positive type cell wall. (B) Dividing coccoid cell in the process of forming crosswalls (black arrows). (C) Coccoid bacterial cell possessing a thinner, gram-negative type cell wall. Black arrow indicates a region where an outer membrane, diagnostic of this type of cell wall, can be seen. (D) Empty gram-positive type cell wall. Ghost cells such as this would not stain with acridine orange but would contribute to the muramic acid content of subsurface samples. Bars = 0.3 μm.

than 1 μm in size. The smallness of these forms could function in survival, as has been suggested for *Azotobacter* (Lopez and Vela, 1981), or it could simply represent an adaptation to low nutrient conditions. Future studies of the life cycles and physiology of subsurface microorganisms could help to clarify these interesting possibilities.

Dividing cells (Figure 20.2B) were observed regularly in the Lula subsurface material, indicating that some of the bacteria were actively growing at the time of fixation. This finding would support our previous conclusion from viable cell counts of Pickett material that a significant portion of the bacteria in subsurface samples is capable of active growth. Cell ghosts (Figure 20.2D) were regularly observed in these samples, suggesting that bacteria in the samples were also dying and that their cell walls may persist in these environments.

Muramic acid analyses of the same samples from the Pickett site that were used for epifluorescence (Table 20.1) and viable cell counts (Table 20.2) indicated that these samples contained approximately 1 μg of muramic acid per gram of oven-dried material. A comparison of this value to published values for muramic acid content of bacterial cells (King and White, 1977; Moriarty, 1977) would predict that 10 to 100 times as many cells were present as were counted by epifluorescence microscopy and plate counts. A 10-fold excess muramic acid content might in part be accounted for by empty muramic acid-containing cell walls such as those seen in the Lula samples (Figure 20.2D). These would not be expected to stain with acridine orange. A predominance of the thicker gram-positive cell wall types might also contribute to the excess muramic acid. However, discrepancies larger than 10-fold are more difficult to explain. Such discrepancies might be resolved by employing measurements of muramic acid in subsurface bacterial isolates as standards.

ACKNOWLEDGMENTS

This work was supported by Subcontract No. 6931-5 under U.S. EPA Cooperative Agreement No. CR806931-02. We are grateful to J. F. McNabb, J. T. Wilson, and the drilling crew at the Robert S. Kerr Environmental Research Laboratory, Ada, Oklahoma for providing the samples of uncontaminated subsurface material. The technical assistance of M. Schowe, W. Hahn, J. Tugel, and K. Whalen is gratefully acknowledged.

NOTE ADDED IN PROOF

Since this manuscript was prepared (nearly 3 years ago), two additional publications have appeared that include more details on enumeration and characterization of subsurface microorganisms and extend our findings to other sampling sites (Ghiorse and Balkwill, 1983; Wilson et al., 1983). In addition, papers presented at two recent symposia on the microbiology of subsurface environments and ground water contamination have also appeared or are to be published imminently (see *Developments in Industrial Microbiology,* Vol. 24, 1983; and C. L. Brierley and J. Brierley (Eds.), *Proceedings of the Sixth International Symposium on Environmental Biogeochemistry, October 1983, Santa Fe, N.M.,* Van Nostrand Reinhold, New York, in press).

REFERENCES

Alexander, M. (1981). Biodegradation of chemicals of environmental concern. *Science* **211**:132–138.

Alexander, M. (1977). *Introduction to Soil Microbiology,* 2nd ed. Wiley, New York.

Allen, M. J. (1980). Microbiology of ground water. *J. Water Pollut. Control Fed.* **52**:132–138.

Bae, H. C. and Casida, L. E., Jr. (1973). Responses of indigenous microorganisms to soil incubation as viewed by transmission electron microscopy of cell thin sections. *J. Bacteriol.* **113**:1462–1473.

Bae, H. C., Cota-Robles, E. H., and Casida, L. E., Jr. (1972). The microflora of soil as viewed by transmission electron microscopy. *Appl. Microbiol.* **23**:637–648.

Balkwill, D. L. and Casida, L. E., Jr. (1973). Microflora of soil as viewed by freeze-etching. *J. Bacteriol.* **114**:1319–1327.

Balkwill, D. L., Rucinsky, T. E. and Casida, L. E., Jr. (1977). Release of microorganisms from soil with respect to electron microscopy viewing and plate counts. *Antonie van Leeuwenhoek J. Microbiol. Serol.* **43**:73–81.

Balkwill, D. L., Labeda, D. P., and Casida, L. E., Jr. (1975). Simplified procedures for releasing and concentrating microorganisms for transmission electron microscopy and viewing as thin-sectioned and frozen-etched preparations. *Can. J. Microbiol.* **21**:252–262.

Bitton, G. and Gerba, C. P. (Eds.) (1984). *Groundwater Pollution Microbiology,* Wiley, New York.

Cholodny, N. G. (1930). Über eine neue Methode zur Untersuchung der Bodenmikroflora. *Arch. Mikrobiol.* **1**:620–652.

Dale, N. G. (1974). Bacteria in intertidal sediments. Factors related to their distribution. *Limnol. Oceanogr.* **19**:509–518.

Daley, R. J. and Hobbie, J. E. (1975). Direct counts of aquatic bacteria by a modified epifluorescene technique. *Limnol. Oceanogr.* **20**:875–882.

Dunlap, W. J. and McNabb, J. F. (1973). Subsurface biological activity in relation to ground water pollution. EPA-6601/273-014, Natl. Environ. Research Ctr., Office Res. and Monitoring, U.S. Environmental Protection Agency, Corvallis, OR.

Dunlap, W. J., McNabb, J. F., Scalf, M. R., and Cosby, R. L. (1977). Sampling for organic chemicals and microorganisms in the subsurface. EPA-600/2-77-176, Natl. Environ. Research Ctr., Off. Res. and Monitoring, U.S. Environmental Protection Agency, Corvallis, OR.

Ehrlich, H. L. (1981). *Geomicrobiology*. Marcel Dekker, New York.

Gray, T. R. G. (1967). Stereoscan electron microscopy of soil microorganisms. Science **155**:1668–1670.

Ghiorse, W. C. and Balkwill, D. L. (1983). Enumeration and morphological characterization of bacteria indigenous to subsurface environments. *Dev. Ind. Microbiol.* **24**:213–224.

Hirsch, P. (1974). Budding bacteria. *Ann. Rev. Microbiol.* **18**:391–444.

Hirsch, P. and Pankratz, S. H. (1972). Study of bacterial populations in natural environments using submerged electron microscope grids. *Z. Allg. Mikrobiol.* **12**:203–218.

Josephson, J. (1980). Safeguards for ground water. *Environ. Sci. Technol.* **14**:39–44.

Karl, D. M. (1978). Occurrence and ecological significance of GTP in the ocean and in microbial cells. *Appl. Environ. Microbiol.* **36**:349–355.

Karl, D. M. (1980). Cellular nucleotide measurements and applications in microbial ecology. Microbiol. Rev. **44**:739–796.

King, J. D. and White, D. C. (1977). Muramic acid as a measure of microbial biomass in estuarine and marine samples. *Appl. Environ. Microbiol.* **33**:777–783.

Lopez, J. G. and Vela, G. R. (1981). True morphology of Azotobacteraceae-filterable bacteria. *Nature* **289**:588–590.

Meyer-Reil, L. A. (1978). Autoradiography and epifluorescence microscopy combined for determination of number and spectrum of actively metabolizing bacteria in natural waters. *Appl. Environ. Microbiol.* **36**:506–512.

Moriarty, D. J. W. (1975). A method for estimating the biomass of bacteria in aquatic sediments and its application to trophic studies. *Oecologia (Berlin)* **20**:219–224.

Moriarty, D. J. W. (1977). Improved method using muramic acid to estimate biomass of bacteria in sediments. *Oecologia (Berlin)* **26**:317–323.

Perfilev, B. V. and Gabe, D. R. (1969). *Capillary Methods of Investigating Microorganisms*. Oliver and Boyd, Edinburgh.

Rosswall, T. (Ed.) (1973). *Modern Methods in the Study of Microbial Ecology*. Bulletins from the Ecological Research Committee, Swedish Natural Science Research Council, Stockholm, **17**.

Staley, J. T. (1971). Incidence of prosthecate bacteria in a polluted stream. *Appl. Microbiol.* **22**:496–502.

Tobin, R. S. and Anthony, D. H. J. (1978). Tritiated thymidine incorporation as a measure of microbial activity in lake sediments. *Limnol. Oceanogr.* **23**:161–165.

Trolldenier, G. (1973). The use of epifluorescence microscopy for counting soil microorganisms. In: T. Rosswall (Ed.), *Modern Methods in the Study of Microbial Ecology*. Bulletins from the Ecological Research Committee, Swedish Natural Science Research Council, Stockholm, **17**:53–59.

Waid, J. S. (1973). A method to study microorganisms on surface films from soil particles with the aid of the transmission electron microscope. In: T. Rosswall (Ed.), *Modern Methods in the Study of Microbial Ecology*. Bulletins from the Ecological Research Committee, Swedish Natural Science Research Council, Stockholm, **17**:53–59.

White, D. C., Davis, W. M., Nickels, J. S. King, J. D., and Bobbie, R. J. (1979). Determination of the sedimentary microbial biomass by extractible lipid phosphate. *Oecologia (Berlin)* **40:**51–62.

Wilson, J. T., McNabb, J. F., Balkwill, D. L., and Ghiorse, W. C. (1983). Enumeration and characterization of bacteria indigenous to a shallow water-table aquifer. *Ground Water* **21:**134–142.

Zimmerman, R. R., Iturriaga, R., and Becker-Birck, J. (1978). Simultaneous determination of the total number of aquatic bacteria and the number thereof involved in respiration. *Appl. Environ. Microbiol.* **36:**926–935.

21

REGIONAL APPROACH TO GROUND WATER INVESTIGATIONS

Wayne A. Pettyjohn

School of Geology
Oklahoma State University
Stillwater, Oklahoma

AN ICONOCLAST'S VIEW OF HYDROGEOLOGIC STUDIES

Ground water investigators love to drill test holes and are never satisfied no matter how many there might be. Once a well is drilled, the hydrologist will probably measure the static level, collect a water sample, and perhaps even conduct an aquifer test, unless he or she wants to test the pump, in which case a pump test is conducted.

Each additional measurement or task creates a new series of problems, which in turn give rise to the ever increasing need for additional test drilling. For example, the well screen may be open to several water-bearing zones, each of which is characterized by a different quality and head. Therefore, the original water level measurement and water sample reflect a composite. Moreover, each water-bearing zone will likely have a permeability that differs from the others, which in turn calls for more observation wells in order to conduct more aquifer tests, which will show that the entire system is as leaky as a sieve.

Then, of course, we cannot be satisfied with the water sample because either its analysis did not conform to our preconceived notions—we could not explain the presence or absence or concentration of some constituents (obviously the chemist messed up)—or it was not collected, shipped, stored, or analyzed in the correct manner in the correct time limit (someone else's fault). Perhaps the well was not pumped long enough or was pumped too long, or perhaps it was not pumped at all (blame this one on the pump). But whatever the reason, it hardly matters anyway because the sample does not actually represent the geochemical conditions that exist in the subsurface. Obviously, with all of these problems, additional drilling is warranted.

Next we need to examine the water level that was originally measured. First, since it reflects a summation of all the water levels of all the water-bearing zones exposed to the screen (assuming no leakage along the casing, which can easily justify at least four more new wells), it is obvious that the measurement is meaningless, worthless, confusing, and inaccurate; and therefore, more wells and measurements are required. After all, if one is going to estimate permeability or ground water recharge, one must have accurate measurements of water levels over a long period (unless the estimate is based on a computer simulation, in which case one needs no information at all).

It is a well-known fact that the cost of many hydrologic studies, particularly those funded by federal agencies, can get rather out of hand because of a lack of confidence, experience, or competence, or because of an abundance of ignorance or stupidity, or because of the desire for an apparent degree of accuracy that exceeds all boundaries of realism, or because we want to protect ourselves from legal action or from sharp-tongued attorneys representing "the other side," or because of the need to provide enough work to last until retirement so that the final report can be written by someone else, or because the project must be dragged out long enough to permit the investigator to stay in the field until the kids grow up and leave home.

There is no doubt but that test drilling is required in nearly every hydrogeologic study, but it is equally obvious that costs are rising so quickly and to such an extent that other sampling methods and techniques to evaluate the subsurface must be considered, developed, and used. Furthermore, for every project or investigation there should be developed, as a first step, an adequate experimental design that incorporates all available data and is based on fundamental hydrogeologic principles. As a matter of fact, it might be a good idea to prohibit all drilling until the final report is prepared; the drilling budget could be used to prove the validity of the already written final report. If the subsurface data indicate that parts of the report are incorrect, then modification and reinterpretation are required. In many cases this actually can be done.

SITE-SPECIFIC VERSUS REGIONAL STUDIES

Various techniques are available for development of at least a general understanding of the hydrogeologic framework, but their usefulness depends on our perspective. The methods can be thought of as point source (test drilling) or non-point source (regional or areal examination) techniques. In far too many investigations point source data are used to the total or nearly total exclusion of non-point data. The usual justification which is valid to some extent, is based on site size, with the smaller site receiving the greater detail. For example, in an examination of a 100 mi^2 area, 10 test holes might be considered adequate because the thickness and permeability of the earth materials are expected to range within fairly broad limits and the water level elevation and gradient are predictable. Conversely, in an examination of a 120-acre potential municipal landfill site, 10 test holes might be considered entirely inadequate because great detail may be required to obtain a permit and to overcome the objections voiced by local landowners, the permitting agency, or environmental groups. In either case, the purpose of drilling is to evaluate or determine permeability, thickness, areal extent of selected rock units, water levels, and water quality in order to predict what will happen when certain stresses are placed on the system.

On the other hand, how much permeability information does a test hole really provide considering the sample return, drilling method, and the exceedingly small area that the hole represents? For example, increasing evidence clearly shows that macropores and fractures play a major role in ground water recharge and quality in addition to well-yield, and this type of secondary permeability is not likely to be evident from test drilling alone.

REMOTE SENSING

It has long been known that water wells drilled on or adjacent to fractures produce higher yields than do other nearby wells. Some data indicate that production of hydrocarbons is also greater adjacent to fracture traces. It is strongly suspected that fracture systems exert a major though subtle control on hydrogeology as they influence permeability, hydraulic gradients, velocity, recharge, and the influx of chemicals to the ground water system. Fracture systems range from tens of miles in length at one extreme to a microscale at the other.

As a first step in mapping fracture systems, satellite imagery can be used since long linear features are readily evident and a map of them provides a framework for determining major trends. Next, low altitude photography can be utilized to key on a specific site. Field checks may be necessary. Commonly, fracture trend and density can be examined in considerable detail on the ground, particularly in surface mines such as quarries.

STREAM HYDROGRAPHS

Another useful method for evaluating regional hydrogeology is analysis of streamflow data, which are abundant, widely available, and easy to acquire.

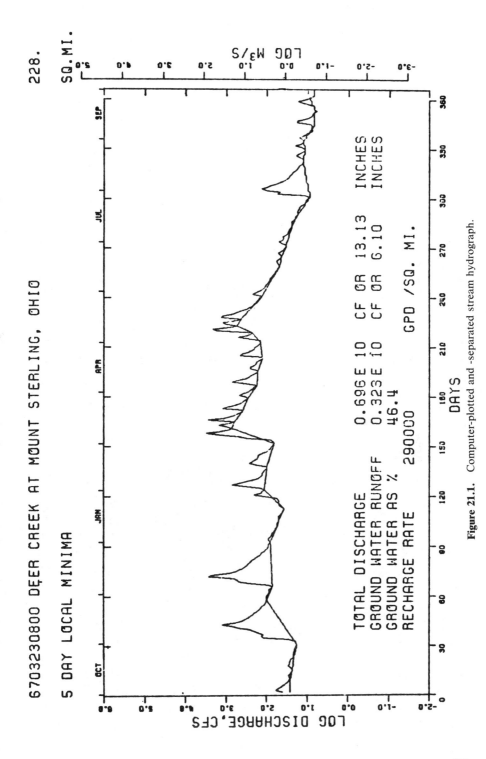

Figure 21.1. Computer-plotted and -separated stream hydrograph.

405

Since a stream is nothing more than an extremely long, very shallow, horizontal well, analysis of its flow will provide clues to the permeability, chemical quality, and potential yield of at least shallow aquifers or those that provide ground water runoff to a stream.

Several stream hydrograph separation techniques are available, including some that are computer-based (Pettyjohn and Henning, 1979). Effective ground water recharge rates, both monthly and annually, can be estimated by relating the volume of ground water runoff to the size of the drainage basin. Granted, the ratios obtained reflect regional conditions, but these may be more realistic than data obtained from rain simulators, infiltration rings, or analyses of ground water hydrographs, all of which are highly site-specific.

A computer-separated stream hydrograph is shown in Figure 21.1. This stream, which lies in a low relief till plain in south central Ohio, has a drainage area of 228 mi^2. Although not evident from the surface, several relatively thin deposits of sand and gravel are incorporated within the till and provide ground water runoff to the tributaries and mainstream of Deer Creek. The calculated effective ground water recharge rate in the basin above the gaging station during a year of normal precipitation (1967) was 290,000 gpd/mi^2 (0.01 gpd/ft^2) or 6.10 in.; thus ground water accounted for 46.4% of the stream's total flow.

Effective ground water recharge rates in different hydrogeologic terrains in Ohio, as indicated by ground water runoff, range from 123,000 to 291,000 gpd/mi^2 (average 291,000) in till covered areas; from 310,000 to 406,000 gpd/mi^2 (average 352,000 in outwash covered areas); and from 160,000 to 198,000 gpd/mi^2 (average 179,000) in the moderate to high relief unglaciated part of Ohio, where alternating layers of sandstone, shale, coal, and limestone crop out.

Regional data of the type described above can be very useful, but they are not site-specific since they represent all of the cause-and-effect relations that occur upbasin. Nonetheless, when used in conjunction with other regional techniques, such as flow-duration curves and ratios and seepage measurements, they prove a good first approximation of hydrogeologic conditions, provide good information on regional differences in permeability, and could significantly reduce drilling costs.

FLOW-DURATION CURVES

As pointed out by Cross and Hedges (1959), flow-duration curves are widely used, and there is considerable technical literature on the subject. From a hydrogeologic viewpoint they are most useful for comparing the flow characteristics of different streams because the shape of the curve is an index of the natural storage within a basin. When plotted on logarithmic probability paper, the more nearly horizontal the curve, the greater is the effect of ground water storage.

During dry weather, the flow of a stream is almost entirely from ground water sources. The lower ends of duration curves therefore indicate in a general way the characteristics of the shallow ground water bodies in the drainage basin above the gaging station. Duration curves thus are useful guides in locating possible sources of ground water (Cross and Hedges, 1959, p. 5).

Despite the fact that changes in the drainage basin, such as regulation, diversion, and variable discharges of effluent, will alter the shape of the flow-duration curve of some streams, records of even these streams can, by judicious comparison with unaffected nearby streams, provide useful approximation of the hydrogeologic system.

A flow-duration curve shows the frequency of occurrence of various rates of flow. It is a cumulative frequency curve prepared by arranging all discharges of record in order of magnitude and subdividing them according to the percentages of time during which specific flows were equaled or exceeded; all chronologic order or sequence is lost (Cross and Hedges, 1959). Flow-duration curves may be plotted on either probability or semilogarithmic paper.

Several flow-duration curves for Ohio streams are shown in Figure 21.2. During low-flow conditions (the flow equaled or exceeded 90% of the time), the curves for several of the streams—such as the Mad, Hocking, and Scioto Rivers and Little Beaver Creek—trend toward the horizontal, while those for Grand River and White Oak and Home Creeks all remain very steep.

Mad River flows through a broad valley filled with very permeable sand and gravel. The basin has a large ground water storage capacity and, consequently, the river maintains a high sustained flow. The Hocking River valley also contains outwash in and along its flood plain, particularly in the upper reaches, which provides a substantial amount of ground water runoff. Above Columbus, the Scioto River crosses glacial till and thin layers of limestone that crop out along the stream channel; ground water runoff in this reach is relatively small. Immediately south of Columbus, however, the Scioto River valley is filled with coarse outwash, and during low flow the discharge increases substantially at succeeding downstream gages.The reason that Mad River has a higher low-flow index than the Scioto River is that the former receives ground water runoff through its length, while ground water runoff to the Scioto River increases significantly only in the area of outwash south of Columbus.

White Oak and Home Creeks originate in bedrock areas where either relatively thin alternating layers of shale and limestone are covered by till (White Oak Creek) or sandstone, shale, and limestone crop out along hillsides (Home Creek). The low permeability of the strata and the greater relief in these basins preclude the storage of large amounts of ground water, and consequently the low flows characteristic of these streams are far less than those of the streams filled or partly filled with outwash.

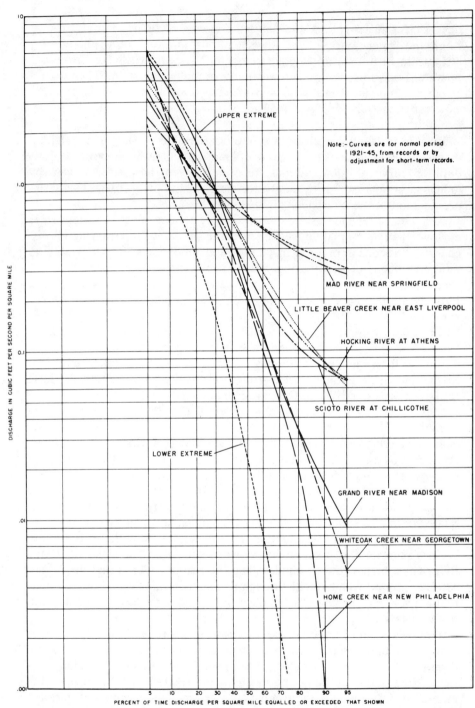

Figure 21.2. Flow-duration curves for selected Ohio streams, showing spread of curves for all non-regulated gaging stations.

FLOW RATIOS

Walton (1970) reported that grain-size frequency-distribution curves are somewhat analogous to flow-duration curves in that their shapes are indicative of water-yielding properties of rocks. He pointed out that a measure of the degree to which all of the grains approach one size (the slope of the grain-size frequency-distribution curve) is the sorting. One parameter of sorting is obtained by the ratio $(D_{25}/D_{75})^{1/2}$, where D_{25} is the grain-size such that 25% of the sample is larger and 75% of the sample is smaller, and D_{75} is the grain-size where 75% is larger and 25% is smaller. Walton modified this equation by replacing the 25 and 75% grain-size diameters with the 25 and 75% flow—$(Q_{25}/Q_{75})^{1/2}$. In this case a low ratio is indicative of a permeable basin that has a large storage capacity. This technique provides another simple and quick method for hydrologic evaluations of drainage basins.

The Q_{25} and Q_{75} data are easily obtainable from flow-duration curves. The data from Figure 21.2 show that the Mad River has a flow ratio of 1.59 and the Scioto River a ratio of 2.64, while Home Creek, typical of a basin of low permeability, has the highest ratio of 5.09.

SEEPAGE OR DRY-WEATHER MEASUREMENTS

Seepage or dry-weather measurements consist of flow determination made at several locations along a stream during a short time interval when there is no surface runoff. Many investigators prefer to initiate seepage runs during the stream's 90% flow.

It is not always possible to conduct an actual seepage survey due to time, manpower, or financial constraints. In these cases, flow-duration curves may serve as valuable substitutes.

Seepage measurements permit a quick evaluation of ground water runoff—how much there is and where it originates—and provide clues to the geology of the basin as well. The flow of some streams increases substantially within a short distance. Under natural conditions the increase is most likely related to increased ground water runoff originating in deposits or zones of high permeability in or adjacent to the stream channel. These gaining reaches may consist of deposits of sand and gravel, fracture zones, solution openings in limestone, or merely local facies changes. In addition, ground water may also discharge through a series of springs or seeps along valley walls or in the stream channel.

In areas where the geology and ground water systems are not well known, streamflow data can provide a means of testing estimates of the ground water system. If the stream-flow data do not conform to the estimates, then the geology must be more closely examined (LaSala, 1968). For example, the northwest corner of Ohio is crossed by the Wabash and Fort Wayne moraines, between which lies the St. Joseph River. As indicated by the

Glacial Map of Ohio (Goldthwait et al., 1961), the St. Joseph Basin consists mainly of till. However, low-flow measurements show that the discharge of the river increases more than 14 cfs along its short reach in Ohio, indicating that the basin contains a considerable amount of outwash, which in this case is covered by a relatively thin layer of till.

The mainstream of the Auglaize River in northwestern Ohio rises from a mass of outwash that lies along the front of the Wabash moraine. The southwest-flowing river breaches the moraine near Wapakoneta and then flows generally north to its confluence with the Maumee River at Defiance. A gaging station is near Ft. Jennings in a till plain area slightly above a reservoir on the Auglaize. This gage measures the flow resulting as an end-product of all causative hydrologic factors upbasin (ground water runoff, surface runoff, slope, precipitation, use patterns, etc.)—it merely shows inflow to the reservoir. Low-flow measurements, however, indicate that nearly all of the baseflow is derived from a small deposit of outwash along the distal side of the Wabash moraine; there is no gain across the wide till plain downstream, which makes up most of the stream's basin.

A number of discharge measurements have been made in the Scioto River basin in central Ohio. The flow measurements in themselves are important because they show the actual discharge—in millions of gallons per day, in this case—at about 90% flow (Figure 21.3). Note that the discharge at succeeding downstream sites on the Scioto River is greater than the flow immediately upstream. This shows that the river is gaining and that water is being added to it by ground water runoff originating largely from the adjacent outwash deposits.

A particularly useful method for evaluating streamflow consists of relating the discharge to the size of the drainage basin (cfs or mgd/mi^2 of drainage basin). A cursory examination of Figure 21.3 shows that it is convenient (and totally arbitrary) to separate the flow into three distinct units: Unit 1 falls in the range of 0.010 to 0.020 mgd/mi^2, Unit 2 includes 0.021 to 0.035 mgd/mi^2, and Unit 3 0.036 to 0.050 mgd/mi^2. The Olentangy River and Alum and Big Walnut Creeks fall into Unit 1, Big Darby and Deer Creeks into Unit 2, and the Scioto River, Walnut Creek, and the lower part of the Big Walnut Creek into Unit 3. Even though the latter watercourses fall into Unit 3, their actual discharges vary widely, from 3.07 to 181 mgd.

Logs of wells drilled along the streams of Unit 1 show a preponderance of fine-grained material that contains only a few layers of sand and gravel; all these wells yield less than 25 gpm and commonly no more than 5 gpm. Logs of wells and test holes along Big Darby and Deer Creek, however, indicate that several feet of sand and gravel underlie 5 to more than 25 ft of fine-grained alluvial material. Adequately designed and constructed wells that tap these buried outwash deposits produce as much as 500 gpm. Glacial outwash, much of it coarse-grained, forms an extensive deposit through which Unit 3 streams and rivers flow. The outwash extends from the surface to depths that in places exceed 150 ft. Industrial wells constructed in these

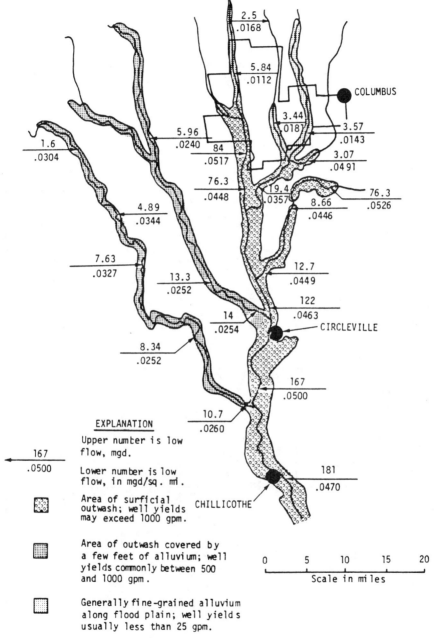

Figure 21.3. Low-flow data coupled with hydrogeologic information can be used to develop a potential well-yield map.

deposits, most of which rely on induced infiltration, can produce more than 1000 gpm. Thus, it is evident that by combining seepage data and well yields with a geologic map, it is possible to develop a potential well yield map. The potential ground water yield map relies heavily on streamflow measurement as well as good judgment, but nonetheless provides, with some geologic data, a good first approximation of ground water availability.

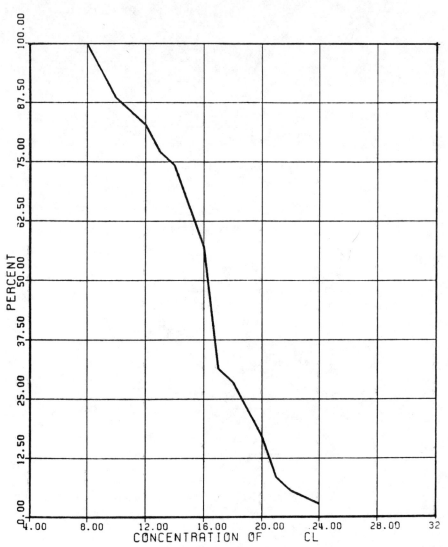

03234000 PAINT CREEK NEAR BOURNEVILLE, OHIO 1967

Figure 21.4.   Chloride duration curve of Paint Creek in 1967.

WATER QUALITY-DURATION CURVES

A water quality-duration curve is similar to its counterpart, the flow-duration curve, and is prepared in the same manner except that the stream's concentration of selected chemical constituents replaces discharge data. The quality-duration curve can be used for three purposes. First, it shows the

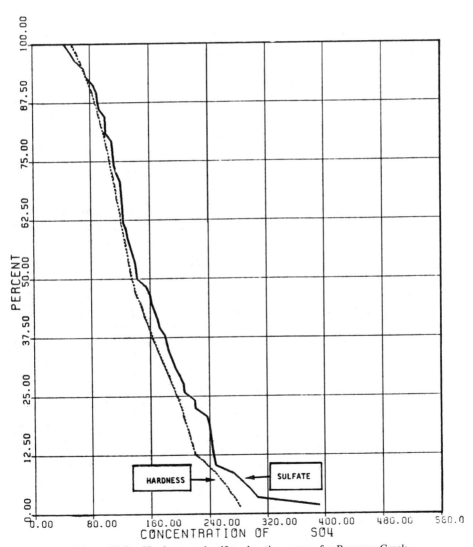

Figure 21.5.   Hardness and sulfate duration curves for Raccoon Creek.

stream's range in concentration of any substance examined; this is useful for water treatment plant designs and dilution studies. Secondly, it can provide a good approximation of the chemical quality of ground water in the zone of active circulation. Finally, in certain instances it can be used to indicate areas of contaminated surface water.

As discussed previously, during a stream's period of low flow, all of the water in the channel consists of ground water runoff, unless the stream receives effluent. Therefore, within limits, at these times a stream's quality closely approximates the chemical quality of shallow ground water in the basin. (This assumption is not entirely correct, however, because some chemical changes may occur at or near the ground water–surface water interface.)

Although more work needs to be done along these lines, it appears that a stream's 10% concentration (the concentration equaled or exceeded 10% of the time) provides a reasonable estimate of the chemical quality of shallow ground water if the stream is not contaminated. It is imperative, however, that values obtained from quality-duration curves be compared with analyses of well water. A quality-duration curve for Ohio's Paint Creek based on data from 1967 is shown in Figure 21.4. This is a typical curve for a stream uncontaminated with respect to chloride. The 10% concentration is about 21 mg/L; shallow well data indicate that the natural concentration of chloride in the basin generally ranges from 3 to 28 mg/L. Thus the estimate of ground water quality based on the quality-duration curve is in close agreement with actual well data.

It is often possible to determine whether a stream is contaminated by means of quality-duration curves, although the method is somewhat subjective and must be based on some prior knowledge. A quality-duration curve of sulfate in Raccoon Creek, a stream contaminated by drainage from coal mines, is shown in Figure 21.5. In this case the 10% concentration of sulfate (250 mg/L) is far greater than the background content (about 75 mg/L), but the concentration of carbonate hardness is within expected limits (220 mg/L). Therefore, despite the fact that a stream is contaminated with respect to one or more constituents, the contamination does not completely invalidate the usefulness of curves depicting other constituents.

QUALITY SEEPAGE MEASUREMENTS

Another method to determine shallow ground water quality is to collect surface water samples during dry-weather conditions from a wide area within a short time interval. In this case emphasis should be placed on small tributaries since they are less likely to be contaminated. These data can be incorporated with seepage measurements, but the latter are not essential for quality evaluations. Once the chemical data are analyzed, concentrations can be plotted on a base map using a color code to represent selected ranges in concentration. In this manner background quality becomes readily evident.

Alum Creek flows southward across a till covered area in central Ohio. Within the basin are scores of abandoned and producing oil wells and even more dry holes. Oil-field brine-holding ponds have been used in this area for years, and leakage from many of them has locally contaminated the ground water with high concentrations of chloride. Contaminated ground water eventually flows into the stream, but during much of the year the chloride-

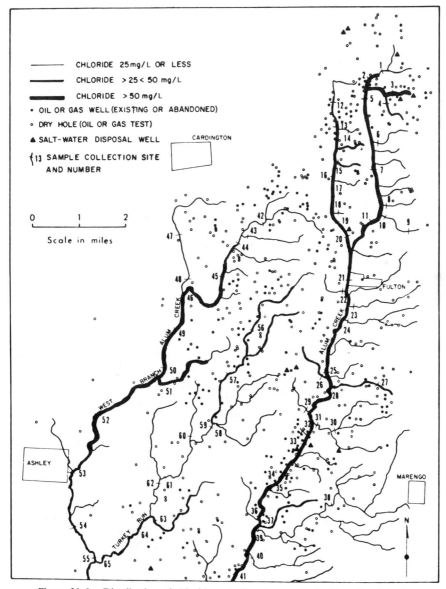

Figure 21.6.   Distribution of chloride and oil and gas wells in Alum Creek basin.

rich ground water is diluted by surface runoff and ground water runoff from
uncontaminated areas.

Nearly 100 stream samples were collected during a single day reflecting
low-flow conditions. Most of the samples were collected from small tributaries near their confluences with the main stream. These data were aug-
mented with samples from field drainage tile. Most of the samples contained
less than 15 mg/L of chloride, which is the background concentration, but

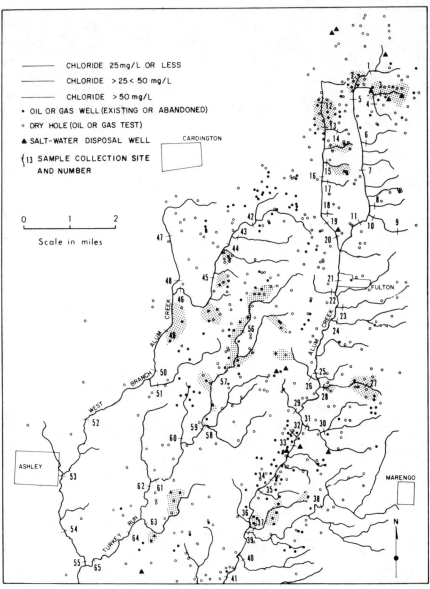

Figure 21.7. Areas of ground water pollution in Alum Creek basin.

others were significantly higher. The small basins that contained higher than background chloride concentrations were assumed to be contaminated (Figure 21.6).

The configuration of each contaminated small basin was delineated on a topographic map, on which were plotted oil and gas wells and tests. The latter provided some control on the point sources of contamination. It was then possible to estimate the general size of each ground water contaminated area (Figure 21.7).

Using the regional approach described above, it is possible to minimize drilling costs for monitoring wells because uncontaminated areas are readily evident and the investigator can then key on selected sites. Once contaminated areas have been located, additional surface water samples can be collected from the sub-basin, and these should permit a more detailed assessment of the area.

SUMMARY AND CONCLUSIONS

Commonly a research plan is not formulated prior to the initiation of hydrogeologic field studies. Such a plan, if based on sound hydrogeologic principles, should lead the investigator down a logical path toward a solution that minimizes costs and maximizes efficiency.

Generally hydrogeologists depend on the very expensive and locally seasonal process of drilling test holes and wells to obtain basic data. Such drilling is usually essential, but a variety of techniques are available to assess hydrogeologic conditions from a regional perspective that can substantially reduce field costs. Generally the data needed for regional studies can be easily and inexpensively obtained.

REFERENCES

Cross, W. P. and Hedges, R. E. (1959). Flow duration of Ohio streams. Ohio Department of Natural Resources, Division of Water, Bulletin 31.

Goldthwait, R. P., White, G. W., and Forayth, J. L. (1961). Glacial map of Ohio. U.S. Geological Survey, Misc. Geol. Inv. Map I-316.

LaSala, A. M. (1967). New approaches to water-resources investigations in upstate New York. *Ground Water* 5(4):6–11.

Pettyjohn, W. A. (1975). Chloride contamination in Alum Creek, central Ohio. *Ground Water* 13(4):332–339.

Pettyjohn, W. A. (1982). Cause and effect of cyclic changes in ground-water quality. *Ground Water Monitoring Rev.* 2(1):43–49.

Pettyjohn, W. A. and Henning, R. (1979). Preliminary estimate of ground-water recharge rates, related streamflow and water quality in Ohio. Water Resources Center, Ohio State University, Project Completion Rept. No. 552.

Walton, W. C. (1970). *Groundwater Resource Evaluation.* McGraw-Hill, New York.

PART FOUR

TRANSPORT AND FATE OF SUBSURFACE CONTAMINANTS

22

OVERVIEW OF SUBSURFACE TRANSPORT AND FATE OF POLLUTANTS

William J. Dunlap

Robert S. Kerr Environmental Research Laboratory
U.S. Environmental Protection Agency
Ada, Oklahoma

HUMAN ACTIVITIES AND GROUND WATER QUALITY

The activities of humankind have always resulted in the release of various contaminants, primarily waste materials, into the terrestrial subsurface. As long as populations remained low and relatively dispersed—and the released contaminants were mostly of natural origin—these activities posed no great threat to ground water quality. However, the discovery during the last half of the nineteenth century that microbial contaminants in ground waters serving as sources of drinking water were responsible for epidemics of diseases such as cholera and typhoid fever clearly indicated that human activities could result in serious pollution of ground water. Nevertheless, a general lack of concern for the potential pollution of ground water has persisted into the last half of the twentieth century, even as the activities of rapidly growing and increasingly affluent populations have been releasing enormous and ever expanding quantities of a vast array of chemical and biological contaminants, including many synthetic substances with no molecular counter parts in nature, into the subsurface. While the environmental movement beginning in the early 1960s has focused great effort and many resources on problems of surface water and air pollution, there has been, until very recently, no concomitant attention to ground water pollution problems.

Indifference to the expanding potential for ground water pollution posed by the various human activities has arisen primarily from the following: a general lack of knowledge on the part of the public and most political entities concerning ground water and the importance of this resource; preoccupation with more readily apparent problems of surface water and air pollution; and, the prevalence of an overly optimistic conception of the capacity of the soil mantle for pollutant assimilation, even among most of those directly concerned with ground water resources management. Within the past few years, however, this indifference has been largely dispelled. Reports of serious incidents of ground water pollution resulting from such activities as land disposal of hazardous wastes and chemical spills have aroused public concern, and field studies have established that many contaminants released into the subsurface are exhibiting sufficient persistence and mobility therein to cause unacceptable deterioration of the quality of valuable and often irreplacable ground water resources. Indeed, this is amply illustrated by many of the papers in this volume, particularly those of Miller (Chapter 4) and Gerba (Chapter 5), presented above. Clearly, human activities releasing contaminants into the subsurface must be managed and controlled much more effectively or eliminated altogether if the quality of ground water is to be preserved.

THE NEED FOR TRANSPORT AND FATE INFORMATION

Source activities releasing contaminants into the subsurface which involve the greatest potential for significant pollution of ground water may be conveniently grouped under four major categories:

1. Waste-disposal activities that utilize the subsurface as a pollutant receptor, such as hazardous-waste landfills, industrial waste ponds and lagoons, waste water land treatment operations, and disposal wells.
2. Industrial and commercial operations involving the handling of large quantities of chemical substances which may be accidentally released into the subsurface in significant amounts as the result of leaks and spills occurring during transport, storage, and utilization activities.
3. Agricultural operations involving intentional application of chemicals to the land.
4. Water reclamation activities entailing either direct or indirect artificial recharge of ground water with contaminated water.

Obviously, these source activities cannot be completely eliminated. Considering the enormous quantities of wastes generated by industrial societies and the very limited possibilities for disposal of this waste to surface media, particularly in view of needs for control of pollution of air and surface water, the subsurface must continue to serve as a receptor of vast and probably increasing quantities of waste. Similarly, although spills and leaks during industrial/commercial operations involving chemical transport, storage, and use can undoubtedly be reduced and better controlled, the sheer magnitude of these activities virtually precludes complete elimination of accidental release without major disruption of commerce in modern industrial societies. Also, increasing demands for food and fresh water seem certain to require continued high levels of utilization of chemicals in agricultural operations and to increase the need for reclamation of waste water by artificial recharge of ground water.

Since the source activities primarily responsible for release of pollutants into the subsurface cannot be eliminated, the goal of ground water protection efforts must necessarily be the control or management of these sources to ensure that released pollutants will be sufficiently attenuated within the subsurface to prevent significant impairment of ground water quality at points of withdrawal or discharge. This goal can be effectively achieved only if control and management options are based on definitive knowledge of the transport and fate of pollutants in the subsurface environment. Such knowledge is required for establishment of criteria for design, location, and operation of new potential sources of pollution, such as new hazardous waste disposal sites and facilities for land treatment of wastewater, in order that these criteria will permit maximum practible use of the subsurface as a pollutant receptor while assuring minimal entry of pollutants into ground water and movement of any pollutants which do enter ground water to points of withdrawal or discharge. Knowledge of transport and fate is also required for assessing the probable impact on ground water quality of existing sources, such as hazardous waste dumps and spill sites, in order to determine a level of remedial action that is both cost-effective and sufficient to prevent serious degradation of ground water quality at points of withdrawal or discharge. Finally, development of improved methods for removing pollutants from and renovating already polluted aquifers is dependent on knowledge of the subsurface behavior of such pollutants. Source control and management options not based on knowledge of pollutant transport and fate are almost certain to result in either undercontrol, with excessive pollution of ground water, or overcontrol, resulting in uneconomical underutilization of the subsurface as a pollutant receptor.

CURRENT KNOWLEDGE AND RESEARCH CONCERNING TRANSPORT AND FATE

Although the need for definitive information pertaining to the transport and fate of pollutants in the subsurface is immediate and pressing, the present state of knowledge on this topic is relatively primitive, reflecting the long prevailing lack of concern for ground water quality. Until the mid-1970s, information concerning subsurface behavior of pollutants was garnered mainly from field monitoring studies conducted in response to observed and usually highly visible cases of ground water pollution. Since these cases were relatively few in number and dealt primarily with inorganic chemicals and bacteria, this information was very limited. More recently, increased interest in ground water, coupled with the availability of more effective analytical methodologies for organic pollutants and viruses, has resulted in rapid expansion of the body of information available from field monitoring studies of ground water quality. These data provide many useful, although limited, insights regarding the transport and fate of various pollutants in the subsurface. For example, the widespread detection of chlorinated ethanes and ethenes in ground water in areas where these substances have been released into the soil profile indicates strongly that they are highly mobile and persistent in the subsurface and, hence, should not be released in appreciable quantities into this environment. For the most part, however, monitoring data are pollutant- and site-specific, and provide little rational basis for extrapolation of this information to other pollutants in different subsurface environmental situations.

Considering the large number and variety of pollutants that may be released to the subsurface and the wide range of environmental situations (geological, hydrological, chemical, and biological) they may encounter, it is apparent that a highly systematic approach must be followed in developing a capability for predicting subsurface transport and fate sufficient to meet the goals of ground water pollution control. Accordingly, the major thrust of transport and fate research now appears to be directed toward the definition and quantitation of the processes governing the behavior of pollutants in subsurface environments, coupled with the development of mathematical models that integrate process descriptions with pollutant properties and environmental (site) characteristics to yield quantitative estimates of subsurface transport and fate. The ultimate objective of this research is to provide methodologies that will permit accurate prediction of the effect specific pollutants released into the subsurface from a particular source activity will have on the quality of ground water at points of withdrawal or discharge. Such research is especially difficult and costly because of the remoteness and relative inaccessibility of the environmental compartments with which it is concerned. It is interdisciplinary in nature and requires the input of a wide range of scientific and technical capabilities. In spite of cost and complexity, however, research in this challenging field is now rapidly accelerating because it is absolutely necessary if ground water pollution control that is both effective and economical is to be achieved. The papers presented in Part Four of this volume describe work that is on the cutting edge of this effort.

23

ADVECTION–DISPERSION–SORPTION MODELS FOR SIMULATING THE TRANSPORT OF ORGANIC CONTAMINANTS

Paul V. Roberts
Martin Reinhard
Gary D. Hopkins
R. Scott Summers

Department of Civil Engineering
Terman Engineering Center
Stanford University
Stanford, California

Pollution of ground water by persistent and potentially hazardous organic chemicals has begun to attract attention as a national problem of grave concern (EPA, 1980). To deal with this emerging problem, it will be necessary to develop and employ mathematical models to simulate the transport of such chemicals in order to plan and interpret field studies. For these purposes, the models employed preferably should be computationally simple, require a less than inordinate amount of input data, and provide output sufficiently accurate to improve our knowledge of the behavior of the contaminants in question.

The purpose of this paper is to describe the types of simple models available, to assess their potential and limitations, and to illustrate their use by applying them to interpret a body of data obtained in a field study of artificial recharge.

ADVECTION–DISPERSION–SORPTION MODELS

Mathematical models of transport incorporating the phenomena of advection, dispersion, and sorption offer a possibility of simulating transport of nondegradable organic solutes. Such models have been reviewed elsewhere (Anderson, 1979; Bear, 1979; Freeze and Cherry, 1979). For organic contaminants that sorb but otherwise react neither chemically nor biologically, the transport equation may be expressed as

$$-u\,\frac{\partial C}{\partial x} + D\,\frac{\partial^2 C}{\partial x^2} - \frac{\rho_b}{\varepsilon}\,\frac{\partial S}{\partial t} = \frac{\partial C}{\partial t} \tag{23.1}$$

where u = average linear velocity (m/sec); C = solute concentration in aqueous phase (g/m^3); x = distance in flow direction (m); D = dispersion coefficient (m^2/sec); ρ_b = bulk density of soil (g/m^3); ε = soil void fraction $(-)$; S = mass of solute sorbed per unit dry mass of soil (g/g); and t = time (sec).

Sorption

Transport Equation

Solutes that sorb strongly onto soil materials are retarded in their movement through an aquifer (Ogata, 1964). If we ignore the second term on the left of Eq. (23.1), the transport equation for a sorbing but nonreacting solute under conditions of ideal plug flow can be simplified to

$$-u\,\frac{\partial C}{\partial x} - \frac{\rho_b}{\varepsilon}\,\frac{\partial S}{\partial t} = \frac{\partial C}{\partial t} \tag{23.2}$$

Further, assume that the mass transfer of solute to sorption sites is rapid relative to the flow velocity (i.e., local equilibrium), and that the sorption equilibrium is linear

$$\frac{\partial S}{\partial C} = K_d \tag{23.3}$$

where K_d = distribution coefficient (m^3/g). For this simple case, the transport equation can be written as

$$-u\,\frac{\partial C}{\partial x} = \left(1 + \frac{\rho_b K_d}{\varepsilon}\right)\frac{\partial C}{\partial t} \tag{23.4}$$

Retardation Factor

The term $(1 + \rho_b K_d/\varepsilon)$ is known as the retardation factor. An advancing front of sorbing solute moves at a linear velocity that is smaller than the velocity of ground water movement by a factor

$$t_r = 1 + \rho_b K_d/\varepsilon = u_r^{-1} \tag{23.5}$$

where t_r = relative residence time, or retardation factor (dimensionless), and u_r = velocity of a sorbing solute divided by the velocity of a conservative tracer.

The value of the retardation factor t_r is influenced primarily by the value of K_d; the values of ρ_b and ε differ to a much lesser extent in natural systems than does K_d. The value of K_d is determined by the strength of solute–soil interactions; the greater the affinity of the solute for the soil phase, relative to its affinity for water, the more strongly the solute will sorb, and the greater the value of K_d. Hence, it is not surprising that solutes exhibiting strongly hydrophobic behavior sorb strongly onto soil materials. In fact, Karickhoff et al. (1979) have shown that sorption of organic solutes by soil materials is governed by a remarkably simple rule: the larger the organic carbon fraction of a soil or sediment, the greater the value of K_d. In experiments with 15 soil materials having organic carbon contents in the range 0.1–3.3%, Karickhoff et al. (1979) found that the value of K_d for a given solute was proportional to the organic carbon content of the soil. Moreover, Karickhoff et al. (1979) found that the value of K_d was approximately proportional to the degree of hydrophobicity of the solute, as measured by the octanol–water partition coefficient K_{ow} (Leo et al., 1971; Hansch and Leo, 1979). These observations are summarized in consistent units:

$$(K_d)_i = 6.3 \times 10^{-7} f_{oc}(K_{ow})_i \tag{23.6}$$

where f_{oc} = fraction of organic carbon in the soil (g organic carbon per g dry soil); K_{ow} = octanol–water partition coefficient (molar concentration basis); and i = solute index. Equation (23.6) represents a best fit of experiments (Karickhoff et al., 1979) with 10 solutes having values of K_{ow} in the range of 100–1,000,000. McCarty et al. (1981) have shown that Eq. (23.6) satisfactorily explains the relative retardation of organic solutes observed in the Palo Alto Baylands pilot study.

Values of the retardation factor calculated for ranges of values of $(K_{ow})_i$ and f_{oc} are plotted in Figure 23.1. The values of ρ_b and ε are assumed in this example to be 2.0×10^6 g/m^3 and 0.2, respectively. Here it is assumed that the inorganic matrix is insignificant as a sorbing phase.

Figure 23.1 can be used to estimate the retardation factor for a solute whose octanol–water partition behavior is known or can be estimated

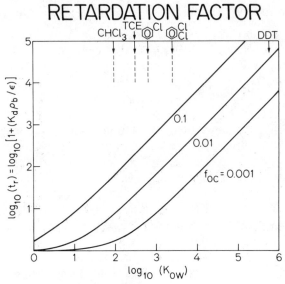

Figure 23.1. Generalized prediction of retardation factor based on octanol–water partition.

(Hansch and Leo, 1979), in an aquifer with known organic carbon content. For example, in an aquifer containing 1% organic carbon, chloroform ($K_{ow} = 90$) will be retarded by a factor of seven, whereas DDT ($K_{ow} = 5.8 \times 10^5$) will be retarded by a factor of 40,000. Clearly, chloroform is expected to migrate much more rapidly than DDT in the subsurface environment; there is little prospect that a strongly sorbing solute such as DDT will be encountered far from its source if the model assumptions are valid.

The Octanol–Water Partition Coefficient

The octanol–water partition coefficient for a solute of interest can be measured simply in the laboratory, or in the absence of experimental data may be estimated using correlation procedures. The development of such approaches has been reviewed recently by Hansch and Leo (1979), who also give an exhaustive compilation of available data. Organic compounds of environmental concern vary by approximately seven orders of magnitude with respect to octanol–water partition. Some polar compounds (e.g., methanol) partition preferentially into the aqueous phase, as evidenced by negative values of log K_{ow}. Some strongly non-polar compounds (e.g., DDT and polychlorinated biphenyl isomers) are concentrated into the octanol phase by a factor of 100,000 or more.

Dispersion

The importance of the dispersion phenomenon can be gauged by the magnitude of the dispersion coefficient D in Eq. (23.1). In most instances, neither laboratory experiment nor theory is of great value in predicting the value of D for a natural aquifer. The value of D is determined primarily by spatial variation of permeability, and hence must be determined by tracer measurements in the field (Anderson, 1979).

Dispersion is reflected in a spreading of the solute concentration front as it moves through the aquifer. The dimensionless Peclet Number ($Pe = ux/D$) is used as a measure of the dispersion tendency: the smaller the value of Pe, the greater the extent of dispersion. Ogata and Banks (1961) and Levenspiel and Bischoff (1963) show that for $Pe > 100$, dispersion can be practically neglected, whereas for $Pe < 5$ the flow regime approaches complete mixing.

In ground water hydrology, the dispersion coefficient is frequently found to be approximately proportional to the velocity

$$D = \alpha\,u \tag{23.7}$$

where α = dispersivity (m). Values of α calculated from field measurements range from 0.1 to 100 m, much larger than the α values (0.0001 to 0.01 m) measured in homogeneous porous media in the laboratory (Anderson, 1979).

How does the response change with increasing distance x? Ogata and Banks (1961) present a general solution for the response to a step change in concentration:

$$\hat{F} = \frac{C}{C_0} = \frac{1}{2}\left\{ \operatorname{erfc}\left[\frac{1-\theta}{2(\theta/Pe)^{1/2}}\right] + \exp(Pe)\ \operatorname{erfc}\left[\frac{1+\theta}{2(\theta/Pe)^{1/2}}\right]\right\} \tag{23.8}$$

where $\theta = ut/x$ and $Pe = ux/D$. The response to a step change is depicted in Figure 23.2 as $\hat{F}$ versus θ for values of $Pe = 0.1, 1, 10, 100,$ and 1000. For small values of Pe ($Pe \leq 10$), pronounced mixing and a broad spectrum of residence times are evident. For large values of Pe ($Pe > 100$), the second

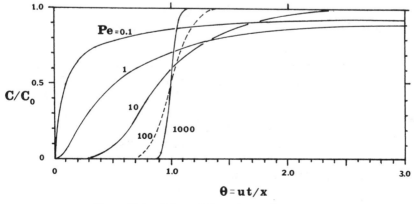

Figure 23.2.   Effect of dispersion on $\hat{F}$-response.

term of Eq. (23.8) can be neglected; the $\hat{F}$-response is nearly symmetric, and the spectrum of residence times is relatively narrow.

Sorption and Dispersion

The behavior of a solute influenced by both sorption and dispersion can be predicted by modifying the transport equation [Eq. (23.1)] as follows:

$$-u_r\, u\, \frac{\partial C}{\partial x} + u_r\, D\, \frac{\partial^2 C}{\partial x^2} = \frac{\partial C}{\partial t} \tag{23.9}$$

where all variables are defined as before.

The sorbing solute travels at an apparent velocity of $u_r \times u$, and exhibits an apparent dispersion coefficient $u_r \times D$. Because the dispersivity α is defined as $\alpha = D/u$, the apparent dispersivity of the sorbing solute

$$\alpha_{\text{app}} = \left(\frac{D_{\text{app}}}{u_{\text{app}}}\right) = \frac{u_r D}{u_r u} = \frac{D}{u} \tag{23.10}$$

is the same as that of a non-interacting solute [Eq. (23.7)]. The remarkable conclusion is that the concentration response of a sorbing solute with retardation factor $t_r = (u_r)^{-1}$ is exactly the same as expected for a non-sorbing solute moving at a velocity equal to u_r times the real average fluid velocity.

Exhaustion of the soil's sorption capacity is a general feature of sorption: the solute concentration front eventually breaks through completely at any given observation point. Thereafter the observed concentration will remain constant. However, for observation points far distant from the contaminant source, and especially for strongly sorbing solutes, the soil may continue to remove solute for many years; the time required for breakthrough can be estimated approximately as $\bar{t}_{\text{tracer}}/u_r$, where $\bar{t}_{\text{tracer}}$ is the average hydraulic or tracer residence time.

Non-Uniform Flow

The simplified summary of advection–dispersion–sorption models above has been limited to the case of uniform, one-dimensional flow. In most real situations the flow field deviates markedly from that simplifying assumption. At first glance, such deviations appear to pose a debilitating limitation to the usefulness of simple advection–dispersion models in real situations. However, the consequences turn out to be not at all severe, as shown by Gelhar and Collins (1971) for several general cases of flow fields characterized by variable velocity. Gelhar and Collins (1971) proved that for Pe > 100 the responses (Figure 23.1) to a step stimulus are identical for markedly different flow fields: uniform, one-dimensional; radial-divergent (two-dimensional); and spherical (three-dimensional). Furthermore, Sauty (1978, 1980), in comparing uniform flow with radial divergent flow, showed that the differences are small for Pe > 3.

Fundamental Limitations of Advection–Dispersion Models

The above discussion of the transport of both non-sorbing and sorbing solutes is based on the suitability of a diffusion-type dispersion model [Eq. (23.1)]. Advection–dispersion models of the simple type represented by Eq. (23.1) have been questioned on various grounds (Fried, 1975, 1981; Smith and Schwartz, 1980; Anderson, 1979).

Smith and Schwartz (1980) concluded that modeling techniques based on advection–dispersion models with large dispersivity values do not truly reflect the physical process about which predictions are being made. Moreover, in field studies the dispersivity has sometimes been found to be scale-dependent, contrary to the premise of the basic dispersion model (Anderson, 1979); investigators have been compelled to hypothesize unexpectedly strong scale dependencies of α to simulate field data, especially to account for the markedly extended "tails" of the observed responses. Increases in dispersivity with increasing scale can be rationalized as expressions of an increasing scale of heterogeneities. Nonetheless, the utility of the advection–dispersion model suffers if it proves necessary to resort to widely varying values of the supposedly constant dispersivity.

These limitations notwithstanding, the advection–dispersion models provide a convenient framework for understanding the transport of solutes in ground water. Stochastic approaches, in which the hydrodynamic properties of the aquifer are treated as distributed parameters (Smith and Schwartz, 1980), offer hope of better simulating field observations, but are conceptually and computationally more difficult and require an inordinate amount of field data.

FIELD STUDY

Ground Water Recharge Project

Ground water recharge by direct injection of reclaimed municipal waste water is the object of a field study in the Palo Alto Baylands begun in 1976 (Roberts et al., 1978). The aims of the program are to rehabilitate a saline aquifer, to prevent intrusion of salt water from nearby San Francisco Bay, and to evaluate the long-term potential for augmenting the potable water supply. The facilities were constructed by the Santa Clara Valley Water District and are operated by the City of Palo Alto under a contract with the Water District.

The facility consists of a 0.09 m³/sec (2 million gal/day) water reclamation plant and a well field, approximately 1 by 3 km, for injection and extraction of the reclaimed water. The reclamation facility is an advanced treatment plant in which secondary effluent is upgraded in a process sequence that includes lime treatment, air stripping, recarbonation, ozonation, filtration, granular activated carbon, and chlorination. The well field is conceived as a set of nine injection/extraction well pairs with attendant monitoring wells. The facility's design is described elsewhere (Roberts et al., 1978, 1980).

Research has been conducted in the Palo Alto Baylands by the Stanford Water Quality Control Research Laboratory since 1976 with the goals of understanding water quality transformations that occur when reclaimed water moves through an aquifer following ground water recharge, and of confirming expectations derived from theory and laboratory studies by comparison to field observations. In 1976–1977, a pilot study was conducted by injecting reclaimed water at a rate of 0.6–1 L/sec (10–15 gal/min) into a test well (Roberts et al., 1978).

A full-scale field study was commenced at an injection well designated I1 in July 1978 and continued for a duration of $2\frac{1}{2}$ years. The objectives were to test the methodology developed during the pilot study by verification with field data corresponding to longer times and distances and differing directions from the injection point; to improve understanding of the behavior of pollutants, especially trace organic contaminants, in the ground water environment; to confirm evidence obtained in the pilot study that the movement of trace organic solutes is retarded significantly by sorptive processes; and to evaluate the feasibility of predicting the arrival of pollutants at distant points based on short-term water quality studies at observation points near the injection point and on an understanding of water movement. The results of the first year of observations (July 1978 to July 1979) have been reported by Roberts et al. (1982). This paper summarizes the results of the field work at injection well I1 from July 1978 through December 1980, with emphasis on the behavior of trace organic contaminants.

EXPERIMENTAL APPROACH

Description of Field Site

The injection well designated as I1 is situated at the northwest end of the Palo Alto Baylands ground water recharge field (Roberts et al., 1978). The aquifer is a permeable stratum of silty sand with some gravel, with a thickness of 1.5 to 3 m, and is found at a depth of 10–15 m below the surface. From examination of observation well logs and core samples, values of average aquifer thickness, porosity, bulk density, and organic carbon content were estimated as 2.0 m, 0.25, 1875 kg/m^3, and 0.45%, respectively. The aquifer is reasonably homogeneous and isotropic in the vicinity of well I1, thus affording more nearly ideal hydraulic conditions than at the previous pilot site. The transmissivity of well I1 was determined in a step-drawdown test to be 100 m²/day (9,500 gal/day/ft), which implies a predicted injection rate of approximately 0.01 m³/sec (150 gal/min) at a wellhead pressure of 100 kPa (15 psig). The aquifer is underlain by an impermeable clay layer and overlain by relatively impermeable, interbedded layers of clay and silty sand.

Six observation wells were available for water quality sampling in the vicinity of injection well I1, as shown in Figure 23.3: P5, P6, P7, and S22 at distances of 11, 20, 40, and 160 m along a straight line, and wells S23 and S24 at 16 m and 43 m in two other directions at angles of approximately 120° to the line I1–S22. The observation wells P5, P6, and P7—from which most of the water quality samples in this work were drawn—are 2-cm (¾-in.) threaded polyethylene tubes connected without adhesive, placed in 20-cm (8-in.) diameter holes drilled by a hollow-stem auger without use of drilling fluids. The polyethylene tubes were slotted to accommodate the full depth of the permeable stratum and placed accordingly. During injection operation, the pressures at the sampling wells were for the most part sufficient to guarantee artesian flow; hence pumping was not necessary to obtain water quality samples.

Analytical Methods

Determination of specific organic compounds was made by gas chromatography (GC). Highly volatile halogenated aliphatic compounds containing one and two carbon atoms were determined by a volatile organic analysis (VOA) procedure (Henderson et al., 1976; Trussell et al., 1979), in which the organic solutes are enriched by pentane extraction in sealed hypovials prior to

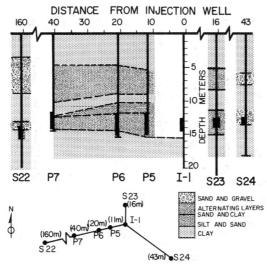

Figure 23.3. Aquifer section and layout of observation wells.

analysis. A 5-μL aliquot of the extract was injected into a packed column gas chromatograph (10% squalane on chromosorb W/AW) equipped with a linearized [63]Ni detector. Chlorobenzene was concentrated by the closed-loop stripping (CLSA) method (Grob and Zürcher, 1976) in preparation for gas chromatography. In the CLSA method, organic solutes are concentrated by circulating air through the sample and thence through an activated-carbon microtrap, which is subsequently extracted with carbon disulfide. The extract is injected splitless onto a glass capillary gas chromatograph equipped with programmed temperature control. Quantitation is achieved by comparison of the flame ionization detector signal with that of the Cl-C$_{12}$ internal standard. The detection limit for the compounds reported herein is approximately 0.1 μg/L using the above methods.

Composition of Injected Water and Formation Ground Water

The approximate composition of the formation ground water near well I1 prior to injection was substantially different from the average composition of reclaimed water during the study period, as shown in Table 23.1. The more than tenfold difference in chloride concentration permitted the use of chloride as a tracer in quantifying the movement of injected water. Concentrations of ammonia and of specific trace organic compounds such as trihalomethanes were below detection limits in the formation ground water, but more than 10 times higher in the injected water. However, the concentrations of total organic carbon (TOC) and chemical oxygen demand (COD) in the reclaimed water were not much higher than those in the formation ground water.

Table 23.1   Composition of Injected Water and Formation Ground Water

	Concentrations	
	Formation Ground Water at I1 Prior to Injection	Injected Water Average for July 1978 to July 1979
COD, mg/L O$_2$	6	10
TOC, mg/L C	2	3
NH$_3$-N, mg/L N	<1	16
Cl$^-$, mg/L Cl	4,000	290
TTHM,[a] μg/L	<0.2	13

[a] TTHM: total trihalomethanes, sum of concentrations measured by gas chromatography.

INTERPRETATION OF FIELD DATA

The methodology used here for analyzing water quality changes occurring during ground water recharge is based on a stimulus–response approach in which the aquifer is considered as a reactor vessel of unknown size, shape, and flow characteristics (Levenspiel and Bischoff, 1963; Roberts et al., 1980). Retardation of solute movement is characterized by comparing the breakthrough for a solute of interest with that for a conservative tracer such as chloride. In this manner, useful quantities can be estimated, such as the rate of transport of a pollutant relative to water and the effective retention capacity of the aquifer with respect to the solute under the conditions of the field experiment. The application of this approach to interpreting solute transport data in connection with artificial recharge of ground water has been described by Roberts et al. (1980).

RESULTS

Breakthrough of Injected Water

The arrival of injected water at the observation wells was estimated from measurements of chloride concentration in the injected water, in the formation ground water, and at the observation wells after injection began. Breakthrough curves for several observation wells are shown in Figure 23.4, based on smoothed values (three-point moving averages) of the chloride concentration response. Comparison of the responses at wells P5, P6, and P7 re-

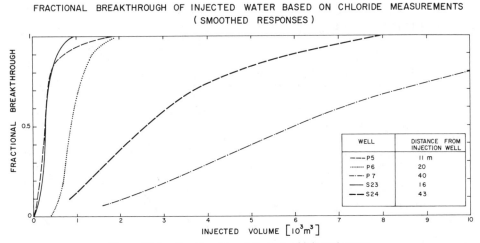

Figure 23.4. Fractional breakthrough of injected water.

veals that a greater volume of water must be injected to achieve breakthrough as the distance from the injection well increases, as is expected. For example, a 50% breakthrough of recharged water is achieved after injection of 300 m^3 at a distance of 11 m (P5), 900 m^3 at 20 m (P6), and 6000 m^3 at 40 m (P7).

The residence time of water in the aquifer before reaching a given observation well was calculated as follows: the volume of pore water displaced was estimated by integrating the area above the chloride breakthrough curve; the pore volume was then divided by the average injection flow rate to estimate the average residence time. The results of these calculations are summarized in Table 23.2. The average injection rate was approximately 0.0063 m^3/sec (100 gal/min) during the breakthrough period. The residence time for travel to S24 was less than to P7, although the two wells are approximately equidistant from the injection well. Similarly, less time was required for recharged water to reach S23 than P5, although P5 is closer to the injection well. This indicates that the aquifer's permeability may be less in the direction P5-P6-P7 than in the directions toward S23 and S24. Approximate residence time contours are shown in Figure 23.5. The contours would be circles under ideal, isotropic flow conditions, but are obviously distended toward S23 and S24. It appears that the flow velocity is approximately half as great in the direction P5-P6-P7 as in the directions S23 and S24. The latter conclusion has been confirmed by Valocchi et al. (1981) using a mathematical model based on the assumption of homogeneous anisotropic flow. This deviation from ideal radial flow is small enough to preclude serious difficulties in interpreting water quality monitoring data.

Evaluation of Longitudinal Dispersivity

The breakthrough responses observed in the field were used to evaluate the Peclet Number and subsequently the longitudinal dispersivity α for the por-

Table 23.2   Hydraulic Parameters

Parameter	Observation Well				
	P5	P6	P7	S23	S24
Distance from injection well, m	11	20	40	16	43
Aquifer pore volume, m^3	400	930	6,300	340	3,100
Average residence time,[a] hr	17	40	270	15	130
Peclet number	2	10	5	4	4
Longitudinal dispersivity, m	5	2	8	4	11

[a] Assuming an average injection rate of 0.0063 m^3/sec.

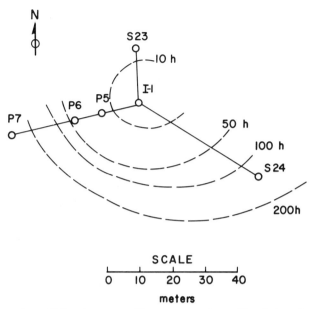

Figure 23.5. Average residence time contours (time in hours).

tions of the aquifer subtended by the various observation wells. The fitting was achieved by visual matching of the observed values with the type curves given by Ogata and Banks (1961). The values of Pe and α so obtained are shown in Table 23.2. The leading portion ($\hat{F} < 0.5$) of the breakthrough response was emphasized where it proved difficult to simulate the entire response with a single value of dispersivity, as with the data from P5 and P6. The values of dispersivity in Table 23.2 range from 2 to 11 m, which is in the same order of magnitude as values reported for two-well tests at similar scale (Anderson, 1979). The apparent variability (a factor of five) of the estimated dispersivities over a relatively small domain (radius of 40 m) is consonant with the notion that dispersion represents spatial variability of permeability (Smith and Schwartz, 1980), and is consistent with previously reported results (Anderson, 1979). The values of the Peclet Number in Table 23.2 exceed 3, except for that of P5; hence, with that exception, deviations from the assumption of uniform flow can be neglected according to the criterion of Sauty (1980).

Transport of Trace Organic Contaminants

Transport of trace organics in the aquifer was studied by observing concentration changes at observation wells following injection. This was possible for a number of halogenated organic compounds whose concentrations in the

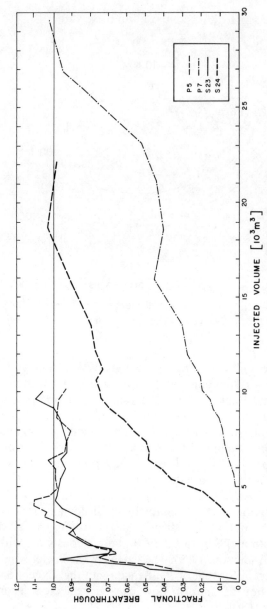

Figure 23.6. Breakthrough of chloroform at several observation wells.

injected water exceed 1 μg/L, compared to background concentrations less than 0.1 μg/L in the formation ground water. Results were expressed as fractional breakthrough of the organic solutes and calculated as the ratio between the observed concentration at the observation well and the average concentration in the injected water. Results were smoothed by calculating three-point running averages to reduce the scatter caused by short-term concentration fluctuations.

Fractional breakthrough data for chloroform at several observation wells are shown in Figure 23.6. The sequential breakthrough at the wells is obviously in accordance with increasing hydraulic residence time. As shown in Table 23.2 and Figure 23.4, the hydraulic residence times are ordered approximately as follows: $t_{P5} \simeq t_{S23} < t_{S24} < t_{P7}$, which corresponds to the order of breakthrough of chloroform (Figure 23.6). Furthermore, it can be seen clearly by comparing the breakthrough curves for injected water (Figure 23.4) with those for chloroform (Figure 23.6) that chloroform appears later than the injected water; its movement is retarded compared to a conservative tracer such as chloride.

Organic compounds differ in the degree to which their movement through the aquifer is retarded, as shown in Figure 23.7. Breakthrough data for chloroform, bromoform, and chlorobenzene at well P5 are compared to that for the injected water. Chloroform appears soon after the injected water, while the appearance of bromoform is retarded to a greater extent and that of chlorobenzene to a still greater degree.

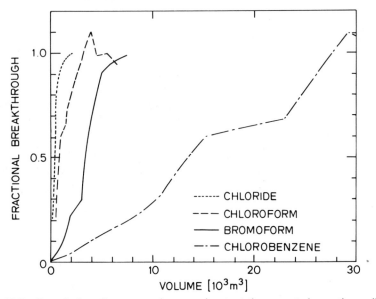

Figure 23.7.   Retardation of trace organic contaminants at the nearest observation well (P5).

DISCUSSION

Retardation Factor

The retardation factors estimated for chloroform at various observation wells (Table 23.3) show good agreement, considering the errors of estimation. The retardation (Table 23.3) of the compounds studied increases in the order

$$CHCl_3 < CHClBr_2 \simeq CHBr_3 < Cl_3CCH_3 < C_6H_5Cl$$

which is in approximate accord with the order of elution of these compounds in gas chromatography. The retardation is attributed to sorption of the compounds on aquifer material and can be related to the hydrophobic nature of the organic solutes. The values of the retardation factor for chloroform and chlorobenzene in the present study agree well with the values previously reported from the pilot study: namely, 5 for chloroform (Roberts, 1979) and 36 for chlorobenzene (Roberts et al., 1981).

Malcolm et al. (1979) have shown that hydrophilic substances such as ethanol and carboxylic acids move through the ground water environment with approximately the velocity of water movement, while the more hydrophobic halogenated compounds are retarded and retained by aquifer material. Karickhoff et al. (1979) have found excellent correlation between the sorption coefficient and the log octanol–water solubility as a measure of

Table 23.3   Retardation Factors and Specific Retention Capacities for Trace Organics

Sample Well	Aquifer Pore Volume[a] (m^3)	Compound	Average Injected Concentration During Breakthrough (mg/m^3)	Retardation Factor t_r	Estimated Specific Retention Capacity $\left(\dfrac{\mu g\ removed}{kg\ aquifer\ material}\right)$
S23	340	$CHCl_3$	7.6	3.5	2.3
P5	400	$CHCl_3$	7.4	3.8	2.5
S24	3400	$CHCl_3$	6.7	2.5	1.5
P7	6300	$CHCl_3$	6.0	3.0	1.5
P5	400	$CHBr_3$	6.6	6	4.0
P5	400	$CHClBr_2$	2.6	6	1.6
P5	400	Cl_3CCH_3	3.3	12	4.4
P5	400	C_6H_5Cl	2.0	33	8.2

[a] Estimated from chloride data.

hydrophobic character. The extent of sorption is approximately proportional to the organic carbon content of the aquifer material, according to Karickhoff et al. (1979). Giger and Molnar-Kubica (1978) have documented the persistence and slow movement in ground water of tetrachloroethylene, a chlorinated hydrocarbon similar in structure to those studied here. Roux and Althoff (1980) observed similar behavior for 1,1,1-trichloroethane at a site where ground water was contaminated by land disposal of industrial wastes.

Prediction of Organic Solute Behavior

The foregoing qualitative analysis suggests that the behavior of trace organic contaminants dealt with in this work may be simulated quantitatively by accounting for advection, sorption, and desorption. In this approach, parameters related to advection and dispersion were evaluated by fitting tracer (chloride) data, and the retardation factor was predicted using the octanol–water partition coefficients for the individual compounds of interest (Hansch and Leo, 1979) in conjunction with Eqs. (23.5) and (23.6).

The values of the advection and dispersion parameters are given in Table 23.2. The same values of dispersivity were used for sorbing solutes as for the chloride tracer, in accordance with Eq. (23.10).

Values of the retardation factor are summarized in Table 23.4 for those organic compounds for which sufficient field data were obtained. The predicted and observed values agree within a factor of 1.5 for five of the six compounds, but for chlorobenzene the observed value exceeds the predicted value by a factor of 2. In view of the difficulties encountered in obtaining sufficiently long time series of data under constant conditions, this order-of-magnitude agreement is considered gratifying, although by no means satis-

Table 23.4   Comparison of Predicted and Observed Retardation Factors

Compound	Octanol–Water Partition Coefficient	Retardation Factor	
		Predicted	Observed
$CHCl_3$	90[a]	3.0	2.5–3.8
$CHBr_2Cl$	300[b]	8	6
Cl_3CCH_3	300[a]	8	12
C_6H_5Cl	690[c]	17	33
$CHBr_3$	417[b]	10	6
$1,2\text{-}C_6H_4Cl_2$	2500[a]	60	~100

[a] Average of values cited by Hansch and Leo (1979).

[b] Calculated using the fragment method (Hansch and Leo, 1979).

[c] Leo et al. (1971).

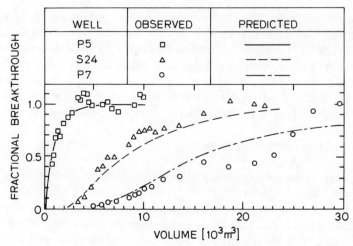

Figure 23.8. Prediction of chloroform response at several observation wells.

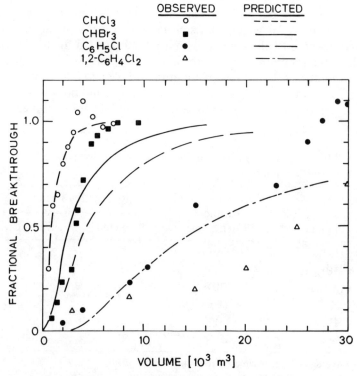

Figure 23.9. Predicted and observed responses of several organic solutes at the nearest observation well (P5).

factory. The breakthrough response is simulated with acceptable accuracy where the predicted and observed values of the retardation factor agree closely, as in the case of chloroform (Figure. 23.8), but not where they disagree markedly, as in the case of chlorobenzene (Figure 23.9). The deviations may be due in part to sorption by soil phases other than organic matter.

For chloroform, the compound for which the data are believed to be most reliable, the interaction between dispersion and sorption is correctly simulated by the model, as demonstrated by the good agreement between observed and predicted values in Figure 23.8. The observed breakthrough responses of some of the organic solutes show a lesser dispersion tendency than predicted (e.g., $CHBr_3$ in Figure 23.9); this deviation probably is due in large part to fluctuations in the respective input concentrations as well as to the use of a moving-average technique to smooth the data.

CONCLUSIONS

There is ample evidence that organic trace contaminants interact with soil solids during artificial recharge. Sorption serves to retard solute transport, whereas both sorption and dispersion attenuate concentration fluctuations.

Advection and dispersion in the subsurface cannot be predicted accurately based on theory or laboratory experiment, but must be measured by means of tracer experiments in the field. Sorption can be predicted with order-of-magnitude accuracy based on knowledge of the octanol–water partitioning of the solute and the organic carbon content of the aquifer material. The combined effect of sorption and dispersion can be simulated satisfactorily if the retardation factor and the dispersivity can be estimated reliably.

Stimulus–response experiments can provide useful information regarding the behavior of organic solutes in the subsurface environment. Such information is valuable in predicting the rate of transport and in identifying solutes that are likely to be transformed under the conditions of the particular environment. Moreover, such knowledge of solute transport rate and dynamic behavior greatly facilitates the rational design of ground water quality monitoring systems.

The hydrophobic interactions of solutes with water appear to have a central significance in determining behavior of those solutes in subsurface transport. As a general rule, hazardous, nondegradable organic chemicals that are only weakly hydrophobic appear to pose the most serious threats of widespread ground water contamination.

ACKNOWLEDGMENTS

This work was supported by the Robert S. Kerr Environmental Research Laboratory of the U.S. Environmental Protection Agency under Grant No.

R-804431. Additional funding was provided by the State Water Resources Control Board and the Department of Water Resources of the State of California. The Santa Clara Valley Water District generously made available their ground water recharge facility. Richard Harnish and Joan Schreiner provided analytical data.

REFERENCES

Anderson, M. P. (1979) Using models to simulate the movement of contaminants through groundwater flow systems. *Crit. Rev. Environ. Control* **9**(2):97–156.

Bear, J. (1979). *Hydraulics of Groundwater*. McGraw-Hill, New York.

Environmental Protection Agency (1980). Proposed ground water protection strategy. Office of Drinking Water, U.S. Environmental Protection Agency, Washington, D.C.

Freeze, R. A. and Cherry, J. A. (1979). *Groundwater*. Prentice-Hall, Englewood Cliffs, NJ.

Fried, J. J. (1975). *Groundwater Pollution*. Elsevier, Amsterdam.

Fried, J. J. (1981). Groundwater pollution mathematical modeling: Improvement or stagnation? In: W. van Duijvenboden, P. Glasbergen, and H. van Lelyveld (Ed.), *Quality of Groundwater,* Elsevier Scientific Publishing Co., Amsterdam, pp. 807–822.

Gelhar, L. W. and Collins, M. A. (1971). General analysis of longitudinal dispersion in nonuniform flow. *Water Resour. Res.* **7**:1511–1521.

Giger, W. and Molnar-Kubica, E. (1978). Tetrachloroethylene in contaminated ground and drinking waters. *Bull. Environ. Contamination Toxicol.* **19**:475–480.

Grob, K. and Zürcher, F. (1976). Stripping of organic substances from water. *J. Chromatogr.* **117**:285–294.

Hansch, C. and Leo, A. (1979). *Substituent Constants for Correlation Analysis in Chemistry and Biology*. Wiley, New York.

Henderson, J. E., Peyton, G. R., and Glaze, W. H. (1976). A convenient liquid–liquid extraction method for the determination of halomethanes in water at the part per billion level. In: L. H. Keith (Ed.), *Identification and Analysis of Organic Pollutants in Water*. Ann Arbor Science, Ann Arbor, MI, pp. 105–111.

Karickhoff, S. W., Brown, D. S., and Scott, T. A. (1979). Sorption of hydrophobic pollutants on natural sediments. *Water Res.* **13**:241–248.

Leo, A., Hansch, C., and Elkins, D. (1971). Partition coefficients and their uses. *Chem. Rev.* **71**:575–616.

Levenspiel, O. and Bischoff, K. B. (1963). Patterns of flow in chemical process vessels. In: T. B. Drew, J. W. Hoopes, Jr., and T. Vermuelen (Ed.), *Advances in Chemical Engineering,* Vol. 4. Academic Press, New York, pp. 95–198.

Malcolm, R. L., Thurman, E. M., Aiken, G. R., and Avery, P. A. (1979). Hydrophilic organic solutes as tracers in groundwater recharge studies. 177th National Meeting of the American Chemical Society, Honolulu, Hawaii.

McCarty, P. L., Reinhard, M., and Rittmann, B. E. (1981). Trace organics in ground-water. *Environ. Sci. Technol.* **15**:40–47.

Ogata, A. and Banks, R. B. (1961). A solution of the differential equation of longitu-dinal dispersion in porous media. U.S. Geological Survey Prof. Paper 411-A, Washington, D.C., p. A-4.

Ogata, A. (1964). Mathematics of dispersion with linear adsorption isotherm. U.S. Geological Survey Prof. Paper 411-H, Washington, D.C.

Roberts, P. V. (1979). Removal of trace organic contaminants from reclaimed water during aquifer passage. In: Kühn and H. Sontheimer (Ed.), *Oxidation Tech-niques in Drinking Water Treatment,* U.S. Environmental Protection Agency, Washington, D.C., EPA-570/9-79-020, pp. 647–672.

Roberts, P. V., Hopkins, G. C., and Schreiner, J. (1982). Field study of organic water quality changes during groundwater recharge. *Water Res.* **16**:1025–1035.

Roberts, P. V., McCarty, P. L., Reinhard, M., and Schreiner, J. (1980). Organic contaminant behavior during groundwater recharge. *J. Water Pollut. Control Fed.* **52**(1):161–172.

Roberts, P. V., McCarty, P. L., and Roman, W. M. (1978). Direct injection of reclaimed water into an aquifer. *J. Environ. Eng. Div., ASCE* **104**(EE5):933–949.

Roux, P. H. and Althoff, W. F. (1980). Investigation of organic contamination of ground water in S. Brunswick Township, New Jersey. *Ground Water* **18**:464–471.

Sauty, J. P. (1980). An analysis of hydrodispersive transfer in aquifers. *Water Re-sour. Res.* **16**(1):145–158.

Sauty, J. P. (1978). Identification des paramètres du transport hydrodispersif dans les aquifères. *J. Hydrol.* **39**:69–103.

Smith, L. and Schwartz, F. W. (1980). Mass transport. I. A stochastic analysis of macroscopic dispersion. *Water Resour. Res.* **16**(2):303–313.

Trussell, A. R., Umphres, M. D., Leong, L. Y. C., and Trussell, R. R. (1979). Precise analysis of trihalomethanes. *J. Am. Water Works Assoc.* **71**(7):385–389.

Valocchi, A. J., Roberts, P. V., Parks, G. A., and Street, R. L. (1981). Simulation of the transport of ion-exchanging solutes using laboratory-determined chemical parameter values. *Ground Water* **19**:600–607.

24

BEHAVIOR AND FATE OF HALOGENATED HYDROCARBONS IN GROUND WATER

René P. Schwarzenbach
Walter Giger

*Swiss Federal Institute for Water Resources
and Water Pollution Control (EAWAG)
Dübendorf, Switzerland*

Halogenated hydrocarbons (halogenated alkanes, olefins, benzenes, etc.) are among the most ubiquitous pollutants in natural waters. Numerous cases of ground water contamination by such chemicals have been reported (e.g., Giger et al., 1978; Zoeteman et al., 1980; Nelson et al., 1981). Since many halogenated hydrocarbons are of great concern with respect to human health (Jolley et al., 1978), it is necessary to extend our knowledge about the processes which determine the movement and fate of such compounds in the ground.

Recently, several studies dealing with the behavior of organic pollutants (including halogenated hydrocarbons) during infiltration of surface waters to ground waters and during artificial ground water recharge have been conducted (Piet and Zoeteman, 1980; Roberts et al., 1980; Sontheimer, 1980; Bouwer et al., 1981a; Schneider et al., 1981; Schwarzenbach et al., 1981). The results of these studies clearly demonstrate the great "pollution potential" of various halogenated hydrocarbons. Some of the individual compounds detected hitherto in Swiss ground waters are listed in Table 24.1. The major causes for these ground water contaminations include spills, infiltra-

Table 24.1 Examples of Halogenated Hydrocarbons Detected in Swiss Ground Waters

Compound	Solubility in Water at 20°C (mg/L)	Log Octanol–Water Partition Coefficient (log K_{ow})	Range of Concentrations (μg/L)	Remarks (source of contamination, etc.)
Chloroform	8200[a]	1.97[b]	0.1–1	?
1,1,1-Trichloroethane	4400[a] 480[e]	2.17[c]	0.02–5	Contaminated aquifer
1-Bromo-3-chloropropane	?	2.21[d]	<0.1	} Leakage from a
1,3-Dibromopropane	?	2.42[d]	<0.1	} chemical plant
Hexachloroethane	50 (22°C)[a]	3.34[c]	15–21	Leachate from a dump site of chemical wastes
Trichloroethylene	1100[e]	2.29[b]	0.02–2	River water infiltrate; contaminated aquifer
Tetrachloroethylene	150 (25°C)[a]	2.88[f]	0.1–150,000	River water infiltrate; contaminated aquifer; spills
Hexachlorobutadiene	2[e]	3.74[c]	0.2–0.3	Leachate from a dump site of chemical wastes
Chlorobenzene	500[a]	2.71[b]	0.1–16	} Leakage from a
Bromobenzene	410[f]	2.99[b]	0.1–8	} chemical plant
1,4-Dichlorobenzene	79 (25°C)[a]	3.38[b]	0.01–0.3	}
1,2,4-Trichlorobenzene	30[g]	4.05[d]	<0.02	}
1,2,3-Trichlorobenzene	?	4.05[d]	<0.02	} River water infiltrate
Hexachlorobenzene	?	6.06[d]	0.001[h]	}

[a] Verschueren (1977).

[b] Hansch and Leo (1979).

[c] Calculated from Tute (1971).

[d] Calculated from Leo et al. (1971).

[e] Pearson and McConnell (1975).

[f] Mackay et al. (1980).

[g] Dow (1978).

[h] Determined by M. Müller, Swiss Federal Research Station for Arboriculture, Viticulture, and Horticulture, 8820-Wädenswil, Switzerland.

tion of polluted surface waters, leachates from dump sites, and leakages from chemical plants.

In this chapter we summarize the results of various laboratory and field studies that have been conducted to elucidate the behavior and fate of halogenated hydrocarbons in ground water and during ground water infiltration. Emphasis will be placed on the sorption behavior (and thus the mobility) of such compounds in aquifers. In addition, chemical and biological transformations of individual halogenated hydrocarbons will be discussed.

EXPERIMENTAL METHODS

Analytical Procedure for the Determination of Individual Halogenated Hydrocarbons

Except for the halomethanes the compounds were concentrated from the water samples by the closed-loop gaseous stripping/adsorption/elution procedure developed by Grob (Grob, 1973; Grob and Zürcher, 1976). The ground water samples (typically 1 L) and the samples from the laboratory studies (typically 0.1 L) were stripped for 90 and 60 min, respectively, at 30°C, and the organic compounds were trapped by adsorption on a filter of 1.5 mg activated charcoal. The filter was then extracted with a total of 20 μL carbon disulfide, and the extract analyzed by high-resolution glass capillary gas chromatography and gas chromatography/mass spectrometry. The gas chromatographic equipment and parameters used have been described elsewhere (Schwarzenbach et al., 1979). The concentration of a compound was calculated by comparing its peak height, as detected by an electronic integrator, with the peak height of a closely eluting internal standard. With this method a relative standard deviation of better than 10% can be obtained for all compounds investigated (Giger et al., 1978; Schwarzenbach et al., 1979).

Halomethanes in ground water samples (typically 0.06 L) were determined by using a micropentane extraction procedure with subsequent gas chromatographic analysis using a thick film SE-54 glass capillary column and an electron capture detector.

Total Purgeable Organo-chlorine Compounds

The total purgeable organo-chlorine compounds (POCl) were determined as chlorine equivalents by the method developed by Zürcher (Zürcher, 1981). The halogenated organic compounds were purged from the water samples (typically 1 L) with oxygen for 30 min at 60°C. The purged compounds were continuously combusted at 950°C, and the produced halides (Cl^-, Br^-, F^-) were trapped and quantified by ion chromatography. The detection limit of this method for organo-chlorine compounds is 0.1 μg Cl^-/L.

Table 24.2 Correlation of the Sorbent–Water Partition Coefficient K_p^z(s) with the Octanol–Water Partition Coefficient K_{ow}^z for Different Sorbents

Sorbent		BET Specific Surface Area (m^2/g_s)	f_{oc} (Fraction Org. Carbon) (g_{oc}/g_s)	Correlation[e] log K_p^z(s) = a log K_{ow}^z + b		
Number	Description			a	b	R^2(n = 10)
1	River sediment A[a]	2.4	0.018	0.73	−1.57	0.97
2	River sediment B[a]	1.2	0.0056	0.69	−1.66	0.95
3	Aquifer material C[a]	4.4	0.0073	0.70	−1.76	0.97
4	Aquifer material[b]	3.2	0.0023	0.67	−1.91	0.98
5	Aquifer material D[c]	4.9[d]	0.0015	0.71	−2.31	0.97
6	Aquifer material E[a]	3.2	0.0008	0.57	−1.96	0.97
7	Aquifer material F[a]	2.1	0.0006	0.55	−1.73	0.96
8	Aquifer material G[a]	2.6	0.0004	0.50	−2.05	0.97
9	Sediment from a eutrophic lake	18	0.019	0.67	−0.86	0.98
10	Sediment from a highly eutrophic lake	n.d.[h]	0.058	0.87	−1.40	0.98
11	Kaolin	12	0.0006	0.43	−1.37	0.96
12	γ-Al_2O_3	120	<0.0001	0.25	−0.83	0.95
13	SiO_2	500	<0.0001	0.12	+0.40	0.29
14	Activated sewage sludge[f]	n.d.[h]	0.33	0.67	+0.40	0.99[g]

[a] Sample from field site in the Glatt Valley (Figure 24.3), prepared by dry sieving; ϕ < 125 μm.

[b] Sample from field site at River Aare (see text); prepared by dry sieving; ϕ < 125 μm.

[c] Sample from field site in the Glatt Valley (Figure 24.3), prepared by wet sieving; 63 μm < ϕ < 125 μm.

[d] Sample degassed at 200°C, value probably too high.

[e] Correlation made for compounds with 2.6 < log K_{ow}^z < 4.7.

[f] Data taken from Matter (1979).

[g] n = 6.

[h] n.d. = not determined.

449

Laboratory Sorption Studies

The sorption of 14 non-polar organic compounds—including tetrachloroethylene and a series of alkylated and chlorinated benzenes—by river sediments, aquifer materials, lake sediments, clay, and pure oxides has been studied in batch and column experiments. The sorbents investigated are summarized in Table 24.2. Sorbent 5 was used in column and batch experiments.

Surface areas of the sorbents were determined on a Carlo Erba Model 1800 Sorptomatic using nitrogen as adsorbing gas (0.16 nm^2/N_2). The samples were outgassed for 48 hr at 50°C and 0.005 ± 0.002 Torr. Adsorption and desorption isotherms were determined at 77 K. With the exception of the SiO_2 sample (sorbent 13, Table 24.2), the measured pore volumes were always smaller than 0.02 cm^3/g_s, indicating that all materials were very nonporous.

The organic carbon content (f_{oc}) of the sorbents was determined by the method of Baccini (1982). Approximately 100 mg of the sample was suspended in 10 mL of $0.5M$ hydrochloric acid. After inorganic CO_2 was removed, a few microliters of the suspension were oxidized with MnO_2 at 850°C. The CO_2 produced from the oxidation of the organic matter was reduced to methane and quantified by a flame ionization detector. The detection limit of this method for sediment samples is 0.1 mg OC/g_s.

The batch experiments were carried out in 12 mL ground glass stoppered test tubes (10 cm $\times$ 1.2 cm I.D.). The sorbent/water ratios were typically between $1:3$ and $1:10$ (g dry weight of sorbent/g water). The initial concentrations of the spiked compounds were in the range between 20 and 100 $\mu g/L$. For equilibration the test tubes were inverted automatically every 2 min. After 18 hr the solids were allowed to settle, and the concentrations of the compounds in the aqueous phase were determined by the procedure described above. The concentrations on the solid phase and the mass-specific partition coefficients ($K_p^z(s)$, concentration on solid phase/concentration in liquid phase, cm_l^3/g_s) were then calculated from mass balances.

In the column experiments the breakthrough behavior of various compounds was studied at two different flow rates (0.001 cm/sec and 0.01 cm/sec). For these experiments a 29 cm $\times$ 1.2 cm I.D. glass column was wet-packed with very fine sand (sorbent 5, Table 24.2). An HPLC pump (Waters Associates Inc., Model M-6000 A) was used to obtain low flow rates.

For a more detailed description of the laboratory studies, the reader is referred to Schwarzenbach and Westall (1981).

RELATIONSHIPS BETWEEN SORBENT–WATER PARTITION COEFFICIENTS AND ORGANIC SOLVENT–WATER PARTITION COEFFICIENTS

The results of various investigations (e.g., Chiou et al., 1979; Karickhoff et al., 1979; Means et al., 1980; Schwarzenbach and Westall, 1981) indicate that for concentrations typically encountered in natural waters the sorption of non-polar organic compounds by natural sorbents is reversible and that a linear sorption isotherm is appropriate to describe sorption equilibrium:

$$S(z) = K_p^z(s)\, C(z) \tag{24.1}$$

where $C(z)$ = concentration of compound z in liquid phase ($\mu g/cm_l^3$)
$S(z)$ = concentration of z on solid phase ($\mu g/g_s$)
$K_p^z(s)$ = equilibrium partition coefficient (cm_l^3/g_s).

The equilibrium distribution of a non-polar solute between an aqueous phase and a natural sorbent can be likened to the distribution of the solute between an aqueous phase and a non-aqueous phase. In a review of the equilibrium partitioning of non-polar solutes between aqueous and non-aqueous phases, Leo et al. (1971) have discussed three topics related to the work presented in this chapter.

1. Linear free energy relationships can be used to relate the partition coefficients of a series of non-polar solutes in different aqueous/non-aqueous solvent systems:

$$\log K_{SW}^Z = a \log K_{RW}^Z + b \tag{24.2}$$

where K_{SW}^Z = the partition coefficient of a solute Z between a solvent S and water
K_{RW}^Z = the partition coefficient of a solute Z between a reference solvent R and water.

n-Octanol is commonly used as the reference solvent. Large amounts of data exist for the octanol–water system, and empirical rules have been developed for the calculation of octanol–water partition coefficients (Leo et al., 1971; Rekker, 1977; Hansch and Leo, 1979).
    Derivation of Eq. (24.2) (compare Leo et al., 1971) shows that the slope a of the correlation line is given by

$$a = \frac{\Delta G_{WS}^X}{\Delta G_{WR}^X} \tag{24.3}$$

where ΔG_{WS}^X (ΔG_{WR}^X) is the change in ΔG_{WS}^Z (ΔG_{WR}^Z) per unit change in lipophilicity of the solute Z, and ΔG_{WS}^Z (ΔG_{WR}^Z) is the free energy of transfer of the solute Z from water to the non-aqueous phase. If the particular sample solvent is more lipophilic than the reference solvent (e.g., n-octanol), the slope is greater than one; if the sample solvent is less lipophilic than the reference solvent, the slope is less than one. The value of the slope a is independent of the units in which K_{SW}^Z is expressed. The value of the intercept b of the linear free energy relationship [Eq. (24.2)], however, depends on the units in which K is expressed. It is difficult to arrive at a universally acceptable set of units for the phase S, since S may include natural sorbents as well as pure liquid solvents. Thus a general interpretation of b is difficult.

2.   The partition coefficients favorable to the non-aqueous phase are primarily the result of the large positive entropy change that occurs upon dehydration of the solute molecules. In the aqueous phase, the non-polar solute is surrounded by an envelope of highly structured water molecules. As the solute is transferred from the aqueous phase to the non-aqueous phase, the structural envelope of water molecules breaks down and the entropy of the system increases. The enthalpy change accompanying the aqueous to non-aqueous phase transfer is only slightly negative or even positive (compare also Chiou et al., 1979). In the case of chlorinated benzenes, for example, each chlorine substituent requires additional structure in the envelope of water surrounding the non-polar molecule in the aqueous phase. The breakdown of this additional structure upon transfer of the non-polar molecule from the aqueous phase to the non-aqueous phase results in an additional increase in the entropy of the system. Thus the partition coefficients of chlorinated benzenes increase with increasing number of chlorine substituents.

3.   The linear free energy relationships have been shown to apply not only to aqueous/non-aqueous solvent systems but also to water-dispersed protein sorption systems.

Recently, linear free energy relationships have been applied to partitioning of non-polar solutes between water and organic matter from the natural environment (e.g., Karickhoff et al., 1979; Rao and Davidson, 1980; Karickhoff, 1981; Schwarzenbach and Westall, 1981).

In summary, Leo et al. (1971) have shown that linear free energy relationships are useful for qualitative interpretation of experimental data and for the representation of an extensive array of data by way of a few adjustable parameters. Linear free energy relationships allow us to estimate partition coefficients and to assess the lipophilicity of a solvent or sorbent. However, one should be aware that linear free energy relationships are not strict thermodynamic functions.

RESULTS AND DISCUSSION

Laboratory Sorption Studies

The most important results of the laboratory sorption studies are summarized and discussed below. For a more detailed discussion of the various experiments the reader is referred to Schwarzenbach and Westall (1981).

For non-polar organic compounds (e.g., halogenated hydrocarbons) of low to intermediate lipophilicity, the following was found to be true.

1.   Over 85% of the sorption by aquifer materials takes place at the size fraction $\phi < 125$ μm. Thus, this fraction can be taken as representative of the whole sample for sorptive properties.

2.   For concentrations typically encountered in natural waters, sorption equilibrium can be described by a linear sorption isotherm [Eq. (24.1)].

3.   The partition coefficient $K_p^z(s)$ [Eq. (24.1)] of a given compound Z between a natural sorbent S and water increases with increasing organic carbon content of the sorbent. As shown in Figure 24.1 for a series of chlorinated benzenes, a highly significant correlation was found between $K_p^z(s)$ and organic carbon content $f_{oc}(s)[g_{oc}/g_s]$ of the sorbent, when $f_{oc}(s)$ was greater than 0.001. The correlation between $K_p^z(s)$ and $f_{oc}(s)$ confirms the findings of others (e.g., Lambert, 1968; Chiou et al., 1979; Karickhoff et al., 1979; Khan et al., 1979) that the organic material in natural sorbents is primarily responsible for sorption and suggests that the lipophilicity of the natural organic sorbents investigated in this study is quite similar. Thus, for sorbents with $f_{oc} \geq 0.001$, the partition coefficient of a particular compound Z can be

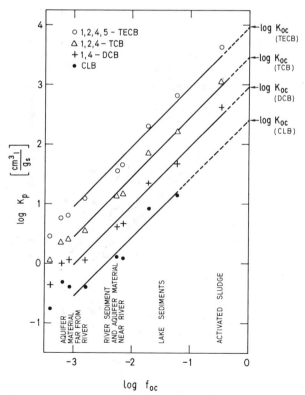

Figure 24.1. $K_p^z(s)$ values for four chlorinated benzenes as a function of the organic carbon content of the sorbents (from Schwarzenbach and Westall, 1981).

expressed in terms of the organic carbon content of the sorbent (Lambert, 1968):

$$K_{oc}^z(s) = \frac{K_p^z(s)}{f_{oc}(s)} \ (cm_l^3/g_{oc}) \tag{24.4}$$

$K_{oc}^z(s)$ is the partition coefficient of Z between water and a hypothetical natural sorbent of 100% organic carbon, which represents the organic material present in the sorbents investigated.

4. For sorbents containing less than 0.1% organic carbon ($f_{oc} < 0.001$), relatively small partition coefficients were found (Table 24.3). In the absence of organic carbon, the specific surface area and the nature of the mineral surface have a greater impact on the degree of sorption. The K_p values for chlorobenzene, for example, are about the same for organic-poor aquifer materials and for γ-Al_2O_3 (see Table 24.3), although the γ-Al_2O_3 used contains virtually no organic carbon. The higher K_p value found for silica is probably due to the much higher specific surface area of this sorbent. For the sorption of more hydrophobic compounds, the organic carbon present in the organic-poor sorbents becomes increasingly important. For example, the partition coefficient for tetrachlorobenzene calculated per square meter of surface area is about 100 times higher for the aquifer material E (Table 24.2) than for Al_2O_3 or SiO_2, whereas that of chlorobenzene is only about 20 times greater for the same materials (Table 24.3). The K_p values found for kaolin are quite similar to those obtained for the organic-poor aquifer material E. In this case the somewhat lower organic carbon content of kaolin is probably balanced by its higher specific surface area.

5. For all sorbents except silica, a highly significant linear correlation was found between the logarithms of the partition coefficients $K_p^z(s)$ of the differ-

Table 24.3 Partition Coefficients of Some Chlorinated Benzenes with Sorbents of Low Organic Carbon Content

			$K_p^z(s)$ (cm_l^3/g_s)			
Sample	BET (m^2/g_S)	f_{oc} (g_{oc}/g_S)	Chloro-benzene	1,4-Di-chloro-benzene	1,2,4-Tri-chloro-benzene	1,2,4,5-Tetra-chloro-benzene
Aquifer material C	4.4	0.0073	1.2	4.4	14.5	37.9
Aquifer material E	3.2	0.0008	0.4	1.1	2.5	6.2
Kaolin	12	0.0006	0.6	1.1	2.4	4.9
γ-Al_2O_3	120	<0.0001	0.6	0.9	1.5	2.2
SiO_2	500	<0.0001	4.2	6.0	7.6	12.1

ent compounds and the logarithms of the corresponding octanol–water partition coefficients K_{ow}^z (Table 24.2). The poor correlation found for silica reflects large differences in the K_p^z values for different isomers. It is suspected that micropores in the surface of the porous silica are the major cause of this effect. It is interesting to compare the slopes a of the correlation lines obtained for the different sorbents (cf. Table 24.2). With the exception of the sediment from an extremely eutrophic lake (Sorbent 10), all sorbents with f_{oc} greater than 0.001 g_{oc}/g_S had slopes of 0.70 ± 0.03, indicating a very similar lipophilicity among these sorbents. This result is consistent with the good linear correlation found between the $K_p^z(s)$ values and the organic carbon contents of the various sorbents (Figure 24.1). The somewhat higher slope obtained for Sorbent 10 can probably be explained by the unusually high lipid content of this sediment (Giger et al., 1980). The smallest slopes were found for the pure oxides. This result is not surprising since, as discussed earlier, the gain in entropy of the system should be small when transferring a lipophilic group from water to a very water-like surface. For the organic-poor sorbents slope values between those obtained for the pure mineral surfaces and those obtained for the organic-rich sorbents can be expected and have been found. Here the observed slope is the net result of (ab)sorption by the organic material and (ad)sorption at the mineral surface.

6.   A highly significant linear correlation was found between the logarithms of the average K_{oc}^z values [Eq. (24.4)] and the logarithms of the K_{ow}^z values of the compounds investigated:

$$\log K_{oc}^z = 0.72 \log K_{ow}^z + 0.5 \ (R^2 = 0.95) \qquad (24.5)$$

Although this linear free energy relationship [Eq. (24.5)] is valid in principle only for the solutes and sorbents used in this study, it is worthwhile to compare this correlation with other similar correlations found in the literature and with experimental results reported for other natural organic sorbents and non-polar solutes. Table 24.4 shows that both Eq. (24.5) and the equation derived by Karickhoff et al. (1979) are consistent with experimental data for a wide range of non-polar compounds. Deviations from both equations are noted for compounds of high lipophilicity. It can be shown that the slope and intercept values of a linear free-energy relationship such as Eq. (24.5) are determined primarily by the type of compounds (i.e., compound class(es), range of hydrophobicity) based on which the relationship is established, and only to a much smaller degree by the type of natural sorbents used. Thus, such relationships with empirically derived parameters are very useful for predicting partition coefficients of a great number of non-polar organic compounds between water and natural sorbents of very different origins.

7.   Partition coefficient values determined from column experiments run at flow velocities $u < 0.001$ cm/sec were similar to those determined from 18 hr equilibrium batch experiments. Column experiments run at $u \sim 0.01$ cm/sec

Table 24.4 Sorbent–Water Partition Coefficients Expressed in Terms of the Organic Carbon Content of the Sorbent: Comparison Between Predicted and Reported K_{oc}^z Values for Compounds and/or Natural Sorbents Not Investigated in This Study

Compound	$\log K_{ow}^z$	Reported (experimental) $\log K_{oc}^z$	Calculated $\log K_{oc}^z$ [Equation (24.5)]	Calculated $\log K_{oc}^z$ ($\log K_{oc}^z$ = 1.00 $\log K_{ow}^z$ − 0.21)[a]
Acetophenone	1.59[b]	1.63[b,c]	1.63	1.42
Benzene	2.11[a]	1.92[a]	2.01	1.90
		1.98[d]		
Chlorobenzene	2.71	2.59[e]	2.44	2.50
Tetrachloroethylene	2.88	2.32[f]	2.57	2.67
Parathion	3.80[g]	3.06[f]	3.23	3.59
β-BHC	3.80[h]	3.46[f]	3.23	3.59
Pyrene	5.09[i]	4.80[i]	4.17	4.88
7,12-Dimethylbenzanthracene	5.98[i]	5.73[i]	4.81	5.77
DDT	6.19[g]	5.14[f]	4.95	5.98
1,2,5,6-Dibenzanthracene	6.50[i]	6.31[i]	5.18	6.29
2,4,5,2',4',5'-PCB	6.72[g]	5.34[f]	5.33	6.51

[a] Karickhoff et al. (1979).
[b] Khan et al. (1979).
[c] Average value (n = 14).
[d] Rogers et al. (1980).
[e] Roberts et al. (1980).
[f] Chiou et al. (1979).
[g] Chiou et al. (1977).
[h] Karickhoff (1981).
[i] Means et al. (1980).

showed the effect of slow sorption kinetics. Comparison of Figure 24.2B with Figure 24.2A shows, qualitatively, that a simple first-order kinetic sorption model is insufficient to describe the observed effect of flow velocity on the elution curve. An excellent paper on this topic has been published by van Genuchten et al. (1974).

From the results of this laboratory study, an equation can be derived for estimation of equilibrium partition coefficients $K_p^z(s)$ for a wide range of

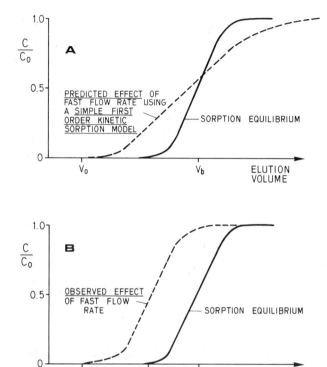

Figure 24.2. Column experiments: effect of fast flow rates on breakthrough curves. (A) Predicted effect by using a simple first-order kinetic sorption model. (B) Observed effect in experiments.

organic compounds and natural sorbents:

$$\log K_p^z(s) = 0.72 \log K_{ow}^z + \log f_{oc}(s) + 0.5 \qquad (24.6)$$

For the range of compounds investigated ($2.6 < \log K_{ow}^z < 4.7$), prediction of $K_p^z(s)$ values within a factor of 2 seems possible, even for sorbents derived from environments very different from the ones investigated (cf. Table 24.5).

Retention Behavior of Halogenated Hydrocarbons in Aquifers

To get a rough quantitative estimate of the retention behavior of halogenated hydrocarbons in aquifers, one can approximate transport in the ground as a one-dimensional process at constant flow in a homogenous porous medium. If the organic carbon content of the size fraction $\phi < 125$ μm of the aquifer material is known, retardation factors (Freeze and Cherry, 1979) can be

Table 24.5   Comparison Between Predicted and Experimentally Determined $K_p^z(s)$ Values of a Series of Compounds for Three Natural Sorbents from Different Environments[a]

Compound	log K_{ow}^z	Conanicut Point Sediment[b] ($f_{oc} = 0.023$) $K_p^z(s)$ [cm$_l^3$/gs] Calc.[e]	Exp.	Fields Point Sediment[c] ($f_{oc} = 0.064$) $K_p^z(s)$ [cm$_l^3$/gs] Calc.[e]	Exp.	Merl Tank Detritus[d] ($f_{oc} = 0.060$) $K_p^z(s)$ [cm$_l^3$/gs] Calc.[e]	Exp.
Chlorobenzene	2.71	6.5	4.6	18	26	17	13
Tetrachloroethylene	2.88	8.5	5.2	24	39	22	28
1,4-Dimethylbenzene	3.15	13	8	38	50	36	29
1,4-Dichlorobenzene	3.38	19	13	55	80	51	43
1,3,5-Trimethylbenzene	3.60	28	16	80	93	74	54
1,2,4,5-Tetramethylbenzene	4.05	60	34	170	216	159	118
1,2,4-Trichlorobenzene	4.05	60	46	170	263	159	154
n-Butylbenzene	4.13	68	47	191	310	178	178
1,2,4,5-Tetrachlorobenzene	4.72	182	148	513	794	478	380

[a] Air-dried samples were received from S. G. Wakeham, Woods Hole Oceanographic Inst., Woods Hole, Mass. 02543, USA.

[b] Sediment from Narragansett Bay (Rhode Island, USA) collected near Jamestown Bridge.

[c] Anoxic sediment from Narragansett Bay (Rhode Island, USA) collected near Providence.

[d] Detritus collected at the bottom of a tank used to simulate an estuary ecosystem (S. G. Wakeham, private communication).

[e] Equation (24.6).

calculated for any halogenated hydrocarbon, provided that its K_{ow} value is available:

$$\text{Retardation factor} = 1 + \frac{\rho(1 - \varepsilon)}{\varepsilon} K_p^{z*}(s) \qquad (24.7)$$

where ρ = density of the aquifer material (g/cm^3)

ε = total porosity

$K_p^{z*}(s)$ = $K_p^z(s)$ × fraction of aquifer material within $\phi < 125$ μm range (Assumption: homogeneous distribution of the size fraction $\phi < 125$ μm.)

$K_p^z(s)$ can be calculated from Eq. (24.6).

Table 24.6 contains such calculated retardation factors of some chlorinated hydrocarbons for typical Swiss aquifers and for a river sediment. The values in Table 24.6 predict that many halogenated hydrocarbons are very mobile in the ground. It should be remembered that at high flow velocities due to slow sorption kinetics, the compounds may be transported even faster than one would assume from equilibrium considerations.

Behavior During Natural Infiltration of River Water into Ground Water

The behavior of organic pollutants during natural infiltration of river water into ground water has been investigated in various field studies (Schneider et al., 1981; Schwarzenbach et al., 1981). Figure 24.3 shows the layout of one of the field study sites situated in the lower Glatt Valley (Switzerland). At this site the River Glatt infiltrates with an average rate of 0.25 m³ water per m² river bed per day. The infiltrated water stratifies in the top layer of the ground water stream and remains there over quite long distances (>100 m). Close to the river the average flow velocity of the water is estimated to be greater than 0.01 cm/sec.

Over the course of one year, water samples from the River Glatt and from the top layer of the ground water at different distances from the river were analyzed for halogenated hydrocarbons. Tetrachloroethylene and 1,4-dichlorobenzene were always among the most abundant compounds detected in the river water samples. During infiltration these two compounds should be rapidly transported from the river to the ground water and even more rapidly in the aquifer itself (compare retardation factors in Table 24.6). Thus at this

Table 24.6 Estimated Retardation Factors of Some Chlorinated Hydrocarbons for Different Aquifers

Compound	Retardation Factors[a]		
	River Glatt Sediment[b]	Aquifer Glattal[c]	Aquifer Close to River Aare[d]
Chloroform	4	1.2	1.5
Trichloroethylene	7	1.3	1.9
Chlorobenzene	13	1.6	2.8
Tetrachloroethylene	17	1.7	3.3
Hexachloroethane	35	2.6	6
1,4-Dichlorobenzene	35	2.6	6
Hexachlorobutadiene	66	4	11
1,2,4-Trichlorobenzene	110	6	17
1,2,4,5-Tetrachlorobenzene	333	17	50

[a] Equation (24.7), $\rho = 2.5$ g/cm³, $\varepsilon = 0.20$. Sorption equilibrium.
[b] $f_{oc} = 0.01$, size fraction $\phi < 125$ μm (evenly distributed) = 42%.
[c] $f_{oc} = 0.001$, size fraction $\phi < 125$ μm (evenly distributed) = 20%.
[d] $f_{oc} = 0.0025$, size fraction $\phi < 125$ μm (evenly distributed) = 25%.

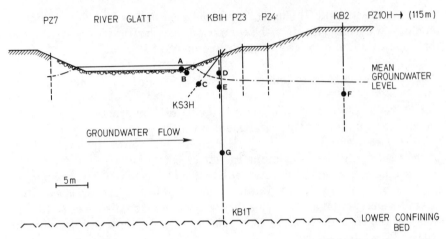

Figure 24.3.   Layout of the field study site in the lower Glatt Valley, Switzerland. Observation wells and locations (A–G) at which soil samples for laboratory sorption studies (see Table 24.2) were collected.

field site, the average concentrations of tetrachloroethylene and 1,4-dichlorobenzene in the top layer of the ground water would be expected to be very similar to those in the river, provided that no chemical or biological transformations of the compounds occur in the ground. In addition, due to dispersion in the ground, the fluctuations in concentration in the river should be increasingly diminished with increasing flow distance (compare Roberts et al., Chapter 23 of this volume). The data presented in Figure 24.4 show that tetrachloroethylene behaves as predicted, whereas the concentration of 1,4-dichlorobenzene decreases significantly with increasing distance from the river. Since 1,4-dichlorobenzene can be considered to be chemically quite inert, it is probable that this compound is removed by biological processes. The ability of microorganisms to transform chlorinated benzenes into chlorinated phenols and catechols has been demonstrated in laboratory cultures (Ballschmitter and Scholz, 1980). However, it is unknown whether such reactions also take place under natural conditions, and whether persistent transformation products are accumulated in the ground water.

The results of the measurements of the total purgeable organo-chlorine compounds (POCl) indicate that a significant part of the chlorinated hydrocarbons are removed during the first few meters of infiltration (Figure 24.4). The results also show that those compounds not removed during infiltration are readily dispersed in the ground.

The findings made at this natural infiltration system in the lower Glatt Valley were confirmed by the results of a similar study at a different field site at the River Aare, Switzerland. In that infiltration system the water has a much longer residence time (weeks) compared to the infiltration system at the River Glatt (hours to days). Unfortunately, detailed information on the

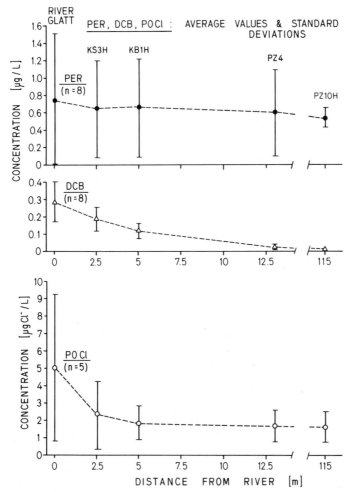

Figure 24.4. Average values and standard deviations of the measured concentrations of tetrachloroethylene (PER), 1,4-dichlorobenzene (DCB), and total purgeable organo-chlorine compounds (POCl) in River Glatt and in the top layer of the ground water at various distances from the river (compare with Figure 24.3). Samples taken approximately monthly (n = number of samplings).

flow direction and on the flow rates of the infiltrated water are not yet available. In general, the same behavior was observed for all halogenated hydrocarbons detected in both rivers (compare examples given in Figure 24.5). It should be noted that in River Aare the concentration of 1,4-dichlorobenzene was always about one order of magnitude lower than in River Glatt. Thus, this compound was almost completely removed during infiltration by biological processes at concentrations of less than 0.05 μg/L (detection limit for 1,4-dichlorobenzene $\simeq$ 0.005 μg/L).

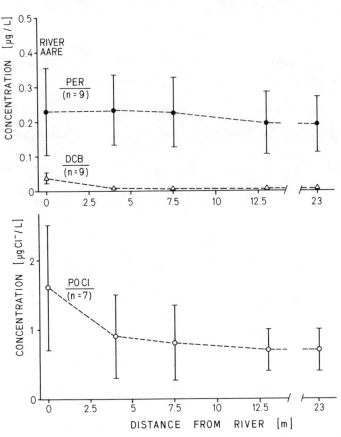

Figure 24.5. Average values and standard deviations of the measured concentrations of tetrachloroethylene (PER), 1,4-dichlorobenzene (DCB), and total purgeable organo-chlorine compounds (POCl) in River Aare and in the top layer of the ground water at various distances from the river. Samples taken approximately monthly (n = number of samplings).

From the results of these field investigations, the following conclusions can be drawn. If a river is continuously charged with persistent volatile halogenated hydrocarbons such as tri- and tetrachloroethylene, large ground water areas may be contaminated. On the other hand, if a ground water in the vicinity of such a river does not contain any of these compounds, it is very unlikely that a significant part of this ground water has been derived from infiltrated river water. Thus, compounds such as tri- and tetrachloroethylene can be used in certain cases as (at least qualitative) tracers for river water infiltration. Such a case is shown in Figure 24.6. Here, the question had to be answered whether—and, if so, which parts of—the ground water beneath a certain area were influenced by the river flowing

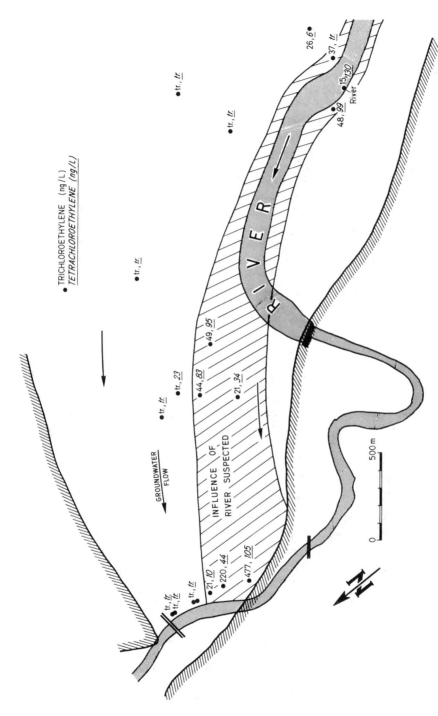

Figure 24.6. Volatile halogenated hydrocarbons as tracers for river water–ground water infiltration: measured concentrations of tri- and tetrachloroethylene in ground water samples from beneath a certain area.

463

through the town. From the results of long-term measurements of "classical" hydrochemical parameters (e.g., hardness), a provisory picture had been drawn (Nänny, 1981). Figure 24.6 shows the measured concentrations of tri- and tetrachloroethylene in the river and in samples from observation wells distributed over the study area.

Although the data presented in Figure 24.6 are based only on single measurements, a clear borderline between influenced and non-influenced areas can be drawn. These results corroborate the conclusions from the previous investigations of the same aquifer (Nänny, 1981).

Ground Water Contamination with Halogenated Alkanes: Hydrolysis and Formation of Sulfur-Containing Compounds Under Anaerobic Conditions

So far, we have dealt primarily with halogenated hydrocarbons, for which chemical transformations are not important under conditions typical for the subsurface. In this section we discuss a case of ground water contamination with halogenated alkanes, a group of compounds which can undergo various chemical reactions in water (i.e., hydrolysis, elimination reactions):

$$RR'R''C\text{---}X \xrightarrow{H_2O/OH^-} RR'R''C\text{---}OH + X^- \qquad (24.8)$$

$$RR'CX\text{---}CHR''R''' \xrightarrow{H_2O/OH^-} RR'C{=}CR''R''' + H^+ + X^- \qquad (24.9)$$

The hydrolysis rate constants and estimated half-lives of some halogenated alkanes under environmental conditions are summarized in Table 24.7. From these data, it can be seen that hydrolysis [Eq. (24.8)] is not an important process for polyhalogenated methanes. However, depending on the constitution of the compound, monochlorinated and monobrominated alkanes hydrolyze with half-lives between a few seconds (e.g., t-butylbromide) and one year (primary chloroalkanes). In general, the hydrolysis rates increase from primary ($R{=}R'{=}H$) to secondary ($R{=}H$) to tertiary halides and are about one order of magnitude greater for alkylbromides than for the corresponding alkylchlorides. The same general sequence also holds for elimination reactions [Eq. (24.9); see March, 1968].

An investigation of ground water contamination caused by a leaking waste water tank at a chemical plant revealed that under anaerobic conditions in ground water, certain alkylhalides can also undergo reactions other than hydrolysis and elimination (Schwarzenbach et al., 1985; Giger and Schaffner, 1981). The chemicals, which in this case had been introduced into the ground over a long period of time, were mostly halogenated hydrocarbons, including a series of predominantly primary alkylbromides (compare examples given in Figure 24.7, "educts"). It was found, however, that most of the compounds detected in the ground water 7 years after the plant operations had ceased contained sulfur. Figure 24.7 ("products") shows some of the structures tentatively elucidated by gas chromatography/mass spectrom-

Table 24.7 Hydrolysis Rate Constants and Estimated Half-Lives of Some Volatile Halogenated Alkanes under Environmental Conditions (pH 7, 298 K)[a]

Compound	Formula	$k_h(\text{sec}^{-1})$[b]	Half-Life
Methyl chloride	CH_3Cl	$2.4 \times (10^{-8})$[c]	334 days
Methyl bromide	CH_3Br	$4.1 \times (10^{-7})$[c]	20 days
Methylene chloride	CH_2Cl_2	$3.2 \times (10^{-11})$[c]	704 yr
Chloroform	$CHCl_3$	$6.9 \times (10^{-12})$[d]	3500 yr
Bromoform	CHB_3	$3.2 \times (10^{-11})$[d]	686 yr
Bromodichloromethane	$CHBrCl_2$	$1.6 \times (10^{-10})$[d]	137 yr
Dibromochloromethane	$CHBr_2Cl$	$8.0 \times (10^{-11})$[d]	274 yr
Ethyl bromide	CH_3CH_2Br	$2.6 \times (10^{-7})$[c]	30 days
Propyl bromide	$CH_3CH_2CH_2Br$	$3.0 \times (10^{-7})$[c]	26 days
Isopropyl bromide	$\begin{matrix} CH_3 \\ \diagdown \\ \quad\ CHBr \\ \diagup \\ CH_3 \end{matrix}$	$3.9 \times (10^{-6})$[c]	2 days
t-Butyl chloride	$\begin{matrix} CH_3 \\ \diagdown \\ CH_3-C-Cl \\ \diagup \\ CH_3 \end{matrix}$	$3.0 \times (10^{-2})$[c]	23 sec

[a] Data taken from Mabey and Mill (1978).
[b] $k_h = k_B[OH^-] + k_A[H^+] + k_N$.
[c] $k_h = k_N$ since k_B is not important below pH 10.
[d] Assume $k_h = k_B[OH^-]$.

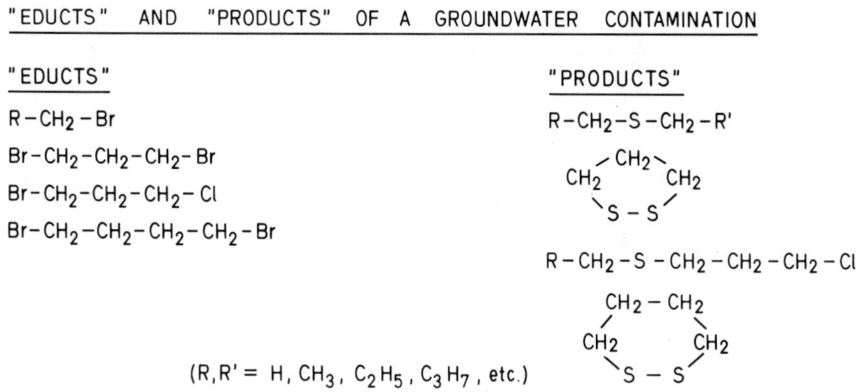

"EDUCTS" AND "PRODUCTS" OF A GROUNDWATER CONTAMINATION

"EDUCTS"	"PRODUCTS"
$R-CH_2-Br$	$R-CH_2-S-CH_2-R'$
$Br-CH_2-CH_2-CH_2-Br$	$\begin{matrix} \diagup CH_2 \diagdown \\ CH_2 \qquad CH_2 \\ \diagdown S-S \diagup \end{matrix}$
$Br-CH_2-CH_2-CH_2-Cl$	
$Br-CH_2-CH_2-CH_2-CH_2-Br$	$R-CH_2-S-CH_2-CH_2-CH_2-Cl$
	$\begin{matrix} CH_2-CH_2 \\ \diagup \qquad\ \diagdown \\ CH_2 \qquad\quad CH_2 \\ \diagdown S-S \diagup \end{matrix}$

$(R, R' = H, CH_3, C_2H_5, C_3H_7, \text{etc.})$

Figure 24.7. "Educts" and "products" of a ground water contamination with halogenated alkanes.

etry. Since hydrogen sulfide was also present in the ground water samples and since it is known that the aquifer has been anaerobic for the last 20 years, it seems safe to assume that these sulfur-containing compounds had been formed by nucleophilic substitution reactions as described in Figure 24.8.

In a series of laboratory experiments, the kinetics of the reactions of 1-bromohexane in water in the presence of hydrogen sulfide has been studied. The rate constants obtained from these experiments (Figure 24.8) show that under natural conditions (i.e., pH = 8, $[H_2S]_{tot}$ ~5 × 10^{-5} M), nucleophilic substitution reactions of primary alkylbromides with hydrogen sulfide can be important. In similar experiments conducted with a secondary alkylbromide (3-bromohexane), no sulfur-containing compounds were detected. A detailed report on this laboratory study is presented elsewhere (Schwarzenbach et al., 1985).

The results of the laboratory experiments are consistent with the general knowledge on nucleophilic substitution reactions of halogenated alkanes (March, 1968). Accordingly, primary alkylbromides should react with HS$^-$ in water by a true SN$_2$ (substitution, nucleophilic, bimolecular) mechanism, whereas tertiary bromides react by a SN$_1$ (substitution, nucleophilic, unimolecular) mechanism. The secondary bromides probably undergo "borderline reactions" (reactions that exhibit properties intermediate between SN$_1$ and SN$_2$ properties—Harris and Wamser, 1976). At the low $[H_2S]_{tot}$ concentrations typical for the environment (<10^{-4}M), significant amounts of alkylmercaptans and dialkylsulfides (compare Figure 24.8) should be formed from alkylbromides only if the reaction proceeds by a SN$_2$ mechanism.

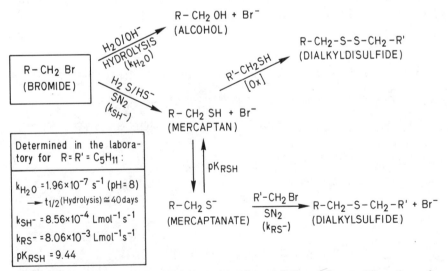

Figure 24.8. Ground water contamination with primary alkylbromides: reaction scheme for the formation of alkylmercaptans and dialkylsulfides under anaerobic conditions.

Thus, predominantly primary alkylbromides should lead to such products in ground water.

With respect to the polyhalogenated methanes (see examples given in Table 24.7), one would expect that these compounds, which are known to be very persistent under aerobic conditions (Bouwer et al., 1981b), should also react very slowly with hydrogen sulfide, since both SN_1- and SN_2-type reactions should be very slow due to the additional halogens bound to the carbon atom at which the reaction takes place. However, Bouwer et al. (1981b) reported that several trihalomethanes disappeared rapidly in seeded cultures in the presence of methanogenic bacteria and hydrogen sulfide at concentrations of $\sim 5 \times 10^{-3} M$. In their control experiments using the same hydrogen sulfide concentrations as in the seeded cultures, the trihalomethanes also disappeared, but at a much slower rate. Bouwer et al. (1981b) suggested nucleophilic substitution reactions [Eq. (24.10)] as a possible explanation for the observed disappearance of the trihalomethanes, but could not explain the differences found between seeded cultures and controls.

$$CHX_3 + HS^- \xrightarrow{SN_2} CHX_2SH + X^- \rightarrow \rightarrow \qquad (24.10)$$

Their comment that the microbial activity of the seeded culture had probably maintained the media in a state more reduced than the controls would agree with the assumption that trihalomethanes do not react with hydrogen sulfide by nucleophilic substitution reactions, but by a radical mechanism involving an electron transfer reaction (Kräutler, 1981):

$$HS^- \rightarrow HS\cdot + e^-$$

$$CHX_3 + e^- \rightarrow CHX_2\cdot + X^- \qquad (24.11)$$

$$CHX_2\cdot + HS\cdot \rightarrow CHX_2SH \rightarrow \rightarrow$$

Further work on this topic is clearly needed to clarify by which mechanisms (chemical and/or biological) trihalomethanes and other polyhalogenated hydrocarbons are removed in ground water under anaerobic conditions.

Persistence of Halogenated Hydrocarbons in the Subsurface

From the results of our field studies of halogenated hydrocarbons in ground water, some conclusions concerning the persistence of some individual compounds can be drawn. It seems that highly halogenated olefins such as trichloroethylene, tetrachloroethylene, and hexachlorobutadiene are extremely persistent in the subsurface under any environmental conditions. These observations are in general agreement with the findings of others (Piet and Zoeteman, 1980; Bouwer et al., 1981b), although Piet and Zoeteman (1980) suspect hexachlorobutadiene to be "decomposed" during dune infiltration. We have also no field evidence that polychlorinated ethanes such as 1,1,1-

trichloroethane and hexachloroethane are affected by any chemical and/ or biological processes. However, there is strong field evidence that 1,4-dichlorobenzene; 1,2,4-trichlorobenzene; and 1,2,3-trichlorobenzene are transformed by microorganisms under mostly aerobic conditions during infiltration of river waters. These findings are corroborated by results of laboratory studies reported in the literature (Marinucci and Bartha, 1979; Ballschmitter and Scholz, 1980). In an anaerobic aquifer, however, appreciable concentrations (up to 15 μg/L) of chlorobenzene, dichlorobenzenes, bromobenzene, and 1,4-dibromobenzene were detected 7 years after the source of contamination had been removed (Giger and Schaffner, 1981). Thus, halogenated benzenes seem to be much more persistent in the ground under anaerobic conditions than under aerobic conditions. The results of laboratory experiments reported by Marinucci and Bartha (1979), who studied the biodegradation of 1,2,3- and 1,2,4-trichlorobenzene in soil and in liquid enrichment culture, are in agreement with this conclusion.

CONCLUSIONS

1.   Halogenated hydrocarbons are among the most ubiquitous ground water pollutants. The major causes for ground water contamination by such chemicals include leachates from chemical waste dumps, spills, polluted surface waters, and use of the chemicals (e.g., as pesticides).

2.   The degree of sorption of a given halogenated hydrocarbon by soil and aquifer materials can be calculated from the organic carbon content of the natural sorbent and from the octanol–water partition coefficient of the compound. Thus, under flow conditions at which sorption equilibrium can be assumed, the retardation factor of a given compound in a given aquifer can be estimated. The results of this study indicate that many halogenated hydrocarbons are very mobile in the ground. Furthermore, at high flow velocities these compounds may be transported faster than one would assume from equilibrium considerations due to slow sorption kinetics.

3.   Many halogenated hydrocarbons (e.g., halogenated olefins, halogenated benzenes, polyhalogenated alkanes) are quite resistent to chemical transformations in ground water. Monohalogenated alkanes undergo hydrolysis and elimination reactions with half-lives between a few seconds and a year. Under anaerobic conditions in the presence of hydrogen sulfide, primary alkylhalides can react with HS^- to yield sulfur-containing compounds such as alkylmercaptans and dialkylsulfides. Polyhalogenated methanes can probably also react with HS^- but by a different reaction mechanism involving an electron transfer reaction.

4.   Presently, very little is known about biological transformations of halogenated hydrocarbons in the ground. From field observations one can conclude that many of these chemicals are quite resistant to biotransformation. However, more work is needed to clarify whether, under which conditions,

and at what rate a given compound is transformed by biologically mediated processes, and whether hazardous metabolites are formed.

ACKNOWLEDGMENT

We thank E. Hoehn, C. Jaques, E. Molnar-Kubica, Ch. Schaffner, J. Schneider, O. Wanner, and J. Westall for their significant contributions to this work. This work was funded by the Swiss National Science Foundation (Nationales Forschungsprogramm "Wasserhaushalt") and by the Swiss Department of Commerce (Project COST 64b bis).

REFERENCES

Baccini, P., Grieder, E., Stierli, R., and Goldberg, S. (1982). The influence of natural organic matter on the adsorption properties of mineral particles in lake water. *Schweiz. Z. Hydrol.* **44**:99–116.

Ballschmiter, K. and Scholz, Ch. (1980). Mikrobieller Abbau von chlorierten Aromaten: VI. Bildung von Dichlorphenolen und Dichlorbrenzkatechinen aus Dichlorbenzolen in mikromolarer Lösung durch *Pseudomonas* sp. *Chemosphere* **9**:457–467.

Bouwer, E. J., McCarty, P. L., and Lance, J. C. (1981a). Trace organic behaviour in soil columns during rapid infiltration of secondary wastewater. *Water Res.* **15**:151–159.

Bouwer, E. J., Rittmann, B. E., and McCarty, P. L. (1981b). Anaerobic degradation of halogenated 1- and 2-carbon organic compounds. *Environ. Sci. Technol.* **15**:596–599.

Chiou, C. T., Peters, L. J., and Freed, V. H. (1979). A physical concept of soil–water equilibria for nonionic organic compounds. *Science* **206**:831–832.

Chiou, C. T., Freed, V. H., Schmedding, D. W., and Kohnert, R. L. (1977). Partition coefficient and bioaccumulation of selected organic chemicals. *Environ. Sci. Technol.* **11**:475–478.

Dow Chemicals U.S.A. (1978). Technical data bulletin for 1,2,4-trichlorobenzene, Organic Chemicals Development, Midland, MI.

Freeze, R. A. and Cherry, J. A. (1979). *Groundwater.* Prentice-Hall, Englewood Cliffs, NJ.

Giger, W., Molnar-Kubica, E., and Wakeham, S. G. (1978). Volatile chlorinated hydrocarbons in ground and lake waters. In: O. Hutzinger, I. H. van Lelyveld, and B. C. J. Zoeteman (Eds.), *Aquatic Pollutants,* Pergamon, Oxford, pp. 101–123.

Giger, W., Schaffner, Ch., and Wakeham, S. G. (1980). Aliphatic and olefinic hydrocarbons in recent sediments of Greifensee, Switzerland. *Geochim. Cosmochim. Acta* **44**:119–129.

Giger, W. and Schaffner, Ch. (1981). Groundwater pollution by volatile organic chemicals. In: W. van Duijvenbooden, P. Glasberger, and I. H. van Lelyveld (Eds.), *Quality of Groundwater,* Elsevier, Amsterdam, pp. 517–522.

Grob, K. (1973). Organic substances in potable water and its precursor. Part I. Methods for their determination by gas-liquid chromatography. *J. Chromatogr.* **84:**255–273.

Grob, K. and Zürcher, F. (1976). Stripping of trace organic substances from water: Equipment and procedure. *J. Chromatogr.* **117:**285–294.

Hansch, C. and Leo, A. (1979). *Substituent Constants for Correlation Analysis in Chemistry and Biology.* Elsevier, Amsterdam.

Harris, J. M. and Wamser, C. C. (1976). *Fundamentals of Organic Reaction Mechanisms.* Wiley, New York, pp. 134–179.

Jolley, R. L., Gorchev, H., and Hamilton, D. H., Jr. (1978). *Water Chlorination Environmental Impact and Health Effects,* Vol. 2. Ann Arbor Science, Ann Arbor, MI.

Karickhoff, S. W. (1981). Semi-empirical estimation of sorption of hydrophobic pollutants on natural sediments and soils. *Chemosphere* **10:**833–846.

Karickhoff, S. W., Brown, D. S., and Scott, T. A. (1979). Sorption of hydrophobic pollutants on natural sediments. *Water Res.* **13:**241–248.

Khan, A., Hasset, J. J., Banwart, W. L., Means, J. C., and Wood, S. G. (1979). Sorption of acetophenone by sediments and soils. *Soil Sci.* **128:**297–302.

Kräutler, B. (1981). Organic Chemistry Department, Swiss Federal Institute of Technology, H, Zürich, Switzerland. Private communication.

Lambert, S. M. (1968). Omega, a useful index of soil sorption equilibria. *J. Agric. Food Chem.* **16:**340–343.

Leo, A., Hansch, C., and Elkins, D. (1971). Partition coefficients and their uses. *Chem. Rev.* **71:**525–616.

Mabey, W. and Mill, T. (1978). Critical review of hydrolysis of organic compounds in water under environmental conditions. *J. Phys. Chem. Ref. Data* **7:**383–415.

Mackay, D., Bobra, A., Shiu, W. Y., and Yalkowsky, S. H. (1980). Relationships between aqueous solubility and octanol–water partition coefficients. *Chemosphere* **9:**701–711.

March, J. (1968). *Advanced Organic Chemistry,* 2nd ed. McGraw-Hill, New York.

Marinucci, A. C. and Bartha, R. (1979). Biodegradation of 1,2,3- and 1,2,4-trichlorobenzene in soil and in liquid enrichment culture. *Appl. Environ. Microbiol.* **38:**811–817.

Matter, C. (1979). Sorptions- und Stoffaustauschprozesse refraktärer organischer Leitsubstanzen in einer Belebtschlammanlage. Ph.D. Thesis No. 6403, Swiss Federal Institute of Technology, Zürich, Switzerland.

Means, J. C., Wood, S. G., Hassett, J. J., and Banwart, W. L. (1980). Sorption of polynuclear aromatic hydrocarbons by sediments and soils. *Environ. Sci. Technol.* **14:**1524–1528.

Nänny, P. (1981). EAWAG, CH-8600 Dübendorf, Switzerland. Private communication.

Nelson, S. J., Iskander, M., Volz, M., Khalifa, S., and Haberman, R. (1981). In: W. van Duijvenbooden, P. Glasbergen, and I. H. van Lelyveld (Eds.), *Quality of Groundwater,* Elsevier, Amsterdam, pp. 169–174.

Pearson, C. R. and McConnell, G. (1975). Chlorinated C_1 and C_2 hydrocarbons in the marine environment. *Proc. Roy. Soc. London, Ser. B,* **189:**305–322.

Piet, G. J. and Zoeteman, B. C. J. (1980). Organic water quality changes during sand bank and dune filtration of surface waters in The Netherlands. *J. Am. Water Works Assoc.* **72**:400–404.

Rao, P. S. C. and Davidson, J. M. (1980). Estimation of pesticide retention and transformation parameters required in nonpoint source pollution models. In: M. R. Overcash, and J. M. Davidson (Eds.), *Environmental Impact of Non-point Source Pollution,* Ann Arbor Science, Ann Arbor, MI, pp. 23–67.

Rekker, R. F. (1977). *The Hydrophobic Fragmental Constant.* Pharmacochemistry Library, Vol. 1. Elsevier, Amsterdam.

Roberts, P. V., McCarty, P. L., Reinhard, M., and Schreiner, J. (1980). Organic contaminant behaviour during groundwater recharge. *J. Water Pollut. Control. Fed.* **52**:161–172.

Schneider, J. K., Schwarzenbach, R. P., Hoehn, E., Giger, W., and Wasmer, H. R. (1981). The behaviour of organic pollutants in a natural river–groundwater infiltration system. In: W. van Duijvenbooden, P. Glasbergen, and I. H. van Lelyveld (Eds.), *Quality of Groundwater,* Elsevier, Amsterdam, pp. 565–568.

Schwarzenbach, R. P., Giger, W., Schaffner, Ch., and Wanner, O. (1985). Groundwater contamination by volatile halogenated alkanes: Abiotic formation of volatile sulfur compounds under anaerobic conditions. *Environ. Sci. Technol.,* in press.

Schwarzenbach, R. P., Giger, W., Hoehn, E., and Schneider, J. K. (1981). Behavior of organic compounds during infiltration of river water to ground water: Field studies. *Environ. Sci. Technol.* **17**:472–479.

Schwarzenbach, R. P., Molnar-Kubica, E., Giger, W., and Wakeham, S. G. (1979). Distribution, residence time, and fluxes of tetrachloroethylene and 1,4-dichloro-benzene in Lake Zürich, Switzerland. *Environ. Sci. Technol.* **13**:1367–1373.

Schwarzenbach, R. P. and Westall, J. (1981). Transport of non-polar organic compounds from surface water to groundwater: Laboratory sorption studies. *Environ. Sci. Technol.* **15**:1360–1367.

Sontheimer, H. (1980). Experience with riverbank filtration along the Rhine River. *J. Am. Water Works Assoc.* **72**:386–390.

Tute, M. S. (1971). Principles and practice of Hansch analysis: A guide to structure-activity correlation for the medicinal chemist. *Adv. Drug. Res.* **6**:1–77.

van Genuchten, M. Th., Davidson, J. M., and Wierenga, P. J. (1974). An evaluation of kinetic and equilibrium equations for the prediction of pesticide movement in porous media. *Soil Sci. Soc. Am. Proc.* **38**:29–35.

Verschueren, K. (1977). *Handbook of Environmental Data on Organic Chemicals,* Van Nostrand Reinhold Company, New York.

Zoeteman, B. C. J., Harmsen, K., Linders, J. B. H. J., Morra, C. F. H., and Slooff, W. (1980). Persistent organic pollutants in river water and groundwater of The Netherlands. *Chemosphere* **9**:231–249.

Zürcher, F. (1981). Simultaneous determination of total purgeable organo-chlorine, -bromine and -fluorine compounds in water by ion-chromatography. In A. Bjørseth and G. Angeletti (Eds.), *Analysis of Organic Micropollutants in Water,* D. Reidel Publishing, Dordrecht, Holland, pp. 272–286.

25

SURVIVAL AND TRANSPORT OF PATHOGENIC BACTERIA AND VIRUSES IN GROUND WATER

G. Matthess
A. Pekdeger

Geological-Paleontological Institute of Kiel University, Kiel, Federal Republic of Germany

For the assessment of ground water protection against pathogenic microorganisms (bacteria and viruses), the mechanisms which control the two main factors for their elimination in the aquifer must be studied:

1. The persistence of bacteria and viruses under the biological and chemical conditions of the ground water.
2. The physical and physical–chemical processes that control the transport of microorganisms in ground water.

The most important pathogenic bacteria and viruses that might be transported in the water path are *Salmonella* sp., *Shigella* sp., *Vibrio cholerae, Yersinia enterocolitica, Y. pseudotuberculosis, Leptospira* sp., *Francisella tularensis, Dyspepsia coli,* enterotoxine forming *E. coli,* Pseudomonades, and viruses—hepatitis, polioviruses, coxsackie viruses, and ECHO viruses (ECHO = enteric cytopathogenic human orphan). The ground water may be contaminated by sewage containing the excrements of carriers or diseased persons. In municipal sewage, the concentration of *E. coli* is of the order of 10^5–10^6/mL and in surface waters about 10^2–10^3/mL. In Central European rivers usually 1–3 *Salmonella*/L are detected.

The virus concentrations in the United States are as high as 10^2–10^4 PFU/L in sewage effluent, 10^0–10^3/L in treated sewage effluent, and 0–10^2/L in surface waters (PFU = plague forming units). According to the EPA (1978), drinking water that has undergone conventional treatment would contain about 1 infectious unit per 10^5–10^8 L. This means the virus concentration in the waste water has to be eliminated by 7 log units before the water is usable as drinking water. About the same order of magnitude holds for the necessary elimination of bacteria: sewage effluent contains about 10^6 *E. coli*/ 100 mL, and drinking water should contain no *E. coli* in 100 mL. The same approach may be used for ground water.

SURVIVAL OF BACTERIA AND VIRUSES IN GROUND WATER

Two groups of microorganisms are to be differentiated when the survival of bacteria and viruses in ground water are considered:

1. Allochthonic pathogenic microorganisms (parasitic bacteria and entero-toxine producing bacteria), which enter the ground water due to contamination.
2. Autochthonic ground water microorganisms.

The autochthonic microbial ground water population flourishes under favorable ecological conditions, developing high population densities ($\gg 10^3$/mL). The allochthonic bacteria are usually eliminated in the ground water environment, but under oligotrophic conditions they may survive without a substantial decrease or with even a slight increase in the germ number during the first 1–7 days. After this period, the elimination of bacteria and viruses may be approximately described by an exponential function (Merkli, 1975):

$$C_{(t)} = C_0 e^{-\lambda(t-t_0)} \qquad (25.1)$$

where $t \geq t_0$ and $t_0 \leq 7$ days
$\quad C_{(t)}$ = concentration at time t
$\quad C_0$ = initial concentration
$\quad \lambda$ = elimination constant = $(\ln 2)/d_{50}$
$\quad d_{50}$ = half-life of microorganism

The equation holds also for viruses (Berg, 1967). Equation (25.1) can be used to estimate the order of magnitude of the number of bacteria and viruses at any time, depending on the initial contamination. Furthermore, it follows from Eq. (25.1) that after a very rapid elimination at the beginning (the half-life of most bacteria and viruses ranges between 1 and 20 days), this process will decrease so that even after longer times, bacteria and viruses may still exist in very small quantities in ground water.

The elimination constant depends on physical, chemical, and biological parameters and is specific for the different microbial species (Matthess and Pekdeger, 1981). It is not yet possible to predict with the necessary accuracy the elimination constant on the basis of controlling factors. Therefore the elimination constant must be measured for each specific species and environment. The published values vary over a broad range (e.g., for a 99.9% elimination of *Salmonella typhi,* time intervals between 2 and 107 days are quoted). Besides the varying parameters mentioned above, different experimental methods lead to deviations in the results. For example, *Salmonella typhimurium* is found in laboratory measurements (Merkli, 1975) to be more resistant in the ground water environment than *S. typhi* or *E. coli* (230, 8, and 23 days, respectively). *In situ* experiments with membrane filter chambers in ground water gave contradictory results (McFeters et al., 1974). The most important ecological factors are the physical (temperature), biological, and chemical conditions. These have a combined effect due to interaction and complementary impacts. The biological factors (concentration of autochthonic bacteria, bacteriophages, etc.) are presumably the most important for the survival of pathogenic bacteria and viruses, although these depend extensively on the physicochemical properties of ground water. In sterilized water or in water with low biological activity, the viruses and bacteria survive longer. The persistence increases under sterilized conditions in the presence of higher contents of organic compounds or in water with favorable pH values (about pH 7) and low oxygen concentrations. In contaminated water or in surface water with high biological activity, the elimination constant is higher under the same physical and chemical conditions due to the activation of the autochthonic bacteria, which can compete very effectively with the pathogenic bacteria. Bacteriophages and parasitic bacteria (e.g., *Bdellovibrio bacteriovorus* and protolytic bacteria), when present in higher concentrations, accelerate the elimination. The persistence of endospore building bacteria is less affected by the biological factors than the other bacteria. Under oligotrophic conditions and at temperatures less than 15°C, pathogenic bacteria and viruses can survive for a long period: *E. coli* > 100 days, *Salmonella typhi* > 100 days, *Salmonella typhimurium* ≤ 230 days, other *Salmonellae* ≤ 70 days, *Yersinia* sp. ≤ 200 days, *Poliovirus* ≥ 250 days.

The published data on the elimination of bacteria and viruses in aquifers show that the survival time cannot be the only criterion for the purifying effect of underground passage. Therefore the physical–chemical transport processes must be considered.

TRANSPORT PROCESSES

In porous aquifers (e.g., sand and gravel) the main transport takes place in the pores. The ground water velocities usually range from less than 1 m/day up to a few meters/day; velocities above 10 m/day are rare. In hard rock aquifers, the ground water flow velocities in fissures are said to be between 0.3 and 8000 m/day, in karstic aquifers up to 26,000 m/day. Therefore it can be concluded that the propagation of pollutants in fissured and karstic aquifers is much faster than in porous aquifers. Furthermore, the larger diameters of the flow paths in fissures and solution canals of hard rocks should allow a better transport of suspended microorganisms. Ground water velocity may be calculated from hydraulic conductivity, hydraulic gradient, and porosity as a mean velocity, or measured by tracer experiments from which the mean velocity and a maximum velocity (from first arrival of measurable tracer amounts at the observation point) can be derived. The maximum velocity is important for the assessment of the vulnerability of ground water systems to pathogenic microorganisms, since it indicates the first possible arrival of microbial pollutants.

Microorganisms are subject to adsorption on underground particles. This may be described for diluted suspensions by the Freundlich isotherm (Merkli, 1975), which defines the equilibrium between the concentration of the suspended (C_s) and adsorbed (C_a) microorganisms:

$$C_a = kC_s^n \tag{25.2}$$

where k and n are assumed to be specific constants for the investigated rock and microorganisms. The Freundlich isotherm shows that the adsorption of microorganisms can be reversible. The adsorption of viruses and bacteria takes place quite rapidly (2 and 24 hr, respectively). Desorption velocity is not as well known as the adsorption velocity and should be measured in further investigations. Another possible description may be the Langmuir isotherm, which possibly is the better mathematical definition of the nature of the adsorption processes:

$$C_a = \frac{KbC_s}{1 + KC_s} \tag{25.3}$$

where K is a constant relating to the bonding energy and b the adsorption maximum when the adsorbent is completely saturated. The continuous adsorption-desorption reactions cause a retardation of the microorganisms with respect to the surrounding ground water. Due to this retardation, more time for the elimination of bacteria and viruses is available. The retardation of the microorganisms is described by the retardation factor R_d, the quotient of mean water velocity v_w to the mean transport velocity of microorganisms v_m. The retardation factor can be calculated if the distribution coefficient K_d of bacteria and viruses is known:

$$R_d = \frac{v_w}{v_m} = 1 + \frac{\rho_b}{n} K_d \tag{25.4}$$

with the bulk density of the aquifer material represented by ρ_b and its porosity by n. The empirical distribution coefficient K_d can be obtained by batch tests in the laboratory. It defines the affinity of the aquifer material for a certain contaminant. K_d is in diluted suspensions equal to the coefficient of the Freundlich or Langmuir isotherm. If the mean velocity of the ground water v_w, of the contaminant v_m, ρ_b, and n are known, the actual K_d values can be calculated by laboratory and field experiments. In field experiments, retardation factors of between 1 and 2 are found for bacteria (*E. coli* and *Serratia marcescens*). The viruses in general, especially the polioviruses, have very high K_d values, depending on water properties and the properties of the respective virus. From equation 25.4, retardation factors of up to 500 can be expected in loamy aquifers with higher cation concentrations. Models using the data of the known adsorption coefficients and elimination rates show that the underground passage can provide a very effective protection against virus contamination. According to other observations, however, the viruses can be desorbed again when cation concentrations decrease (e.g., by means of a very intensive rainfall), and are thus enabled to travel further (Duboise et al., 1976). Model calculations, using data of Gerba and Goyal (1978) for virus adsorption and the elimination constant of viruses from Akin et al. (1971), show that the transport of viruses can be very different, depending on the water chemistry, for the same elimination constant (Figure 25.1).

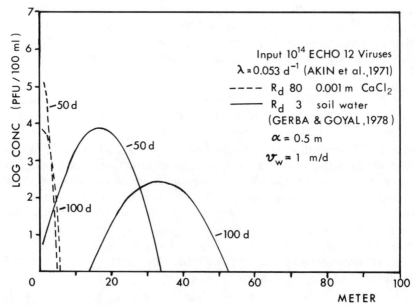

Figure 25.1.   Model calculation of virus transport in ground water (single impulse, point source).

Bacteria can attach themselves actively onto the surface of the solid aquifer materials. The attached bacteria are protected against other influences and find higher nutrient concentrations, which decrease the elimination rate. The investigations with autochthonic bacteria show that the attachment is most intensive in the phase of exponential growth. The enteric bacteria show hardly any growth in ground water; thus the active attachment should be at a minimum.

Polluted ground water plumes undergo dispersion, which causes a distribution of the pollutants in time and space so that in the contaminated ground water body the concentration decreases with time and with transport distance. This process may be described by the general transport equation in vector form (Bear, 1972):

$$\frac{\partial C}{\partial t} = \text{div} \left(\frac{D}{R_d} \cdot \text{grad } C \right) - \frac{v_w}{R_d} \cdot \text{grad } C - \lambda C \qquad (25.5)$$

where the coefficient of hydrodynamic dispersion $D = D' + D_d + D_e$, ($D' =$ coefficient of dispersion, $D_d =$ coefficient of diffusion—important for viruses—and $D_e =$ coefficient of active mobility of the bacteria), grad C is the concentration gradient, v_w the average ground water velocity, λ the elimination constant, and R_d the retardation factor. The active mobility of the bacteria decreases with decreasing temperatures (e.g., for *E. coli* 0.1 m/day at 20°C). The hydrodynamic coefficient D depends on ground water flow velocity (Scheidegger, 1974):

$$D = \alpha v_w + b \qquad (25.6)$$

The dispersivity coefficient α is a function of the inhomogenity of the aquifer; thus α increases with the scale of the experiments. Porous media used in laboratory experiments have dispersivity coefficients α on the order of 0.01–1 m, in field experiments on the order 0.01–100 m, and fissured and karstic rocks in field experiments 10–1000 m (Bertsch, 1978; Matthess and Pekdeger, 1981). For given boundary conditions, numerical models are available. Inserting the above-mentioned dispersivities shows that in fissured and karstic aquifers a "purification" of ground water occurs due to dilution below detection level even though the residence time in these aquifers may be too small for an effective elimination and the filtering qualities are generally poor.

Model calculations of bacteria transport using laboratory data indicate that under certain conditions bacteria can be transported over large distances. In fact, coliform bacteria were observed to be transported in loamy sand aquifers for more than 1 km, and in fissured karstic aquifers for more than a few kilometers. Thus, depending on all the factors mentioned above, very different transport times and distances can occur. Many of these controlling factors need further intensive laboratory and field investigations, and further theoretical model calculations with different parameters should be done.

FILTRATION AND SUFFUSION

The transport of microorganisms may be limited if the pore size of a porous medium is smaller than the size of the microorganisms. The pore size of a porous medium is not homogenous but is heterogenously distributed. Some of the pore diameter is greater and some smaller than the bacteria. Calculations of suffusion can be used for determination of the mechanical filtering criteria (Busch and Luckner, 1974). The geometrical suffusion security η_{SG} defined as in Eq. (25.8) must be greater than 1.5 to limit bacteria transport.

$$\eta_{SG} = \frac{d_m}{F_s d_K} \geq 1.5 \tag{25.8}$$

where d_m = diameter of microorganism
   F_s = empirical transit factor for suffusion—numerically 0.6 is used; this is a factor for the heterogeneity of a porous media
   d_k = hydraulic equivalent diameter of pore canals
       = $0.2d_{10}$ or $0.455 \sqrt[6]{U}(e)d_{17}$
   U = uniformity coefficient (ration of d_{60} to d_{10})
   e = void ratio (ratio of void volume to solid volume)
       d_{10}, d_{17}, d_{60} = grain size, with 10, 17, and 60% of the soil finer.

The mechanical filter processes in gravelly aquifers cannot be very effective due to the small diameters of bacteria (0.2–5 μm) and viruses (0.25–0.02 μm). For example the critical (or limiting) pore diameter for particle transport in medium sand is calculated to be 72 μm (Table 25.1). For the bacteria a mechanical filtration should be expected below the grain size of coarse loam with uniform grain size distribution. In natural sediments the grain size distribution is more heterogeneous, and some percentage of the pore diameters can interfere with bacteria transport (>10 percent in sand—Figure 25.2). The particle accumulation on solid substance surfaces is affected by sedimentation, flow processes, diffusion and interception (Rolke, 1971; see also Figures 25.3 and 25.4).

The importance of sedimentation for bacteria and viruses has apparently been overestimated in the past; the sedimentation is very important for the accumulation of inorganic mineral suspension (density about 2.5 g/cm^3), but not for microorganisms (density about 1 g/cm^3). The kinetic energy of a small particle transported by the ground water flow to the surface of a grain is not high enough to overcome the repulsing surface forces.

As a filtering process for particles with diameters of less than 1 μm (e.g., viruses), the diffusion is very important, becoming increasingly effective

**Table 25.1   Relation Between Grain Size
of Uniform Soil and the Critical Pore Size**

Soil	Grain Size (mm)	$F_s d_k$ (μm)
Silt		
Fine silt	0.002–0.006	0.72
Medium silt	0.006–0.020	2.4
Coarse silt	0.02–0.06	7.2
Sand		
Fine sand	0.06–0.2	24
Medium sand	0.2–0.6	72
Coarse sand	0.6–2	240
Gravel		
Fine gravel	2–6.3	720
Medium gravel	6.3–20	2400
Coarse gravel	20–63	7200

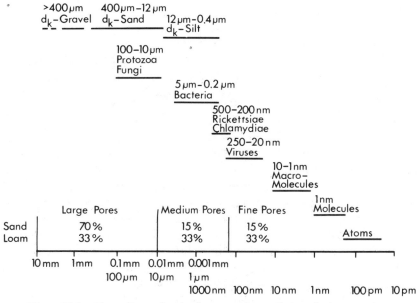

Figure 25.2.   Comparison of grain size, pore size, and size of microorganisms.

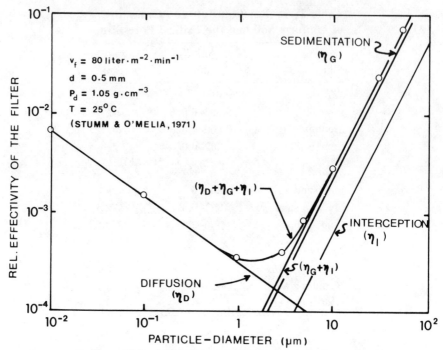

Figure 25.3. Mechanisms of the filtration processes.

with decreasing particle size. The interception is the most effective process for the accumulation of bacteria although this process has its minimum at particle diameters of about 1 μm. The most important bonding forces which can retain microorganisms are van der Waals forces (mass) and Coulomb (electrostatic) forces. The effects of van der Waals forces are restricted to very short distances ($r \leq$ few nm), and their power decreases rapidly with increasing distances ($\sim 1/r^7$), whereas the Coulomb forces decrease much less rapidly with increasing distances ($\sim 1/r^2$).

The solid particles of an aquifer are usually negatively charged. Exceptions are the iron and manganese hydroxides and organic substances at low pH values. The generally negatively charged bacteria and viruses are strongly adsorbed by anionic adsorbents and only slightly by cationic adsorbents. It is well known from filtering processes that the negatively charged particles stay in suspension in sand filters, as the repulsing electrostatic forces are stronger than the van der Waals forces. The dissolved cations in water decrease the repulsing forces of the grain surfaces. Monovalent cations are adsorbed by the solid substance and decrease their charge deficiency on the particle surfaces. Under these conditions, the mass forces are more effective, and an accumulation of particles can take place. This can be demonstrated by the dependence of virus adsorption on the solute concentration of the water (Gerba and Goyal, 1978).

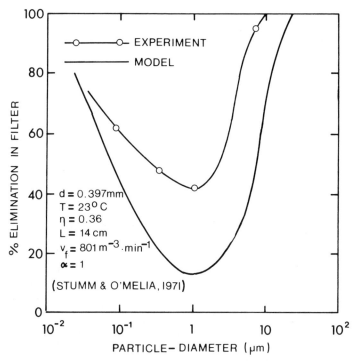

Figure 25.4. Model calculation and experimental results of filtration effectivity.

Bivalent cations can also cause a positive charge deficiency so that the electrostatic forces can be more efficient. Thus higher calcium concentrations increase bacteria and virus adsorption.

CONCLUSIONS

The purifying characteristics of river bank filtration and slow sand filtration can be only partly compared with the processes in ground water. The biologically active layer at the boundary of water and sediment is very effective due to the high content of sorptive small particles and microbial slimes. If this layer is destroyed, bacteria and virus transport may take place, contaminating the aquifer. It needs a certain time to build up this layer again (few days to few weeks).

Disturbances such as erosion or drying lead to a breakthrough into the deeper layers that will decrease after some time. During a continuous contamination of the ground water by organic substances and microorganisms, the contamination plume becomes smaller with time. At very high bacteria concentrations flocculation and aggregation can occur at the source of contamination so that only a limited transport into the aquifer can take place.

Thereafter, intermediate levels of contaminants, which initially enter a ground water biotop of moderate biological activity, are potentially the worst.

REFERENCES

Akin, E. W., Benton, W. H., and Hill, W. F., Jr. (1971). Enteric viruses in ground and surface waters—a review of their occurrence and survival. Proc. 13th Water Quality Conference, *Univ. of Illinois Bull.* **65**:55–74.

Bear, J., 1972. *Dynamics of Fluids in Porous Media*. Elsevier, New York, 1972.

Berg, G. (Editor) 1967. *Transmission of Viruses by the Water Route*. Wiley-Interscience, New York.

Bertsch, W., 1978. Die Koeffizienten der longitudinalen und transversalen hydrodynamischen Dispersion—ein Literaturüberblick. *Deutsch. Gewässerk. Mitt.* **22**:42–45.

Busch, K. F. and Luckner L., Geohydraulik. 1974, 2nd ed., Stuttgart, Enke.

Duboise, S. M., Moore, B. E., and Sagik, B. P., 1976. Poliovirus survival and movement in a sandy forest soil. *App. Environ. Microbiol.* **31**:536–543.

Environmental Research Center (ORD) (1978). Human Viruses in the Aquatic Environment. (Report to Congress) for the Office of Drinking Water, U.S. E.P.A. Washington, D.C. 20460, EPA–57019-78/006, Cincinnati, Ohio, 37 pp.

Gerba, C. P. and Goyal, S. M. (1978). Health considerations. Adsorption of selected enteroviruses to soils. In: *State of Knowledge in Land Treatment of Wastewater,* Vol. **2**, pp. 225–232, U.S. Government Printing Office, Washington, D.C., 1978-700-171/123.

Matthess, G. and Pekdeger, A. (1981). Erfordernisse u. Konzeptionen der *Ermittlung der engeren Trinkwasserschutzzone.* In: *Proceedings of Bewertung chemischer Stoffe im Wasserkreislauf, Tagung Langen* October 8–10, 1979, Institut für Wa.-Bo.-Lufthygiene Berlin pp. 74–98, Schmidt Verlag, Berlin.

McFeters, G. A., Bissonette, G. K., Jezeski, J. E., Thomson, C. A., and Stuart, D. G. (1974). Comparative survival of indicator bacteria and enteric pathogens in wellwater. *Appl. Microbiol. Biotech.* **27**:823–829.

Merkli, B. (1975). Untersuchungen uber Mechanismen und Kinetik der Elimination von Bakterien and Viren im Grundwasser, Ph.D. thesis, *ETH* 5400, Zürich.

Rolke, D. (1971). Vergleichende Untersuchungen an Trocken- und Überstaufiltern zum Mechanismus der Partikelablagerung in Kiesfiltern. Veröfftl. Lehrstuhl Wasserchem. **5**:21–42.

Scheidegger, A. E. (1958). *Can. Mining Metall. Bull.* **561**:26–30.

Scheidegger, A. E. (1974). *The Physics of Flow Through Porous Media*. University of Toronto Press, Toronto.

Stumm, W. and O'Melia, C. (1971). Chemische Vorgänge bei der Filtration. Veröfftl. Abt. Lehrstuhl Wasserchem. **5**:42–67.

26

BIODEGRADATION OF CONTAMINANTS IN THE SUBSURFACE

John T. Wilson

Robert S. Kerr Environmental Research Laboratory
Ada, Oklahoma

Michael J. Noonan

Lincoln College
Canterbury, New Zealand

James F. McNabb

Robert S. Kerr Environmental Research Laboratory
Ada, Oklahoma

Widely occurring contamination of ground water by synthetic organic pollutants has created a need for information about the biotransformation of these contaminants in aquifers and associated regions of the unsaturated subsurface.

Recent data on the occurrence of microbes in the subsurface environment indicate that there are biogeochemically significant densities of bacteria in deeper regions of the unsaturated subsurface and the shallow regions of water table aquifers. Ghiorse and Balkwell (Chapter 20) and White et al. (Chapter 14) report biomass estimates and plate counts equivalent to 10^6 organisms/g dry soil. If a sizable proportion of these microorganisms is metabolically active, they can transform organic pollutants at a significant rate. For example, the rate of biotransformation expected from an active population of 10^6 cells/g dry soil is 0.1 mg pollutant/L/day, based on typical values for water content of the soil, dry biomass of the cells, and the fraction of organic pollutant converted to cells (0.33 mL/g, 10^{-13}g, and 0.33 g dry biomass/g pollutant) and a very slow growth rate of 0.1/day. Because the concentrations of organic pollutants in ground water rarely exceed 1.0 mg/L, and the residence time of water in the active zone entends from months to decades, there is ample opportunity for biodegradation if the subsurface microflora is active and competent to degrade the pollutant.

This study examined the biodegradation of selected organic pollutants by the indigenous microflora of uncontaminated samples of the subsurface environment.

ACQUISITION OF UNCONTAMINATED SAMPLES OF THE SUBSURFACE

A modification of procedures described by Dunlap et al. (1977) was used to acquire samples of the deeper subsurface that were not contaminated with surface microbes. A bore hole of the desired depth was drilled with an auger. Then the auger was removed and the sample taken with an autoclaved thin-wall core barrel 9.5 cm I.D. × 45 cm long (Figure 26.1). The core was extruded 3 to 5 cm, then broken off to produce an aseptic face. The remainder of the core was extruded past a device that pared away the outer 1.2 cm of core material (Figure 26.2). As a result, the core material that was in contact with the core barrel, and thus might be contaminated with surface microbes, was discarded. The last 4 cm of the core was not extruded to prevent contamination from the extruder.

Material was collected from positions immediately above and below the

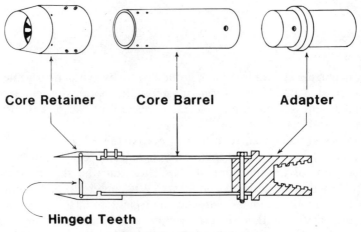

Core Retainer   **Core Barrel**   **Adapter**

Hinged Teeth

Figure 26.1. Coring device.

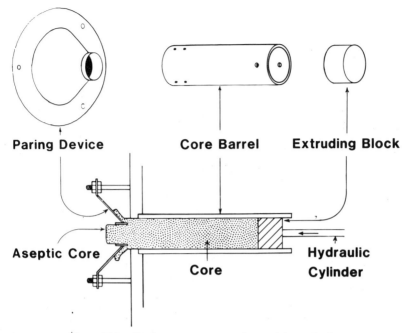

Figure 26.2.   Device to recover cored material aseptically.

water table at sites in Lula, Oklahoma; Pickett, Oklahoma; and Fort Polk, Louisiana. The actual depths were 2.1 and 3.6 m at Fort Polk, 3.0 and 4.9 m at Lula, and 3.6 and 4.8 m at Pickett. At the Pickett site, a special Teflon well was constructed to provide water from the same lens of sand that was sampled by the core barrel. The water was filter-sterilized, then dosed with a suite of organic chemicals and stirred in a sealed container until all compounds were dissolved.

ANALYSIS FOR ORGANIC COMPOUNDS

Water samples were analyzed in essential accordance with EPA protocol Method #601 (EPA, 1979). A 10 mL sample was purged with N_2 at 40 mL/min for 12 min, using a Hewlett Packard 7675A Purge and Trap Unit. Volatile organic compounds were collected on a Tenax resin trap, then thermally desorbed into a gas chromatograph for analysis.

CONSTRUCTION OF THE STATIC MICROCOSMS

Uncontaminated samples of the subsurface were obtained as described above for material from Pickett and Fort Polk. Sterile water was mixed into

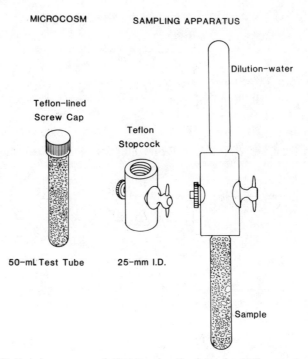

Figure 26.3. Static microcosm used to determine the fate of volatile organic pollutants in subsurface materials.

the sample until it became a slurry, which was then transferred to 35-mL screw-cap test tubes (Figure 26.3). Space was left for approximately 1 mL of dose solution. The remaining space in the microcosms was filled with ground water dosed with organic chemicals prepared as described above. The tip of the tube was wiped clean with a sterile piece of paper, and the tube sealed with a 25-mm Teflon-lined screw cap. Then the dose solution was dispersed through the slurry with a vortex mixer. A portion of the subsurface material was autoclaved prior to dispensing into the test tubes to provide an abiotic control. For material from Lula, the dosed ground water was added to the slurry of the subsurface material prior to the transfer to test tubes.

The microcosms were sampled by diluting the pore water. The cap was removed and the microcosm screwed into one end of a large Teflon stopcock with a 10 mm bore. A 35-mL test tube filled with water was screwed into the other end. Then the cock was opened and the contents of the two tubes mixed together with a vortex mixer. Finally the cock was closed, and the tubes were separated, sealed with Teflon-lined screw caps, and centrifuged gently until the diluted pore water was clear. Microcosms were analyzed after 0, 1, 3, and 9 weeks of incubation. At each time period, at least two microcosms plus an autoclaved microcosm were analyzed.

BEHAVIOR OF ORGANIC POLLUTANTS IN STATIC MICROCOSMS REPRESENTING THE SATURATED ZONE AND DEEPER REGIONS OF THE UNSATURATED ZONE

The behavior of selected organic pollutants added at trace concentrations ($\sim$1 mg/L) to uncontaminated samples of the subsurface is presented in Table 26.1.

Although all the compounds examined in this study have been reported to be biodegradable to at least some extent (Tabak, et al., 1981), all of the chlorinated alkanes and alkenes, with the exception of bromodichloromethane, were stable in material from the deeper subsurface. This finding is consistent with the observed persistence of these compounds in ground water (Zoeteman et al., 1981). Bromodichloromethane degraded slowly (2.4 $\pm$ 0.8% of initial concentration/week) in material from below the water table at Lula, but did not degrade in material that had been autoclaved, suggesting a biological process. However, this process may not have been direct metabolic attack; Bouwer et al. (1981) have shown that HS$^-$, which could have been produced by the respiration of sulfate in the microcosm, can degrade brominated hydrocarbons in ground water.

Toluene and styrene are readily degradable in surface environments, while chlorobenzene is more recalcitrant (based on BOD$_5$ data tabulated by

Table 26.1 Fate of Organic Pollutants in Uncontaminated Samples of the Subsurface[a]

	Lula, OK		Pickett, OK		Fort Polk, LA	
Compound	Above[a]	Below[a]	Above	Below	Above	Below
Bromodichloromethane	S[b]	D[c]				
Chloroform	S	S	S	S	S	S
1,1-Dichloroethane			S	S	S	S
1,2-Dichloroethane	S	S				
1,1,1-Trichloroethane			S	S	S	S
1,1,2-Trichloroethane	S	S				
Trichloroethene	S	S	S	S	S	S
Tetrachloroethene	S	S	S	S	S	S
Toluene	D	D	S	S	D	D
Chlorobenzene	D	S	S	S	S	S
Styrene			D	D	D	D

[a] Sample taken immediately above or below the water table.

[b] S means stable; degradation was not detectable at $P = 0.95$. The average detectable degradation was 3% of the initial concentration per week.

[c] D means degraded, significant at $P = 0.95$.

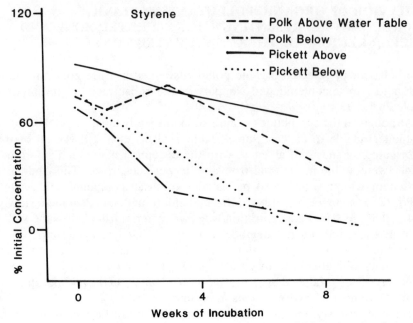

Figure 26.4.   Biodegradation of styrene in uncontaminated material from the subsurface environment.

Verschueren, 1977). These compounds did degrade to some extent in the subsurface environment. Toluene degraded rapidly in subsurface material from Lula (>91%/week), and more slowly in material from Fort Polk (<3%/week); however, degradation was not detectable in material from Pickett. In contrast, the rate of degradation of styrene showed a surprising uniformity, with respect to both position and site (Figure 26.4). Finally, degradation of the more recalcitrant compound, chlorobenzene, was detected only in material from above the water table at Lula (Table 26.1).

CONSTRUCTION OF THE COLUMN MICROCOSM

The column microcosm (Figure 26.5) is described in detail by Bengtsson (Chapter 15). Aseptic core material from below the water table at Pickett was extruded directly into glass columns 7.0 cm I.D. × 40 cm long. The column had a sampling port at 20 cm and at the effluent end.

The feed solution for the columns was stored in flexible Teflon bags supported in a plexiglass box filled with water. The bags had virtually no headspace, so that volatile organic compounds from the feed solution would not be lost. Pressure from a reservoir forced the feed solution onto three replicate columns. Flow through the columns was maintained at 2 cm/day by a peristaltic pump downstream of the column effluent port.

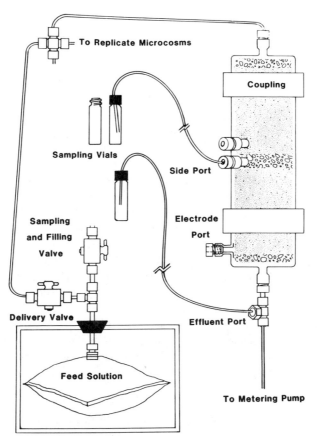

Figure 26.5. Column microcosm used to determine the transport and fate of volatile organic pollutants in sandy aquifer material.

Samples for volatile constituents were drawn by connecting a sterile, gas-tight syringe to the sample port. Samples for non-volatile constituents were drawn by opening the port and allowing the sample to drip into a vial. All materials touching the feed solution, aquifer matrix material, or samples were either Teflon or borosilicate glass. All experimental manipulations were performed aseptically. The microcosm was incubated at 17°C, the *in situ* temperature.

This microcosm was part of a larger project concerning the fate of volatile organic pollutants during land treatment of municipal waste water. The feed solution for the microcosm was prepared from primary municipal waste water that had been renovated by passage through 1.5 m of unsaturated sandy soil. The renovated waste water had dissolved oxygen >4 mg/L and total organic carbon <5 mg/L. The water was filter-sterilized, then dosed with a suite of organic chemicals and stirred in a sealed container until all compounds were dissolved.

BEHAVIOR OF ORGANIC POLLUTANTS IN A COLUMN MICROCOSM REPRESENTING A SANDY AQUIFER

The feed solution to the column microcosm contained eight organic pollut-ants, each at a concentration of approximately 1 mg/L. The feed solution also had a fairly high conductivity, due primarily to chloride salts in the renovated waste water used to prepare the feed solution. This allowed the use of conductivity as a conservative tracer for the infiltrating feed solution. Typical behavior of an organic pollutant in this microcosm is illustrated in Figure 26.6. Chloroform broke through in the column effluent simultane-ously with the increase in conductivity, and showed the same dispersion. There was no detectable sorption or degradation.

Schwarzenbach and Westall (1981) have shown a strong correlation be-tween the extent of sorption of a series of hydrocarbons to aquifer material and the octanol–water partition coefficient of the compound, which is an index of the relative affinity of the compound for hydrophobic environments. They also showed that the extent of sorption of several hydrocarbons to aquifer material and a series of natural sediments was directly related to the organic carbon content of the subsurface material.

The aquifer material from the Pickett site had very little organic matter— 0.02% organic C (personal communication from Marvin Piwoni, Ground Water Research Branch, R. S. Kerr Laboratory, Ada, OK). As a result, the expected retardation of the organic compounds was very low. In fact, all of

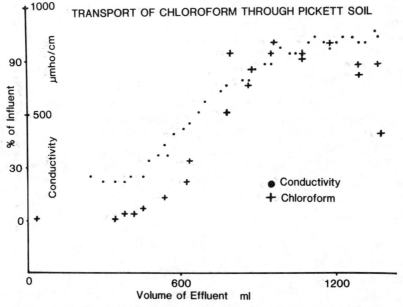

Figure 26.6.   Transport of chloroform and electrolytes through a sandy aquifer.

Table 26.2   Transport and Fate of Organic Pollutants in a Sandy Aquifer (Column Study)

Compound	Highest Concentration in Column Effluent (% of avg. influent conc.)	Retardation relative to Water[a]	Predicted Retardation[b]
Chloroform	100	0.9	1.1
1,1-Dichloroethane	110	0.9	1.1
1,1,1-Trichloroethane	100	0.9	1.1
Trichloroethene	80	1.2	1.1
Tetrachloroethene	60	1.8	1.4
Toluene	120	1.0	1.3
Chlorobenzene	190	<1	1.3

[a] Retention volume of the organic compound ÷ retention volume for conductivity.

[b] Calculated after Schwarzenbach and Westall (1981).

the compounds essentially moved with the water (Table 26.2). As was the case in the test tube microcosm, there was little evidence of degradation of chloroform; 1,1-dichloroethane; 1,1,1-trichloroethane, toluene, or chlorobenzene during the 20 days required for passage through the column. Chlorobenzene slowly degraded in the reservoir of feed solution for the columns. This degradation was probably prevented once the compound entered the microcosm, which would explain the anomalously high recovery of chlorobenzene. Tri- and tetrachloroethene may have degraded in the column microcosm, even though they did not degrade in the test tube microcosms. However, it is more likely that a portion of these highly volatile compounds was lost during sampling and analysis.

CONCLUSIONS

1.   The deeper regions of the unsaturated subsurface and shallow regions of water table aquifers may harbor a microflora which can transform some, but not all, of the organic pollutants commonly encountered in the subsurface environment.

2.   Biodegradation of toluene and chlorobenzene was highly variable between sampling sites, even though these compounds are generally degraded in surface materials. This suggests that biodegradation in the subsurface should not be predicted from the behavior of compounds in material from the surface but must be verified by direct experiment for each compound and each site of interest.

3.   In this study, low concentrations of most halogenated hydrocarbons did not undergo detectable biotransformation in subsurface matrix material obtained from both unsaturated and saturated zones at three different sites.

REFERENCES

Bouwer, E. J., Rittman, B. E., and McCarty, P. L. (1981). Anaerobic degradation of halogenated 1- and 2-carbon organic compounds. *Environ. Sci. Technol.* **15**(5):596–599.

Dunlap, W. J., McNabb, J. F., Scalf, M. R., and Cosby, R. L. (1977). Sampling for organic chemicals and microorganisms in the subsurface EPA-600/2-77-176, Natl. Environ. Research Ctr., Off. Res. and Monitoring, U.S. Environmental Protection Agency, Corvallis, OR.

Schwarzenbach, R. P. and Westall, J. (1981). Transport of nonpolar organic compounds from surface water to groundwater: Laboratory sorption studies. *Environ. Sci. Technol.* **15**(11):1360–1367.

Tabak, H. H., Quave, S. A., Mashni, C. I., and Barth, E. F. (1981). Biodegradability studies with organic priority pollutant compounds. *J. Water Pollut. Control Fed.* **51**(10):1503–1518.

United States Environmental Protection Agency (1979). Guidelines establishing test procedures for the analysis of pollutants. *Fed. Reg.* **44**(223):69464–69575.

Verschueren, K. (1977). *Handbook of Environmental Data on Organic Chemicals.* Van Nostrand Reinhold, New York.

Zoeteman, B. C. J., De Greef, E., and Brinkmann, F. J. J. (1981). Persistency of organic contaminants in groundwater: Lessons from soil pollution incidents in The Netherlands. *Sci. Total Environ.* **21**:187–202.

27

ANAEROBIC TRANSFORMATION, TRANSPORT, AND REMOVAL OF VOLATILE CHLORINATED ORGANICS IN GROUND WATER

Paul R. Wood
Russell F. Lang
Iris L. Payan

Drinking Water Research Center
School of Technology
Florida International University
Tamiami Campus
Miami, Florida

Our research over the last 5 years has shown that ground water may contain from low levels (less than 1 μg/L) to high levels (over 1 g/L) of a family of chemical compounds that can be classified as low-solubility volatile organics. A partial list is presented in Table 27.1.

Many of these compounds are included in the EPA Priority Pollutant List. The list in Table 27.1 is by no means complete. This chapter covers our research work on (1) anaerobic transformation of parent compounds introduced in the aquifer, (2) transport of parent and biodegraded daughter compounds in the aquifer as measured at actual contaminated aquifer sites, and (3) decontamination of an aquifer site with resulting reclamation and use of the contaminated water.

Table 27.1   Low Solubility Volatile Organics Found in Many Contaminated Waters

Aliphatics:	Propane up to C_{10} straight or branched chain hydrocarbons
Aromatics:	Benzene, toluene, ethylbenzene, etc.
Chlorinated aromatics:	Mono and Dichlorobenzenes, etc.
Chlorinated (Halogenated) alkanes:	Chloroethane; methylene chloride; 1,1- and 1,1,1- and other chloroethanes; trihalomethanes; and higher molecular weight halogenated alkanes.
Chlorinated ethenes:	Vinyl chloride, vinylidene chloride, *cis-* and *trans*-1,2-dichloroethylene, trichloroethylene, and tetrachloroethylene.

SOURCE OF CONTAMINATION

We have studied actual contaminated aquifer sites to a depth of approximately 200 to 340 ft and have found the compounds listed in Table 27.1. The presence of many of the compounds listed is the result of dumping or accidental spilling of the compounds onto the ground. For example, hydrocarbon fuels contain the aliphatic and aromatic hydrocarbons listed in Table 27.1. Also, some of the chlorinated methane, ethane, and ethylene compounds originate from usage of the common cleaning and degreasing compounds listed in Table 27.2.

Our initial research work was prompted by the presence of compounds in the aquifer supplying our drinking-water plants. These compounds, which seemed to have no logical source, included vinyl chloride; 1,1-dichloroethene; *cis-* and *trans*-1,2-dichloroethene; 1,1-dichloroethane; and chloroethane. These chemicals are either not actually produced or are not in wide use across the whole country as are the parent compounds in Table 27.2. We have shown that these compounds in the aquifer are the result of anaerobic biodegradation of the three latter parent compounds listed in Table 27.2 (Wood et al., 1981a).

Table 27.2   Annual U.S. Production (1979) of Four Chlorinated Solvents (lb)

Methylene chloride	625,000,000
1,1,1-Trichloroethane	700,000,000
Trichloroethylene	325,000,000
Tetrachloroethylene	750,000,000
Total	2,400,000,000

ANAEROBIC TRANSFORMATION OF PARENT COMPOUNDS

Anaerobic bacteria found in ground water were able to transform parent compounds into the following daughter compounds:

Carbon tetrachloride → chloroform → methylene chloride

$$\begin{array}{l}\text{Tetrachloro-} \\ \text{ethylene}\end{array} \rightarrow \begin{array}{l}\text{trichloro-} \\ \text{ethylene}\end{array} \rightarrow \left\{\begin{array}{l}cis\ 1,2\text{-dichloroethene} \\ trans\ 1,2\text{-dichloroethene} \\ 1,1\text{-dichloroethene}\end{array}\right\} \rightarrow \begin{array}{l}\text{vinyl} \\ \text{chloride}\end{array}$$

1,1,1-trichloroethane → 1,1-dichloroethane → chloroethane

The laboratory work resulted in biodegradation half-life values for the parent and daughter compounds (based on our particular laboratory conditions) as shown in Table 27.3. A half-life value of "long" represents no detectable reduction of the compound under the test conditions over a time period of observation averaging 30–60 days.

In some of the laboratory work using different actual muck-water samples, we found bacteria profiles that seemed to result in loss by biodegradation of an injected parent compound (tetrachloroethylene), with only trace amounts of daughter compounds detected. Thus, as the aquifer bacteria profile changes, different end results may occur. Anaerobic bacteria were isolated from ground water and muck-water samples, and cultured in laboratory media in the presence of tetrachloroethylene. Tentative conclusions on biodegradation of tetrachloroethylene by specific bacteria were as follows:

1. *Clostridium cadaveris* and/or *Clostridium limosum* and/or G+ *cocci* (tetrads) may, in the course of biodegrading tetrachloroethylene, favor the production of trichloroethylene and *cis*-1,2-dichloroethene, with some methylene chloride and/or 1,1-dichloroethene.

2. Big G+ rods and filaments (2 × 10 Trichome) may, alone or in the presence of G+ *cocci* (tetrads), result in tetrachloroethylene biodecay with minor formation of chlorinated by-products.

3. *C. limosum* alone may result in tetrachloroethylene biodecay with minor formation of chlorinated by-products. The same applies to a mix of *Pseudomonas maltophilia* and *Pseudomonas fluorescens* and to *Ps. fluorescens* alone.

4. *Ps. maltophilia* alone may favor heavy growth of chlorinated by-products.

5. G-short, wide rods alone may result in the fastest biodecay of tetrachloroethylene of all the bacteria—singly or in combinations—tested (an honor probably equally shared by *Ps. fluorescens* alone).

6. *Proteus vulgaris* alone seems to result in biodecay of tetrachloroethylene with minor growth of chlorinated by-products.

7. *Enterobacter cloacae, Escherichia coli*, and *Pseudomonas aeruginosa* alone allow growth of chlorinated by-products, including some vinyl chloride in our limited 17 day test.

8. Large G+ rods alone seem to favor minor growth of chlorinated organic by-products as they biodecay tetrachloroethylene.

Examination of actual aquifer contamination sites supported the findings in our laboratory work.

In wells downstream from a site where only trichloroethylene was spilled underground were found the parent compound and the expected daughter compounds *cis-* and *trans-*1,2-dichloroethene, 1,1-dichloroethene, and vinyl

Table 27.3   Biodegradation and Physical Properties of Interest on the Volatile Organics Found in Ground Water

	Bio Half-Life (days)	Solubility (ppm)	H_{iPC}	% Removal per Four Series Aeration Stage
Vinyl chloride	long	2700?	5.2	94
trans-1,2-Dichloroethene	long	6300	0.16	87
cis-1,2-Dichloroethene	long	3500	0.29	85
1,1-Dichloroethene	53	8000	0.62	88
Trichloroethylene	43	1100	0.48	86
Tetrachloroethylene	34	130	1.2	89
Methyl chloride	est.<11	—	—	—
Methylene chloride	11	19400	0.1	82
Chloroform	36	8200	0.15	87
Carbon tetrachloride	14	800	0.97	89
Bromodichloromethane	—	6060	0.099	82
Chlorodibromomethane	—	5190	0.043	80
Bromoform	—	4240	0.023	78
Chloroethane	10	—	—	—
1,2-Dichloroethane	long	8700	0.05	80
1,1-Dichloroethane	long	5100	0.24	84
1,1,2-Trichloroethane	24	—	—	—
1,1,1-Trichloroethane	16	720	1.2	89
Benzene	—	1780	0.23	85
Chlorobenzene	—	488	0.13	85
p-Dichlorobenzene	—	79	0.11	82
o-Dichlorobenzene	—	145	0.083	81

chloride. In the laboratory we had previously found that biodegradation of either tri- or tetrachloroethylene resulted in production of *cis-* over *trans-* 1,2-dichloroethylene by a factor of perhaps 25 to 1 or more. This was confirmed in the above spill site. This was also confirmed in other actual spill sites where tri- and/or tetrachloroethylene was present. In other sites, we also confirmed that 1,1,1-trichloroethane biodegrades to 1,1-dichloroethane, chloroethane, and methylene chloride.

TRANSPORT OF PARENT AND DAUGHTER COMPOUNDS IN THE AQUIFER

In actual contaminated aquifer sites, we were involved in detailed mapping of the original spill area and establishing of the boundaries of the downstream contamination plume. This work extended to an approximate maximum depth of 200 to 340 ft and a downstream distance of 2 miles. It was estimated that some of the initial spills were perhaps 15 or more years old. In this work, it soon became apparent that as investigations were made of the plume further from the initial spill site, patterns were developing in the types and ratios of specific compounds found. Analysis of these patterns suggested that there might be some predictability in what might be found based on what was actually found in the sites investigated. Our observations suggested that the biodegradation half-life values reported in Table 27.3 might assist in explaining and thus predicting what compounds have been and might be found at progressively further distances from the initial spill site. Also, assuming the compound initially spilled was known, the half-life values along with our findings on the favored ratio of *cis-* over *trans-*1,2-dichloroethene might allow prediction of what daughter compounds would form and perhaps their ratios. Also, perhaps, using the compounds found, one might project backward to the initial parent compounds spilled, even though they perhaps no longer existed at the site studied.

The biodegradation half-life values reported in Table 27.3 represent our findings in the laboratory under our particular test conditions. We do not suggest that they define an actual spill condition. From actual sites where spills occurred perhaps more than 15 years ago, we have found extremely high concentrations of the parent compounds spilled. For example, concentrations of tri- and tetrachloroethylene and 1,1,1-trichloroethane in monitor well water samples were in the 500,000 μg/L range, with combined contaminant levels of over 1,200,000 μg/L. However, samples of water taken from monitoring wells progressively further from the spill site and progressively deeper in the aquifer showed a marked difference in chemical profile, both in types of compounds and concentrations. At such sites as described above, the further and deeper from the initial spill site, the level of the parent compound decreased—actually decreased to zero levels in some cases— and the ratio of long half-life daughter compounds increased. Daughter compounds with a short half-life were not found at distances furthest from the spill site.

Parent compounds may persist in a spill site for many years, even though their half-life value is short. This may occur because above certain concentrations the compound might inhibit growth of bacteria. Natural dilution as the plume moves downstream may then allow biodegradation to occur.

In actual spill sites, where tri- and/or tetrachloroethylene was spilled, we observed that with increased distances from the initial spill site, both downstream and in depth, the concentration of the parent compounds decreased, in some cases reaching zero, while the ratios of vinyl chloride, *cis*- and *trans*-1,2-dichloroethene, and 1,1-dichloroethene to the parent compounds increased. At the furthest distances, vinyl chloride and *cis*-1,2-dichloroethene were found with decreasing levels of *trans*-1,2-dichloroethene and 1,1-dichloroethene. These findings appeared to be consistent with the half-life ratios reported in Table 27.3 and with the favored biodegradation to *cis*- over *trans*-1,2-dichloroethene.

In sites where 1,1,1-trichloroethane was spilled, the concentration of the parent compound decreased with distance while the concentration ratio of 1,1-dichloroethane increased. Chloroethane was found when the concentration of the parent compound was appreciably high, but further from the spill site where only 1,1-dichloroethane was present, chloroethane was not

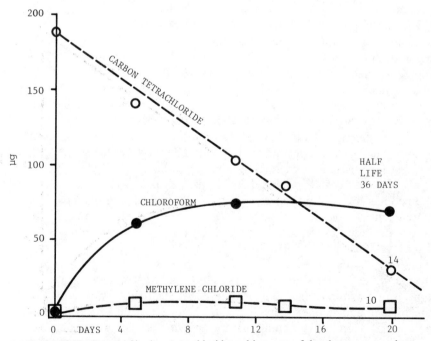

Figure 27.1. Decay of carbon tetrachloride and increase of daughter compounds.

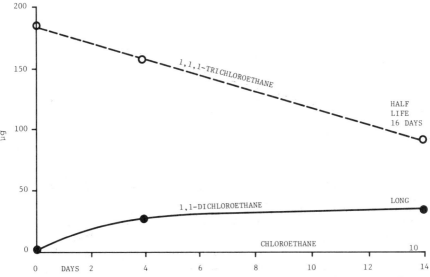

Figure 27.2.   Decay of 1,1,1-trichloroethane and increase of daughter compounds.

found. Again, these results appeared to be explainable by the half-life values in Table 27.3.

No actual sites involving spills of methylene chloride alone or chloroform or carbon tetrachloride were studied. However, our laboratory work with these compounds suggests that the reported half-life values may be of value in actual spill cases. In Figure 27.1, the biodegradation of carbon tetrachloride in a muck-water sample is shown. Chloroform, with a half-life of 36 days, built up in concentration from day zero as the carbon tetrachloride (half-life 14 days) steadily decreased. Methylene chloride, with a half-life of only 11 days, did not build up to a very high concentration. No methyl chloride was detected, suggesting that its half-life was probably less than the 11 days for methylene chloride. From these data, one might be able to explain or predict the compounds found at a carbon tetrachloride, chloroform, or methylene chloride spill site. Similar laboratory degradation curves are shown for 1,1,1-trichloroethane in Figure 27.2, and for tetrachloroethylene in Figure 27.3.

DECONTAMINATION OF AQUIFER AND WATER RECLAMATION

It was recognized that it would be desirable to have a practical method of decontaminating both the initial spill site and the downstream aquifer both to prevent the spread of the contamination plume and to reclaim the contaminated water (Wood et al., 1981b; Wood et al., 1980).

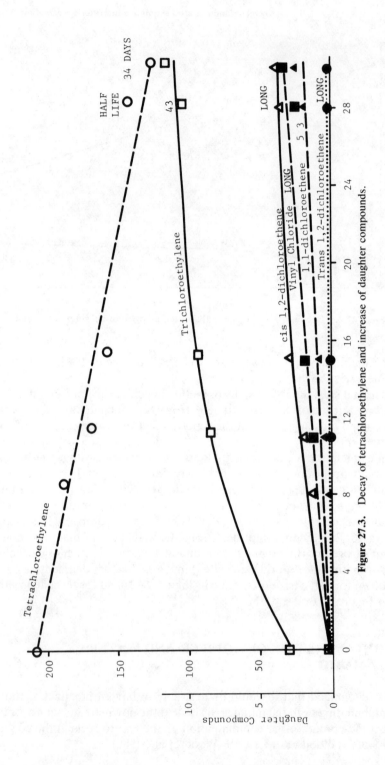

Figure 27.3. Decay of tetrachloroethylene and increase of daughter compounds.

500

Table 27.4   Spray Nozzles Evaluated for Removing Chlorinated Solvents from Ground Water[a]

Nozzle No.	Spray Pattern	Water Flow Rate (gal/min)
4CRC250	Hollow cone (45°)	300 at 10 psi
$1\frac{1}{2}$H20	Full cone (74°)	24 at 10 psi
2H35	Full cone (75°)	42 at 10 psi
2H47W	Full cone (124°)	55 at 10 psi
2H151150	Full cone (15°)	58 at 10 psi
2H50	Full cone (83°)	59 at 10 psi
4H154500	Full cone (15°)	225 at 10 psi
4RR65160	Full cone (65°)	279 at 10 psi
$\frac{3}{4}$ FF-18	Fog jet	13 at 20 psi
$1\frac{1}{2}$ 29F-35	Fog jet	25 at 20 psi
$1\frac{1}{4}$ FF-70	Fog jet	50 at 20 psi

[a] Spray Systems Company, Wheaton, IL, Industrial Catalog 27.

An actual spill site was chosen, but before a spray head aeration system could be designed for the specific site, it was necessary to determine the parameters affecting the rate of loss of the volatile compounds due to spray head aeration. Spraying Systems Co. of Wheaton, IL (Industrial Catalog 27) supplied a series of spray nozzles. The series covered a wide range of water flow rates and spray pattern types (Table 27.4). The nozzles were tested individually at well sites with varying levels of volatile organic chemical concentrations, ranging from combined levels of contamination of over 10^5 μg/L to very low levels of less than 10 μg/L. The nozzles were positioned 8 ft above the ground and sprayed either up or down at varying water flow rates and pressures. When the spray pattern was directed upward, the average distance of water droplet travel was estimated.

In Figure 27.4, a plot of log H_{iPC} versus percent removal for six contaminants at total contaminant levels of 11, 131, 4648, and 130,170 μg/L shows that for a spray head aeration system, the rate loss is the same regardless of concentration, as long as contaminants are completely dissolved (Wood et al., 1981b). These data also show that in a spray head aeration system log H_{iPC} versus percent removal is approximately a straight line plot. This approximate straight line relationship applies to any type of aeration system. For any aeration system, once this plot is established with data points for at least two compounds, the rate loss in the system for any volatile compound can be predicted if its H_{iPC} is known. Also, from such a plot if the initial concentration for any compound in water is known, we can predict the final concentration after one or more passes through the spray head system and

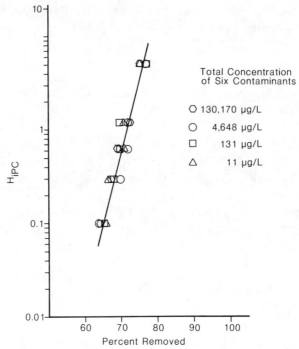

Figure 27.4.   Removal data for six contaminants through spray nozzle 2H35 (all three runs sprayed 8 ft down, 42 gal/min at 10 psi).

thus can design a system to achieve any final concentration desired, including zero.

Figures 27.5, 27.6, and 27.7 show the removal results through three spray heads, sprayed down 8 ft and then up starting at the same level of 8 ft. The water flow rate and psi (energy input) was the same when sprayed either up or down. In each case the average distance of water droplet travel was estimated. For the same energy input, spraying upward resulted in a much greater percent removal in all three cases. The average water droplet travel distance for the upward spray was 24 ft, 16 ft, and 32 ft in Figures 27.5, 27.6, and 27.7, respectively, compared to 8 ft in the downward spray. In Figures 27.5, 27.6, and 27.7, the total contaminant concentration in μg/L is shown for each series of tests, but the concentration of each chemical was determined to obtain each point on its plot. From all the data collected, it became apparent that the droplet travel distance was the controlling factor in the removal rate. This is illustrated with data for vinyl chloride in Figure 27.5. In the downward spray test, the initial concentration of vinyl chloride was 232 μg/L, and spraying 8 ft down reduced the concentration to 53 μg/L, a removal rate of 77%. In the upward spray test run, where the average droplet travel distance was 24 ft, the initial concentration was 220 μg/L and the final

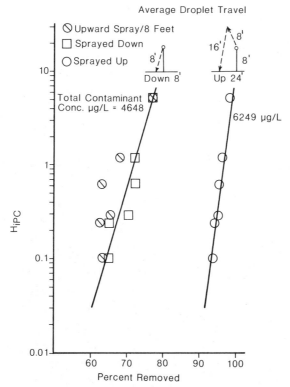

Figure 27.5. Percent reduction of six contaminants through spray head 2H35 sprayed up and down (full cone 75°, 42 gal/min, 10 psi).

concentration 2.3 μg/L, a removal rate of 99%. We can consider the data in two ways: (1) if the 8 ft down test were performed three times in series with a removal rate of 77% per pass, the initial concentration of 232 μg/L would be reduced to 2.8 μg/L, which is very close to the 2.3 μg/L obtained in the upward spray test, which was equal to a series of three 8 ft droplet travel times (24 ft ÷ 8 ft = 3); or (2) we can take the upward spray data and calculate the percent removal for three 8 ft passes in series. This calculates to 78% removal per 8 foot section:

$$\text{initial conc.} = 220 \quad \mu\text{g/L} \times (0.22) = 48 \quad \mu\text{g/L}$$

$$48 \quad \mu\text{g/L} \times (0.22) = 10.6 \,\mu\text{g/L}$$

$$10.6 \,\mu\text{g/L} \times (0.22) = 2.3 \,\mu\text{g/L}$$

This 8 ft section removal value for each compound in the upward spray test is plotted in Figure 27.5, and all the compounds fall near the linear curve for the 8 ft downward spray test. The same calculations are shown in Figures

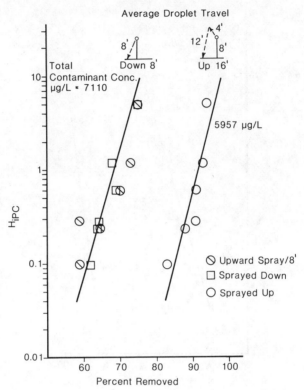

Figure 27.6. Percent reduction of six contaminants through spray head 2H47W sprayed up and down (full cone 124°, 55 gal/min, 10 psi).

27.6 and 27.7. Thus, it appears that the rate of loss is proportional to droplet travel distance. If the rate of loss is determined for any unit distance, 8 ft for example, we can calculate the rate of loss for any other travel distance.

Early in the spray head research program, it became obvious that a hollow cone spray pattern was undesirable. In a full cone spray pattern, the same volume of water would be broken up into smaller water droplets, resulting in higher percent removals. For comparison, even when nozzle 4CRC250 was sprayed upward, the removal rate was almost identical to the removal rate for nozzle 2H35 sprayed downward (Figure 27.5). Of course, the two tests cannot be directly compared since the water flow rates at 10 psi vary. Water droplet size also greatly influences rate loss. The droplet size in the 4CRC250 run was much larger than in the 2H35 run. The effect of droplet size can be seen in Figure 27.8, where a study of a series of full cone spray nozzles is depicted. In all three runs at 10 psi, the average droplet travel distance was approximately equal. The major difference was in droplet size, which was larger as water volume increased. In Figure 27.8, the 24 and 42 gal/min heads gave much higher percent removals. Data points based on this one series of data do not show much difference between the 24 and 42 gal/

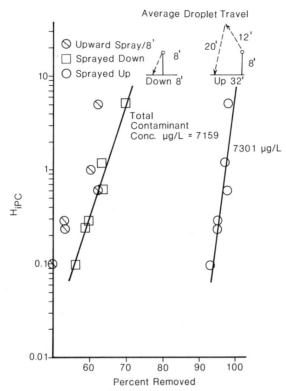

Figure 27.7. Percent reduction of six contaminants through spray head 1¼ FF-70 sprayed up and down (fog jet, 50 gal/min, 20 psi).

min flow rates. Repeated runs would probably indicate a preference for the lower flow rate.

Fog nozzles produce the smallest water droplets. Results obtained with three fog nozzles are shown in Figure 27.9. The fog nozzles require a minimum pressure of 20 psi for fogging to occur. However, the removal curve in Figure 27.9 for the 25 and 50 gal/min runs is almost identical to the curve in Figure 27.8 for the 24 and 42 gal/min runs using full cone spray nozzles. In Figure 27.9, the three data points for the 13 gal/min run appear to fall on a line to the right of the other two runs. In this run, 99+% was being removed, making accuracy of data plots questionable. From all of the data, it appears that in general we desire a small droplet, but must weigh droplet size with other requirements. For our case, in which the water was to be purified and then used, maximum recovery of water was essential. Fog nozzles would result in losses due to particle drift, so for maximum recovery it was more efficient to use full cone nozzles.

Additional tests showed that nozzle 2H50 gave consistently better removal rates than 2H35. Therefore, 2H50 was chosen for the spray head aeration system.

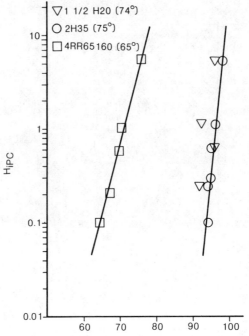

Figure 27.8. Effect of water flow rate on percent removal (full jet, 10 psi up).

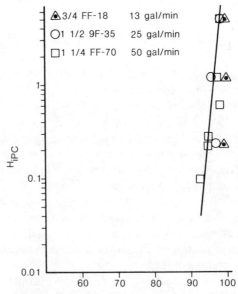

Figure 27.9. Removal data using fine particle spray fog nozzles (fog jets, 20 psi up).

SPRAY HEAD AERATION SYSTEM DESIGN

The finished four stage series unit is shown in Figure 27.10. The first stage built was the elevated stage to the right. It was originally a cascade aerator with a single column approximately 6 ft tall, in the center of a 17 ft × 17 ft basin reservoir. The basin reservoir capacity was increased and a 14 ft cypress wall box constructed over the basin to retain all the water spray. A 9 in air gap was maintained between the top of the concrete basin and the bottom of the cypress wall. Twenty 2H50 spray heads were equally spaced just above the water level in the basin. Spraying upward at 10 psi, each head could handle a flow of approximately 60 gal of water/min. The requirement was 1000 gal/min through the unit; 20 spray heads would handle 1200 gal/min.

The top of the spray pattern was approximately 2 ft from the top of the cypress wall. The maximum droplet travel distances were approximately 26 ft (13 ft up, and 13 ft down). We had no idea how the individual nozzle spray patterns would interact with closely spaced neighboring nozzles. However, the performance of the unit was excellent and prompted design and construction of three more units in series. These units have a concrete reservoir base 20 ft × 25 ft × 6 ft deep (approximately 20,000 gal capacity). The cypress walls are 14 ft high, with a 1 ft air gap between the bottom of the cypress wall and top of the concrete wall. Again, 20 2H50 spray heads were evenly spaced in each unit at just about the surface of the water in the basin. The pump for each unit handled 1500 gal/min, and the unit received 1000 gal/min from the wells. Thus, the pump draws 500 gal/min from the reservoir. A view down into the cypress walls is shown in Figure 27.11. The four stage

Figure 27.10. Four series spray head aeration system installation for aquifer decontamination.

Figure 27.11. View of spray in cypress wall units (no forced air) of four series spray head aeration installation.

aeration unit was tested after construction with water from one well with high levels of contaminants. Results are shown in Table 27.5.

A plot of removal data for the first three stages is shown in Figure 27.12. Fourth stage data are not included because accuracy of data in the 0 to 0.1 μg/L range is doubtful. In Figure 27.12, it is apparent that the percent removal is approximately equal for all three stages, as expected. For example, approximately 89% of 1,1,1-trichloroethane was removed by each stage. The

Table 27.5 Concentration of Four Main Volatile Organic Compounds in Water on Initial Test Run of Four Stage Aeration Unit

	Concentration (μg/L)	
	Entering First Stage	Fourth Stage Effluent
Vinyl Chloride	665	0
1,1,1-Trichloroethane	1220	0.2
cis-1,2-Dichloroethene	360	0.1
1,1-Dichloroethene	3150	2.0

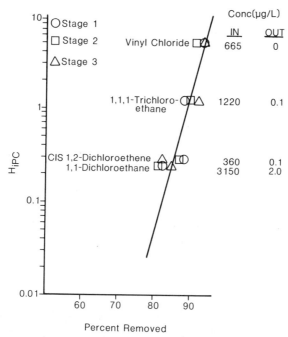

Figure 27.12.   Removal data for first three stages in four stage spray head aeration system.

first stage received 1220 μg/L and reduced the value by 89%, to 134 μg/L. The second stage reduced this value to 15 μg/L and the third stage to 1.6 μg/L. The fourth stage reduced the value to 0.2 μg/L.

The water droplet size from the 2H50 nozzles was large enough to result in minimal loss of water through the system by spray drift.

The data in Table 27.6 illustrate what the four stage unit would do if higher influent levels were introduced.

Table 27.6   Projected Four Stage Spray Head Aeration Unit Performance on High Concentration Volatile Organic Contaminated Water

	Concentration (μg/L)				
	Inlet	First Stage Effluent	Second Stage Effluent	Third Stage Effluent	Fourth Stage Effluent
Vinyl chloride	10,000	600	36	2.2	0.1
1,1,1-Trichloroethane	100,000	10,000	1,000	100	10
cis-1,2-Dichloroethene	20,000	2,900	440	65	10
1,1-Dichloroethane	15,000	2,440	390	63	10

Table 27.3 shows the percent removed per stage for a wide range of volatile organic contaminants based on H_{iPC}.

If recovery of the maximum amount of water through a spray head aeration system were not important, much greater removal rates could be achieved than are shown in Tables 27.3, 27.5, and 27.6. For example, using a $\frac{3}{4}$ FF-18 fog jet nozzle (Table 27.4), the rate loss for 1,1-dichloroethane (the hardest compound in Table 27.6 to remove) would be 15,000 μg/L to 150 μg/L to 1.5 μg/L in just two passes. Even higher removal rates would be achieved if the fog nozzle were positioned high off the ground.

After 6 months of operation, the spray aeration system is not only producing water of potable quality, but also decontaminating the ground water. Because water is being pumped from the wells in the contamination plume, downgradient movement of contaminants is prevented. Consequently, the well field aeration system is acting as an effective contaminant containment and clean-up scheme.

CONCLUSIONS

1. This study indicates that the presence of the highly volatile chloroethene compounds—vinyl chloride, 1,1-dichloroethene, *cis-* and *trans-*1,2-dichloroethene—in our raw ground water is probably a result of biodegradation of trichloroethylene and/or tetrachloroethylene, which are found widely spread in the environment as a result of our widespread use of these compounds.

2. A simple laboratory method was developed for assaying biodegradation of highly volatile chlorinated organic compounds in water, soils and sediments, and bacteria culture media.

3. All of the chlorinated methane, ethane, and ethene compounds studied appear to be susceptible to biodegradation in the environment.

4. A rate-of-biodegradation technique, based on the assumption that the degradation slope observed was constant in the test conditions, assigned biodegradation half-life values for all the compounds tested under anaerobic conditions favoring daughter compound formation.

5. The bacterial population profile of a given system, which varies seasonally, appears to determine the intermediate biodegraded chlorinated organic compound profile in the system. Certain bacterial profiles, for example, will reduce all the tetrachloroethylene present in a given time, with the formation of only intermittent trace quantities of lower chlorinated compounds. Other bacterial profiles will reduce tetrachloroethylene producing all possible lower chlorinated compounds.

6. The resulting concentration of the lower chlorinated compounds may be dependent on the individual half-life of each product and the favored end-product of biodegradation in a given system. For example, biodegradation of tri- and/or tetrachloroethylene favors the production of *cis-* over *trans-*1,2-

dichloroethene by a ratio of approximately 25 : 1. Therefore, while the bio-degradation half-life of the two compounds (*cis* and *trans*) in a bacterial profile system where they are formed is long and perhaps equal, the concentration of *cis* will always be much greater than that of *trans*.

7.   Using a technique for estimation of biodegradation half-life for the families of chlorinated compounds studied, we may be able to understand and predict the chlorinated organic profile and transport in an environmental system after introduction of any single member of the family.

8.   Field data from actual below-ground accidental spills of these halogenated parent compounds in Florida and other states has confirmed our laboratory biodegration findings.

9.   Tentative conclusions are presented for specific bacterial activity in the biodegradation of tetrachloroethylene under laboratory conditions.

10.   Spray head aeration is very effective for removal of a wide range of volatile organic contaminants in water. No forced air is necessary. Spray heads are available for a wide range of applications.

11.   Designing and predicting the performance of spray head aeration systems is now possible based on our research, design, and application.

12.   Merits of such a system are water reclamation and reuse, prevention of a contaminated plume's spreading in the aquifer, and aquifer decontamination.

REFERENCES

Wood, P. R., Parsons, F. Z., Lang, R. F., and Payan, I. L. (1981a). Introductory study of the biodegradation of the chlorinated methane, ethane, and ethene compounds. Presented at the American Water Works Association Annual Conference and Exposition, June 7–11, St. Louis, MO.

Wood, P. R., Curtis, F. W., Lang, R. F., and Payan I. L. (1981b). Removal of organics from water by aeration. Presented at the American Water Works Association Annual Conference and Exposition, June 7–11. St. Louis, MO.

Wood, P. R. DeMarco, J., Curtis, F. W., Harween, H. J., and Lang, R. F. (1980). Removal of organics from water by aeration. A progress report for Experimental Design No. 11, of Pilot Plant Project for Removing Organic Substances from Drinking Water, Cooperative Agreement #CR 806890-01, Water Supply Research Div., Municipal Environment Research Laboratory, Office of Research and Development, U.S. Environmental Protection Agency, Cincinnati, OH.

28

BASIC CONCEPTS FOR GROUND WATER TRANSPORT MODELING

Philip B. Bedient
Robert C. Borden
David I. Leib

National Center for Ground Water Research
Rice University
Houston, Texas

Many articles have been written over the past several years reviewing the science and art of ground water transport modeling. These reviews have generally been written at a fairly sophisticated level, so audiences with little formal training in transport theory may have difficulty in interpreting them. While a great deal of accessible literature exists for ground water flow models in standard ground water hydrology texts, the concepts of transport modeling have been reported in many different sources and at many different levels of difficulty. Very few articles exist to explain transport models at a basic level for hydrologists, hydrogeologists and ground water resource managers concerned mainly with the migration of contaminants from various waste sources.

Recent efforts to bridge this gap include two excellent books by Mercer and Faust (1981) and Wang and Anderson (1982). Freeze and Cherry (1979) also offer clear descriptions of many transport mechanisms and discuss some of the governing equations. However, a source still does not exist that describes transport concepts beginning with one- and two-dimensional solutions and including advection, dispersion, adsorption, and decay for the audience identified above. The simple concepts presented all in one place in this chapter serve as a compilation of existing theory and application for predicting migration and transport from waste sources in ground water.

TRANSPORT PROCESSES

Transport processes of concern in ground waste include advection, dispersion, adsorption, decay, and chemical reaction. The first two mechanisms have been analyzed in some detail for both laboratory and field conditions, while the latter three processes are the focus of current research efforts. The incorporation of these transport mechanisms into ground water model formulations is described in more detail by Bredehoeft and Pinder (1973), Fried (1975), Anderson (1979), Bear (1979), and Freeze and Cherry (1979).

Advection represents the movement of a contaminant with the bulk fluid according to the seepage velocity in the pore space. Figure 28.1 shows the resulting breakthrough curve in one dimension (1-D) for solute transport in the absence of dispersion. There are certain cases in the field where an advective model provides a useful estimate of contaminant transport. Some models include the concept of arrival time by integration along known

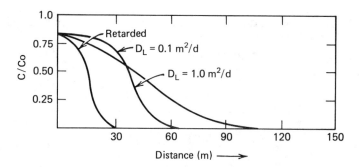

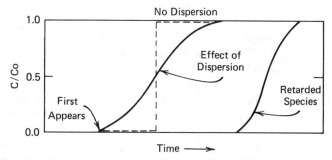

Figure 28.1. Breakthrough curves in 1-D showing effects of dispersion and retardation.

streamlines (Nelson, 1977). Others set up an induced flow field through injection or pumping and evaluate breakthrough curves by numerical integration along flow lines. Dispersion is not directly considered in these models, but results from the variation of velocity and arrival times in the flow field (Charbeneau, 1981, 1982).

The dispersion process is described in detail in Bear (1979), Anderson (1979), and Freeze and Cherry (1979). Dispersion is due mainly to heterogeneities in the medium that cause variations in flow velocities and flow paths. Laboratory column studies yield dispersion estimates on the order of centimeters, while values in field studies may be on the order of meters. Figure 28.1 shows that spreading out of the contaminant front due to the dispersion of the contaminant in the porous media.

Dispersive flux in a flow field with average velocity components V_x and V_y can be written in terms of the statistical fluctuations of velocity about the average, V_{x^*} and V_{y^*}. For the case where $V_x = \bar{V}_x$, a constant, and $V_y = 0$, dispersive flux is assumed proportional to the concentration gradient in the x direction:

$$f_{x^*} = nCV_{x^*} = -nD_L \frac{\delta C}{\delta x} \tag{28.1}$$

where D_L is the longitudinal dispersion coefficient, n is porosity and C is concentration of contaminant tracer. Similarly, the dispersive flux f_{y^*} is assumed proportional to the concentration gradient in the y direction:

$$f_{y^*} = nCV_{y^*} = -nD_T \frac{\delta C}{\delta y} \tag{28.2}$$

For a uniform flow field with average velocity $\bar{V}_x$

$$D_L = \alpha_L \bar{V}_x \tag{28.3}$$
$$D_T = \alpha_T \bar{V}_x$$

where α_L and α_T are the longitudinal and transverse dispersivities, respectively.

Dispersivity values have usually been set constant in transport models, but recent efforts by Smith and Schwartz (1980) and Gelhar et al. (1979) indicate that dispersivity depends on the distribution of heterogeneities and the scale of the field problem. Many investigators are presently working on the complex problem of estimating dispersivity from field tracer studies and pump tests, and both statistical and deterministic models have been postulated (Anderson, 1979).

It should be noted that dispersion coefficients become more complex in a nonuniform flow field characterized by V_x and V_y. The dispersion coefficient relates the mass flux vector to the gradient of concentration and can be represented as a second rank tensor (Bear, 1979). Through careful definition of the coordinate system, relationships can be developed between D_L, D_T and the components of the tensor $[D]$ (Wang and Anderson, 1982). Bear (1979) defines the coefficient of hydrodynamic dispersion D_h as the sum of coefficients of mechanical dispersion D and of molecular diffusion, D_m, although molecular diffusion is often several orders of magnitude less than D.

While there exist many reactions which can alter contaminant concentrations in ground water, adsorption onto the soil matrix appears to be one of the dominant mechanisms. The concept of the isotherm is used to relate the amount of contaminant adsorbed by the solids S to the concentration in solution, C. One of the most commonly used forms is the Freundlich isotherm,

$$S = K_d C^b \qquad (28.4)$$

where S is the mass of solute adsorbed per unit bulk dry mass of porous media, K_d is the distribution coefficient, and b is an experimentally derived coefficient. If $b = 1$, Eq. (28.4) is known as the linear isotherm and is incorporated into the 1-D advective–dispersion equation in the following way:

$$D_L \frac{\partial^2 C}{\partial x^2} - \overline{V} \frac{\partial C}{\partial x} - \frac{\rho_b}{n} \frac{\partial S}{\partial t} = \frac{\partial C}{\partial t} \qquad (28.5)$$

where ρ_b is the bulk dry mass density, n is porosity, and

$$- \frac{\rho_b}{n} \frac{\partial S}{\partial t} = \frac{\rho_b}{n} \frac{dS}{dC} \frac{\partial C}{\partial t}$$

For the case of the linear isotherm, $(dS/dC) = K_d$, and

$$D_L \frac{\partial^2 C}{\partial x^2} - \overline{V} \frac{\partial C}{\partial x} = \frac{\partial C}{\partial t} \left(1 + \frac{\rho_b}{n} K_d \right)$$

or finally,

$$\frac{D_L}{R} \frac{\partial^2 C}{\partial x^2} - \frac{\overline{V}}{R} \frac{\partial C}{\partial x} = \frac{\partial C}{\partial t} \qquad (28.6)$$

where $R = [1 + (\rho_b/n) K_d]$ = retardation factor, which has the effect of retarding the adsorbed species relative to the advective velocity of the ground water (Figure 28.1). The retardation factor can be a useful tool for the case of linear isotherms with fast, reversible adsorption. Retarded fronts can be derived from conservative fronts by adjusting $\overline{V}/R$ and D_L/R in 1-D. Typical values of R for organics often encountered in field sites range from 2 to 10. The effect can be seen in 2-D by solving the governing equations in x and y (see next section). Pickens and Lennox (1976) present a graphic display of the effects of K_d and α_L and α_T values on observed front locations.

The use of the distribution coefficient assumes that partitioning reactions between solute and soil are very fast relative to the rate of ground water flow. Thus, it is possible for nonequilibrium fronts to occur that appear to migrate faster than retarded fronts, which are at equilibrium. These complexities involve other rate kinetic factors beyond the scope of simple models discussed in this paper.

GOVERNING FLOW AND TRANSPORT EQUATIONS

The differential equation for simulating ground water flow in two dimensions is usually written

$$\frac{\partial}{\partial x_i}\left(T_{ij}\frac{\partial h}{\partial x_j}\right) = S\frac{\partial h}{\partial t} + W \tag{28.7}$$

where T_{ij} = transmissivity tensor $(L^2/T) = K_{ij}b$
K_{ij} = hydraulic conductivity tensor (L/T)
S = storage coefficient
W = source or sink term (L/T)
x_i = cartesian coordinates (L)
h = hydraulic head (L)
b = aquifer thickness (L)

The dispersion equation is usually written

$$\frac{\partial}{\partial x_i}\left(D_{ij}\frac{\partial C}{\partial x_j}\right) - \frac{\partial}{\partial x_i}(CV_i) - \frac{C_0 W}{nb} + \Sigma R_k = \frac{\partial C}{\partial t} \tag{28.8}$$

where C = concentration of solute (M/L^3)
V_i = seepage velocity (L/T) averaged in the vertical direction
D_{ij} = coefficient of dispersion (L^2/T)
C_0 = solute concentration in source or sink fluid (M/L^3)
R_k = rate of addition or removal of solute $(\pm)$ (M/L^3T)
n = porosity

The various terms in Eq. (28.8) are usually referred to as dispersive transport, advective transport, source or sink, and reaction—all of which sum to produce a change in species concentration. At the present time, there are difficulties in attempting to use the mass transport equation to describe an actual field site in 2-D. Dispersivities in the x and y direction are difficult to estimate from tracer tests due to the presence of spatial heterogeneities and other reactions in the porous media. Estimation of hydraulic conductivity, and associated velocities, is also related to the presence of field heterogeneities that are often unknown. The source or sink concentrations that drive the model are usually assumed constant in time, but may actually have varied through historical time. A particularly serious problem appears to be the reaction term, which may represent adsorption, ion exchange, or decay. The assumption of equilibrium conditions and the selection of rate coefficients are both subject to some error and may create difficult prediction problems at many field sites.

Despite all the above mentioned problems, mass transport models still offer the most reliable approach to organization of field data, prediction of plume migration, and ultimate management of waste-disposal problems. While a complete three-dimensional scenario with all rate coefficients included will probably never be achieved, presently existing 1-D and 2-D solute transport models have much to offer in simplifying and providing insight into complex ground water problems.

ONE-DIMENSIONAL MODELS

The governing mass transport Eq. (28.8) is difficult to solve in cases of practical interest due to boundary irregularities and variations in aquifer characteristics, so numerical methods must generally be employed. There are however, a limited number of relatively simple, 1-D problems for which analytical solutions exist. Some of these cases are presented here in order to gain insights into the effect of dispersion and adsorption on the overall patterns produced.

The governing assumptions include the following: (1) the tracer is ideal, with constant density and viscosity; (2) the fluid is incompressible; (3) the medium is homogeneous and isotropic; and (4) only saturated flow is considered. For the case of a nonreactive tracer in 1-D flow in the $+x$ direction,

$$\frac{\partial C}{\partial t} = D_h \frac{\partial^2 C}{\partial x^2} - \frac{q}{n} \frac{\partial C}{\partial x} \qquad (28.9)$$

where n = porosity, q = specific discharge, and D_h = coefficient of hydrodynamic dispersion (Bear, 1979).

Several different solutions can be derived for Eq. (28.9) depending on initial and boundary conditions and whether the tracer input is a slug input or a continuous release. For an infinite column with background concentration C_1 ($0 \leq x \leq +\infty$) and input tracer concentration C_0 at $-\infty \leq x \leq 0$ for $t \leq 0$,

Bear (1960) solves the problem using the Laplace transform:

$$\frac{C(x,t) - C_0}{C_1 - C_0} = \frac{1}{2} \text{ erfc} \left[\frac{x - (q/n)t}{2\sqrt{D_h t}} \right] \tag{28.10}$$

where erfc = complementary error function = $(2/\sqrt{\pi}) \int_x^\infty e^{-u^2} du$. The point where $(C - C_0)/(C_1 - C_0) = 0.5$ travels with the mean flow $V = q/n$, and $\sigma^2 = 2Dt$ is proportional to the total path traveled.

The corresponding solution can be derived for the injection of a tracer slug at $x = 0$ with background concentration $C_1 = 0$ in the column. As the slug moves downstream with specific discharge q in the $+x$ direction, it spreads out according to

$$C(x,t) = \frac{M/n}{(4\pi D_h t)^{1/2}} \exp \left[-\frac{(x - (q/n)t)^2}{4D_h t} \right] \tag{28.11}$$

where M is the injected mass.

A graph comparing Eqs. (28.10) and (28.11) is shown in Figure 28.2 and indicates the obvious difference between slug and continuous release transport problems in 1-D.

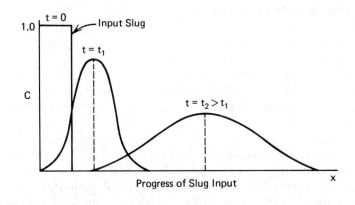

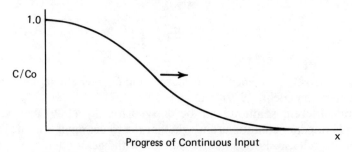

Figure 28.2.   Comparison of slug and continuous tracer input.

A case of more practical interest is that of a semi-infinite column where the column ($x > 0$) is initially at $C = 0$, and is connected to a reservoir containing tracer at $C = C_0$ ($x = 0$). The tracer continuously moves down the column at specific discharge q in the $+x$ direction. It is assumed that $C = 0$ at $x = \infty$. For the case of linear adsorption, Eq. (28.9) becomes

$$\frac{\partial C}{\partial t} = \frac{D_h}{R}\frac{\partial^2 C}{\partial x^2} - \frac{q}{nR}\frac{\partial C}{\partial x} \tag{28.12}$$

where R = retardation factor described earlier for adsorption. The solution in this case becomes (Bear, 1972)

$$C(x,t) = \frac{C_0}{2}\left[\text{erfc}\left(\frac{Rx - (q/n)t}{2\sqrt{RD_h t}}\right)\right.$$

$$\left.+ \exp\left(\frac{qx}{nD_h}\right)\text{erfc}\left(\frac{Rx - (q/n)t}{2\sqrt{RD_h t}}\right)\right] \tag{28.13}$$

Ogata and Banks (1961) showed that the second term in Eq. (28.13) can be neglected where $D_h n/qx < 0.002$; this condition produces an error of less than 3%. Equation (28.13) then reduces to a form similar to Eq. (28.10).

Equation (28.10) or (28.13) can be used in laboratory studies to determine dispersion coefficients for nonreactive and adsorbing species. Figure 28.1 shows several graphs depicting the effect of dispersion and adsorption on the observed breakthrough curve. Larger values of D_h tend to spread out the fronts while large values of R tend to slow the velocity of the center of mass ($C/C_0 = 0.5$) and reduce D_h by a factor of $1/R$. Conceptually, in 1-D, these transport mechanisms are relatively well understood for laboratory scale experiments.

TWO-DIMENSIONAL MODELING

Two-dimensional (2-D) modeling is generally the most useful tool for analyzing and predicting solute transport in a field situation. This type of model is considerably more versatile than the 1-D models previously discussed.

Horizontal 2-D modeling will be most useful where (1) the flow field is nonuniform, or (2) observation wells or other monitoring points provide data in more than one direction. Nonuniformities in the flow field may be the result of variations in topography and permeability, of natural or artificial sinks such as wells and springs, or of recharge points such as injection wells or leakage from lagoons. Probably the most common reason for using a horizontal 2-D analysis is that monitoring wells are typically distributed about the horizontal plane. Most wells are fully penetrating, and therefore any samples collected represent an approximate vertical average in that portion of the aquifer.

Sampling in the vertical direction may be warranted where there are significant differences between the contaminant density and water density or where detailed information is available on vertical variations in aquifer hydraulic conductivity or porosity. Profile 2-D models are available for such cases.

To study the variation in contaminants in 2-D, it is first necessary to solve the governing transport Eq. (28.8) in the x and y directions. The techniques available for developing a solution include analytical and numerical methods.

Analytical Models

Analytical models are developed by solving the transport equation for certain simplified boundary and initial conditions. Numerous solutions are available in the literature (Bear, 1979; Hunt, 1978; Wilson and Miller, 1978; Cleary and Adrian, 1973; Shen, 1976) for pulse and continuous contaminant sources with boundary conditions ranging from homogeneous confined aquifers to infinite-thickness water table systems. Processes that may be included in these models are advection, dispersion, adsorption, and decay. Analytical solutions generally require simple geometries and boundary conditions, but do provide useful insights to many ground water contaminant problems.

Ground water generally moves slowly, and there are a limited number of locations where humans come into contact with ground water. The adverse consequences of contamination can be analyzed if the concentration of a contaminant can be reliably estimated for a given location and time. Two very useful tools for summarizing the variation of a contaminant in space and time are location and arrival time distributions. In a series of papers, Nelson (1977) introduced the concept of arrival distribution and presented a technique for developing those distributions based on determining the travel time along individual flow lines. Variation in the arrival distribution is the result of convergence and divergence of flow lines. Nelson's procedure does not account for dispersion directly, but the arrival distribution concept is useful because it uses the actual flow field to predict transport processes.

The 2-D analytical model developed by Wilson and Miller (1978) is one of the simplest to use and can account for lateral and transverse dispersion, adsorption, and first order decay in a uniform flow field. Concentration C at any point in the s,y plane can be predicted using Eq. (28.14) for an instantaneous impulse or Eq. (28.15) for continuous injection. Contaminants are assumed to be injected uniformly throughout the vertical axis. The flow velocity must be obtained from a flow model or from detailed field monitoring. The x axis is oriented in the direction of flow.

$$C(x,y,t) = \frac{m'}{4\pi nt\sqrt{D_xD_y}} \exp\left[-\frac{(x - vt)^2}{4D_xt} - \frac{y^2}{4D_yt} - \lambda t \right] \quad (28.14)$$

$$C(x,y,t) = \frac{f'_m \exp (x/B)}{4\pi n\sqrt{D_xD_y}} \, W\left(u, \frac{r}{B}\right) \quad (28.15)$$

where m' = injected contaminant mass per vertical unit aquifer (M/L)

$\quad\quad f'_m$ = continuous rate of contaminant injection per vertical unit aquifer (M/LT)

$\quad\quad D_x$ = longitudinal dispersion coefficient (L²/T)

$\quad\quad D_y$ = transverse dispersion coefficient (L²/T)

$\quad\quad n$ = porosity

$\quad\quad t$ = time since start of injection (T)

$\quad\quad \lambda$ = decay coefficient (l/T)

$\quad\quad V$ = seepage velocity (L/T)

$\quad\quad B = 2 D_x/V$

$\quad\quad \gamma = 1 + (2B\lambda/V)$

$\quad\quad u = \dfrac{r^2}{4\gamma D_xt}$

$\quad\quad r = \left[\left(x^2 + \dfrac{D_x\,y^2}{D_y}\right)\gamma\right]^{1/2}$

$W(u, r/B)$ is the leaky well function described in papers by Hantush (1956) and Wilson and Miller (1978) and tabulated in many ground water texts (Bear, 1979; Freeze and Cherry, 1979). When analyzing the effects of adsorption, the retardation coefficient R is used to redefine V, D_x and D_y as

$$V' = V/R$$

$$D'_x = D_x/R$$

$$D'_y = D_y/R$$

Equation (28.15) was verified by estimating the size and shape of a plume of chromium contamination and comparing with previous numerical modeling and field monitoring (Wilson and Miller, 1978). In general the predicted contaminant plume compared very well with the measured extent of contamination. Some deviation from the prediction did occur below a small stream that caused the natural flow field to deviate from the idealized uniform flow field used in the analytical model.

Numerical Methods

Numerical or computerized solutions of the transport equation in two dimensions are among the most plentiful and commonly used techniques.

These solutions are generally more flexible than analytical solutions because the user can approximate complex geometrics and combinations of recharge and withdrawal wells by judicious arrangement of grid cells.

The general method of solution is to break up the flow field into small cells, approximate the governing partial differential equations by differences between the values of parameters over the network at time t, then predict new values for time $t + \Delta t$. This continues forward in time in small increments Δt. The most common mathematical formulations for approximating the partial differential equation of solute transport are the methods of finite difference, finite element, and characteristics.

The earlier finite-difference methods operate by dividing space into rectilinear cells along the coordinate axes. Homogeneous values within each cell are represented by values at a single node. Partial differentials can then be approximated by differences and the resulting set of equations solved by iteration (Mercer and Faust, 1981; Carnahan et al., 1969; Prickett, 1975). Approximating the differentials by a difference requires neglecting remaining terms, which results in truncation error. Finite-difference models have been developed for a variety of field situations including saturated and unsaturated flow, and for transient and constant pollutant sources. The primary disadvantage of these methods is that the truncation error in approximating the partial differential equations can result in error of the same order of magnitude as does the physical dispersion process (Anderson, 1979).

The finite element method also operates by breaking the flow field into elements, but in this case the elements may vary in size and shape. In the case of a triangular element, the geometry would be described by the three corner nodes where heads and concentrations are computed. The head or concentration within an element can vary in proportion to the distance to these nodes. Sometimes complex interpolating schemes are used to predict parameter values accurately within an element and thereby reduce the truncation errors common in finite difference procedures. Some numerical dispersion may still occur but is usually much less significant. The use of variable size and shape elements also allows greater flexibility in the analysis of moving boundary problems such as occur when there is a moving water table or when contaminant and flow transport must be analyzed as a coupled problem. A disadvantage of the finite element methods is the need for formal mathematical training to understand the procedures properly and generally higher computing costs (Pinder and Gray, 1977; Pinder, 1973; Wang and Anderson, 1982).

The method of characteristics (MOC) is most useful where solute transport is dominated by convective transport. The most common procedure is to track idealized particles through the flow field. In step one a particle and an associated mass of contaminant is translated a certain distance according to the flow velocity. The second step adds on the effects of longitudinal and transverse dispersion. This procedure is computationally efficient and minimizes numerical dispersion problems (Konikow and Bredehoeft, 1978).

Description of the USGS Ground Water Model

This section presents a more detailed description and application of one of the most widely used 2-D ground water transport models (Konikow and Bredehoeft, 1978). The USGS method of characteristics (MOC) model simulates solute transport in flowing ground water and can be applied to a wide range of 1- and 2-D problem types involving steady-state or transient flow. The model computes changes in concentration over time caused by mixing (or dilution) from fluid sources. The original model assumes that the solute is nonreactive and that the gradients of fluid density, viscosity, and temperature do not effect the velocity distribution. However, the aquifer may be heterogeneous and/or anisotropic.

The computer program solves two simultaneous partial differential equations to simulate transport in the ground water. The program uses an iterative alternating direction implicit (ADI) procedure to solve a finite-difference approximation to the ground water flow equation, and it uses the methods of characteristics to solve the solute transport equation. The latter uses a particle-tracking procedure to represent convective transport and a two-step explicit procedure to solve a finite-difference equation that describes the effects of hydrodynamic dispersion, fluid sources and sinks, and divergence of velocity.

The computer program first solves the equation describing the transient 2-D flow of a homogeneous fluid through a nonhomogeneous anisotropic aquifer [Eq. (28.7)] with the implicit finite-difference approximation presented in Eq. (28.16). After the head distribution is computed, the velocity of the ground water flow is computed for each node using an explicit finite-difference form of Darcy's equation:

$$
T_{xx[i-1/2,j]} \frac{h_{i-1,j,k} - h_{i,j,k}}{(\Delta x)^2} + T_{xx[i+1/2,j]} \frac{h_{i+1,j,k} - h_{i,j,k}}{(\Delta x)^2}
$$

$$
+ T_{yy[i,j-1/2]} \frac{h_{i,i-1,k} - h_{i,j,k}}{(\Delta y)^2} + T_{yy[i,j+1/2]} \frac{h_{i,j+1,k} - h_{i,j,k}}{(\Delta y)^2}
$$

$$
= S \frac{h_{i,j,k} - h_{i,j,k-1}}{\Delta t} + \frac{q_{w(i,j)}}{\Delta x \Delta y} - \frac{K_z}{m} [H_{s(i,j)} - h_{i,j,k}] \quad (28.16)
$$

where $h_{i,j,k}$ = hydraulic head (L) at point i, j, k
$\quad$ $H_{s(i,j)}$ = hydraulic head in source node i, j (L)
$\quad$ i, j, k = indices in the x, y, and time dimensions, respectively
$\Delta x, \Delta y, \Delta t$ = increments in the x, y, and time dimensions, respectively
$\quad$ q_w = volumetric rate of withdrawal or recharge at the i, j node (L³/T)

Next, the solute transport equation is solved, which is used to describe the 2-D areal transport and dispersion of a given nonreactive dissolved chemical species in flowing ground water [Eq. (28.8)]. This equation is solved through a three-step procedure. The first step involves the use of the particle mover that solves for the change in concentration over distance. This in turn is used by a two-step explicit finite difference approximation to the dispersion equation. The new nodal concentrations are then calculated at the end of the time period as:

$$C_{i,j,k} = C_{i,j,k} + \Delta C_{i,j,k} \tag{28.17}$$

Input parameters for the model include:

1. Effective porosity.
2. Storage coefficient.
3. Longitudinal dispersivity.
4. Transmissivity (as a tensor).
5. Ratio of transverse to longitudinal dispersivity.
6. Ratio of T_{yy} to T_{xx}.
7. Setting up the time steps, grid spaces and options for output.
8. Initialization of the aquifer with size and shape parameters, locations of sources and sinks, location of observation wells, boundary conditions, background concentrations, potentiometric or water table levels and stability criteria.

Constant-head and constant-flux boundary conditions are allowed.

The model output can include the head distributions, the x- and y- direction velocity distributions, the drawdown depths, and the concentration values at each grid location.

After application, the accuracy of the model was evaluated for two idealized problems for which analytical solutions could be obtained. The agreement was nearly exact for 1-D flow, but yielded some numerical dispersion for the case of plane radial flow. Mass-balance errors were generally less than 10 percent, with largest errors during the first several time steps. The model has been applied to several field examples by the USGS and other groups involved in predicting ground water pollutant patterns.

COMPARISON OF ONE- AND TWO-DIMENSIONAL MODELS

Two-dimensional (2-D) models can more accurately represent contaminant transport than a simple 1-D approach but do require a significant effort to run and calibrate. To compare the two approaches quantitatively, the idealized field site shown in Figure 28.3 was modeled using a 1-D analytical solution to the advection-dispersion Eq. (28.13) and the USGS method of characteristics model. Flow-through velocity was assumed to be 0.15 ft/day. Adsorption and decay of the contaminant was assumed to be zero.

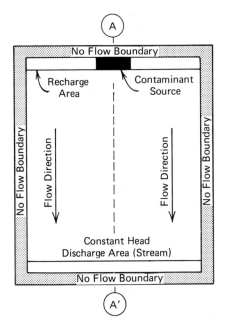

Figure 28.3. Idealized field site.

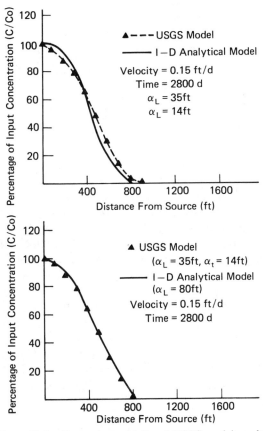

Figure 28.4. Comparison of 1-D and USGS model results.

To speed calculations, the 1-D model was written into a short program (175 lines) and loaded into an HP-86 microcomputer. The USGS model had already been loaded onto an IBM main frame computer.

Figure 28.4 illustrates the predicted variation in concentration for the two models along line A-A' after 2800 days ($\alpha_L = 35$ ft; USGS model $\alpha_t/\alpha_L = 0.4$). The maximum difference between the two models was 11% of the input concentration at 600 ft. As transverse dispersivity increases, this difference will increase. In a simple field situation where dispersivity is being adjusted to match observed concentrations, this difference is small since the 1-D model can be adjusted to match the USGS model by increasing longitudinal dispersivity to 80 ft (Figure 28.4). This illustrates that dispersivity is a fitted parameter specific to the model scale being used.

PARAMETER ESTIMATION

To calibrate any 1-D or 2-D transport model properly to an actual site, it is first necessary to define the input parameters. For modeling a non-adsorbing solute such as chloride, the parameters of primary concern are water table elevation, hydraulic conductivity K, and dispersivity α. For small project sites, accurate delineation of the water table will depend on the number of wells that can be drilled. Porosity can easily be determined by laboratory testing of a few samples. While this limited sampling may result in errors of 10 to 20%, these errors will probably not be significant compared to the problems inherent in measuring hydraulic conductivity and dispersivity.

Hydraulic Conductivity

Hydraulic conductivity is probably the single most important parameter for accurately predicting contaminant transport and at the same time one of the most difficult to measure. Conventional pump tests were developed to determine the capacity of large water-supply wells and are generally not suited for determining K in the shallow aquifers, where most contamination problems occur. Because of the tremendous variation in K common to most aquifers, laboratory testing of core samples is a very expensive proposition at best, and may not accurately represent heterogeneities that may be present across the field site.

One feasible approach where there is an existing contaminant plume of conservative tracer is to intensively monitor the plume and define the breakthrough front. Hydraulic conductivity can then be calculated from flow distance, water table gradient, and time since release. The use of artificial tracers is not usually feasible because of the low velocities common in ground water. A disadvantage of this approach is the very high monitoring cost.

The approach most useful in smaller, low-budget sites is to use so-called bail tests on single wells. The general procedure is to measure the steady-state water table, quickly pump out the well, then measure the rise in water level in the well with time. The exact procedure for calculating K from this data depends on type of well screen, depth of confining layer, and depth below water table. Details of these procedures are provided by Hvorslev (1951), Cooper et al. (1967), Boast and Kirkham (1971), and most ground water texts.

Bail tests generate hydraulic conductivities that are horizontal and vertical averages of the area affected by the well drawdown. Various researchers (Freeze, 1975; Gillham and Farvolden, 1974) have indicated that this procedure of identifying an equivalent porous medium is unsatisfactory since to model dispersion properly it is necessary to define the variations in hydraulic conductivity. This is undoubtedly true but also impractical for most field sites. In most field problems, the only practical approach is to identify an average K and then adjust dispersivity and adsorption to account for any resulting spreading of the front.

Dispersivity

Dispersivity α is defined to be the characteristic mixing length and is a measure of the spreading of the contaminant. Unfortunately, this parameter has little physical significance and varies with the scale of the problem and sample method. Laboratory values range from 1 to 10 cm, while field studies often obtain values from 10 to 100 m.

Dispersivity can be determined experimentally by injecting a tracer into a well and monitoring concentrations at observation wells. Experimental data are then fitted using analytical or numerical solutions to the dispersion equation to obtain α. Fried (1975) defined four levels to measure field dispersivities, dependent on distance traveled:

1. Local scale: 2–4 m.
2. Global scale 1: 4–20 m.
3. Global scale 2: 20–100 m.
4. Regional scale: >100 m.

Different experimental techniques are recommended for each scale. An excellent review of experimental values for α is provided by Anderson (1979).

In a field problem, experimental determination of α is often impractical due to the very long flow times. In these cases the most common approach is to run the transport model for a range of dispersivities and then adjust α until the predicted plume matches observed concentration data. A typical starting range for longitudinal dispersivity (α_L) is 10 to 100 ft. Transverse dispersivity typically ranges from 10 to 30% of α_L.

The procedure of fitting α to predict observed concentration patterns is admittedly crude but can lead to useful descriptions of contaminant transport. Bouvette (1983) performed a fairly extensive sensitivity analysis for a waste site at Conroe, Texas with eight surrounding wells. Variation in the storage coefficient S and dispersivity α estimates were not nearly as important as variations in K and waste pond height. Dispersivities ranged from 30 to 100 ft (9 to 30 m) and had a small effect (~20%) on plume migration patterns after 25 years of simulation. The ratio of α_T/α_L was selected as 0.1 with $K = 0.86$ ft/day (3×10^{-4} cm/sec) for the best simulated pattern. Increasing K to a value of 4.3 ft/day (1.5×10^{-3} cm/sec) caused the 50% breakthrough contour to move twice the distance from the waste source.

Adsorption

Freeze and Cherry (1979) briefly discuss chemical adsorption and methods of measurement. Several methods exist to describe the process, including (1) use of computational models based on equilibrium constants, (2) laboratory batch or column experiments in which contaminated water is allowed to react with soils under very controlled conditions, (3) field experiments in which the degree of adsorption is measured during passage of contaminant solutions through aquifer material, and (4) studies of existing sites where adsorption has already occurred.

While batch tests and column tests are routinely used to determine adsorption isotherms and retardation constants, considerable uncertainty exists regarding the application of these types of results to field sites. Field heterogeneities and possible nonequilibrium conditions create serious problems for prediction.

Pickens et al. (1981) report one of the very few field experiments for the *in situ* determination of dispersive and adsorptive properties of a well-defined sandy aquifer system. The technique involves the use of a radial injection dual-tracer test with ^{131}I as the nonreactive tracer and [85]Sr as the reactive tracer. Tracer migration was monitored at various radial distances and depths with multilevel point-sampling devices. In the analysis of curves, nonequilibrium adsorption effects were incorporated into the dispersion terms of the solute transport equation rather than introducing a separate kinetic term. Effective dispersivity values (α) obtained for [85]Sr averaged 1.9 cm, typically a factor of 2–5 larger than those obtained for [131]I (0.8 cm average). The K_d values obtained by various laboratory and field techniques were within a factor of 4 of the mean values of K_d for [85]Sr based on a separate analysis of sediment cores in another part of the aquifer. Thus, the usefulness of the dual-tracer injection test has been demonstrated along with the existence of nonequilibrium conditions at a field site.

Charbeneau (1982) and Bedient et al. (1983) propose the use of a three-well injection-production test at the Conroe, Texas waste site to allow the *in situ* estimation of adsorption and decay parameters within a reasonable time frame. Charbeneau (1982) shows that there is a one-to-one correspondence between points on the breakthrough curve and the streamlines of the flow field. The exact correspondence need not be determined, but the existence of the correspondence allows one to consider transport along the separate

streamlines even with the configuration of the streamlines remaining unknown. The transport problem along the individual streamlines may be non-
dimensionalized so that a single dimensionless streamline applies to all
streamlines in the flow field. The transport problem along this single dimensionless streamline is solved using the method of characteristics. Charbeneau (1981) has shown how the method of characteristics may simply be
used in the analysis of linear or nonlinear sorption and multicomponent ion
exchange. Finally, the transport solution from the dimensionless streamline
and the breakthrough curve are combined to yield the pollutant breakthrough curve. The mathematical details are outlined in the referenced
papers.

CONCLUSIONS

Basic concepts in ground water transport modeling include problems in both
one and two dimensions with advection, dispersion, adsorption, and decay.
The 1-D solutions to the advective dispersion equation allow simple predictions of breakthrough with adsorption or decay. Analytical expressions allow reasonable estimates for travel time and spreading of contaminant fronts
in 1-D. Two-dimensional models can be solved analytically for simple
boundary conditions, but numerical models are generally required for actual
waste problems. Various numerical techniques are available and are briefly
reviewed in the chapter, but a more detailed description is included for the
USGS method of characteristics model. The USGS model is a well-documented 2-D model capable of handling many typical field geometrics with
waste ponds or pumping wells. Recent changes to the model allow consideration of adsorption.

The final section reviews the serious problems of parameter estimation at
field sites, including estimation of hydraulic conductivity, dispersivity, and
adsorption coefficients. Extrapolation of results from laboratory column experiments to field sites has proven to be quite difficult, and therefore methods that can be directly applied to the field are being developed by researchers. Several of these methods, including the dual-tracer injection-production
test, for determining dispersion and adsorption coefficients at actual field
sites are described in detail.

REFERENCES

Anderson, M. P. (1979). Using models to simulate the movement of contaminants
    through groundwater flow systems. *Crit. Rev. Environ. Control* **9**:97–156.
Bear, J. (1960). The transition zone between fresh and salt waters in coastal aquifers.
    PhD dissertation, University of California, Berkeley.
Bear, J. (1972). *Dynamics of Fluids in Porous Media*. Elsevier, New York.

Bear, J. (1979). *Hydraulics of Groundwater.* McGraw-Hill, New York.

Bedient, P. B., Rodgers, A. C., Bouvette, T. C., and Tomson, M. B. and Wang, T. H. (1984). Ground water quality at a creosote waste site. *Ground Water,* **22**(3):318–329.

Boast, C. W. and Kirkham, D. (1971). Auger hole seepage theory. *Soil Sci. Soc. Am. Proc.* **35**:365–373.

Bouvette, T. (1983). Characterization of hazardous waste sites with analytical and numerical models. M.S. Thesis, Rice University, Houston, TX.

Bredehoeft, J. D. and Pinder, G. F. (1973). Mass transport in flowing groundwater. *Water Resour. Res.* **9**:194–209.

Carnahan, B., Luther, H. A., and Wilkes, J. O. (1969). *Applied Numerical Methods.* Wiley, New York.

Charbeneau, R. J. (1981). Groundwater contaminant transport and adsorption and ion exchange chemistry: Method of characteristics for the case without dispersion. *Water Resour. Res.* **17**:705–713.

Charbeneau, R. J. (1982). Calculation of pollutant removal during groundwater restoration with adsorption and ion exchange. *Water Resour. Res.* **18**:1117–1125.

Cleary, R. W. and Adrian, D. D. (1973). New analytical solutions for dye diffusion equations. *J. Environ. Eng. Div., ASCE* **99**:213–227.

Cooper, H. H., Jr., Bredehoeft, J. D., and Papadopoulos, I. S. (1967). Response of a finite-diameter well to an instantaneous charge of water. *Water Resour. Res.* **3**:263–269.

Freeze, R. A. (1975). A stochastic-conceptual analysis of one-dimensional groundwater flow in nonuniform homogeneous media. *Water Resour. Res.* **11**:725.

Freeze, R. A. and Cherry, J. A. (1979). *Groundwater.* Prentice-Hall, Englewood Cliffs, NJ.

Fried, J. J. (1975). *Ground Water Pollution.* Elsevier, Amsterdam.

Gelhar, L. W., Gutjahr, A. L., and Naff, R. L. (1979). Stochastic analysis of macrodispersion in a stratified aquifer. *Water Resour. Res.* **15**:1387–1397.

Gillham, R. W. and Farvolden, R. N. (1974). Sensitivity analysis of input parameters in numerical modeling of steady state regional ground water flow. *Water Resour. Res.* **10**:529.

Hantush, M. S. (1956). Analysis of data from pumping tests in leaky aquifers. *Trans. Am. Geophys. Union* **37**:702–714.

Hunt, B. (1978). Dispersive sources in uniform ground-water flow. *J. Hydrol. Div. ASCE* **104**:75–85.

Hvorslev, M. J. (1951). Time lag and soil permeability in groundwater observations. U.S. Army Corps Engrs. *Waterways Exp. Sta. Bull.* **36,** Vicksburg, MS.

Konikow, L. F. and Bredehoeft, J. D. (1978). Computer model of two-dimensional solute transport and dispersion in ground water. In: Automated Data Processing and Computations, Techniques of Water Resources Investigations of the U.S. Geological Survey, Book 7: *Automated Data Processing and Computations,* U.S. Geological Survey, Washington, D.C.

Mercer, J. W. and Faust, C. R. (1981). *Ground-Water Modeling.* National Water Well Association, Worthington, OH.

Nelson, R. W. (1977). Evaluating the environmental consequences of groundwater contamination. #1. An overview of contaminant arrival distributions as general evaluation requirements. *Water Resour. Res.* **14**:409–415.

Ogata, A. and Banks, R. B. (1961). A solution to the differential equation of longitudinal dispersion in porous media. U.S. Geological Survey Professional Paper 411-A, Washington, DC.

Pickens, J. F. and Lennox, W. C. (1976). Numerical simulation of waste movement in steady groundwater flow systems. *Water Resour. Res.* **12**:171–180.

Pickens, J. F., Jackson, R. E., and Inch, K. J. (1981). Measurement of distribution coefficient using a radical injection dual tracer test. *Water Resour. Res.* **17**:529–544.

Pinder, G. F. (1973). A Galerkin finite element simulation of groundwater contamination on Long Island, N.Y. *Water Resour. Res.* **9**:1657–1669.

Pinder, G. F. and Gray, W. G. (1977). *Finite Element Simulation in Surface and Subsurface Hydrology.* Academic Press, New York.

Prickett, T. A. (1975). Modeling techniques for groundwater evaluation. In: *Advances in Hydroscience* Vol. 10. Academic Press, New York, pp. 1–143.

Shen, H. T. (1976). Transient dispersion in uniform porous media flow. *J. Hydrau. Div. ASCE* **102**:707–716.

Smith, L. and Schwartz, F. W. (1980). Mass transport: 1. A stochastic analysis of macroscopic dispersion. *Water Resour. Res.* **16**:303–313.

Wang, H. F. and Anderson, M. P. (1982). *Introduction to Groundwater Modeling.* Freeman, San Francisco.

Wilson, J. L. and Miller, P. J. (1978). Two-dimensional plume in uniform groundwater flow. *J. Hydrau. Div. ASCE* **104**:503–514.

INDEX

NEW TECHNOLOGY OF PEST CONTROL
  Carl B. Huffaker, Editor

THE SCIENCE OF 2,4,5-T AND ASSOCIATED PHENOXY HERBICIDES
  Rodney W. Bovey and Alvin L. Young

INDUSTRIAL LOCATION AND AIR QUALITY CONTROL: A Planning Approach
  Jean-Michel Guldmann and Daniel Shefer

PLANT DISEASE CONTROL: Resistance and Susceptibility
  Richard C. Staples and Gary H. Toenniessen, Editors

AQUATIC POLLUTION
  Edward A. Laws

MODELING WASTEWATER RENOVATION: Land Treatment
  I. K. Iskandar, Editor

AIR AND WATER POLLUTION CONTROL: A Benefit Cost Assessment
  A. Myrick Freeman, III

SYSTEMS ECOLOGY: An Introduction
  Howard T. Odum

INDOOR AIR POLLUTION: Characterization, Prediction, and Control
  Richard A. Wadden and Peter A. Scheff

INTRODUCTION TO INSECT PEST MANAGEMENT, Second Edition
  Robert L. Metcalf and William H. Luckman, Editors

WASTES IN THE OCEAN—Volume 1: Industrial and Sewage Wastes in the Ocean
  Iver W. Duedall, Bostwick H. Ketchum, P. Kilho Park, and Dana R. Kester, Editors

WASTES IN THE OCEAN—Volume 2: Dredged Material Disposal In the Ocean
  Dana R. Kester, Bostwick H. Ketchum, Iver W. Duedall and P. Kilho Park, Editors

WASTES IN THE OCEAN—Volume 3: Radioactive Wastes and the Ocean
  P. Kilho Park, Dana R. Kester, Iver W. Duedall, and Bostwick H. Ketchum, Editors

LEAD AND LEAD POISONING IN ANTIQUITY
  Jerome O. Nriagu

INTEGRATED MANAGEMENT OF INSECT PESTS OF POME AND STONE FRUITS
  B. A. Croft and S. C. Hoyt, Editors

PRINCIPLES OF ANIMAL EXTRAPOLATION
  Edward J. Calabrese